Greek Mathematical Thought and the Origin of Algebra

Jacob Klein

Translated by Eva Brann

Dover Publications, Inc.
New York

Published in Canada by General Publishing Company, Ltd., 30 Lesmill
Road, Don Mills, Toronto, Ontario.
Published in the United Kingdom by Constable and Company, Ltd., 3 The
Lanchesters, 162–164 Fulham Palace Road, London W6 9ER.

This Dover edition, first published in 1992, is an unabridged and unaltered
republication of the edition published by The M.I.T. Press, Cambridge,
Massachusetts, in 1968. The original German text was published as "Die
griechische Logistik und die Entstehung der Algebra" in *Quellen und Studien
zur Geschichte der Mathematik, Astronomie und Physik*, Abteilung B: *Studien*, Vol.
3, fasc. 1 (Berlin, 1934), pp. 18–105 (Part I); fasc. 2 (1936), pp. 122–235 (Part II).
The translation of Vieta's *Introduction to the Analytical Art* which appears in the
Appendix is by the Reverand J. Winfree Smith. This edition is published by
special arrangement with The M.I.T. Press, 55 Hayward Street, Cambridge,
Massachusetts 02142.
Manufactured in the United States of America
Dover Publications, Inc., 31 East 2nd Street, Mineola, N.Y. 11501

Library of Congress Cataloging-in-Publication Data

Klein, Jacob, 1899–1978.
[Griechische Logistik und die Entstehung der Algebra. English]
Greek mathematical thought and the origin of algebra / by Jacob Klein ;
translated by Eva Brann.
p.   cm.
"Unabridged and unaltered republication of the edition published by the
M.I.T. Press, Cambridge, Massachusetts, in 1968"—T.p. verso.
Includes bibliographical references (p.   ) and indexes.
ISBN 0-486-27289-3
1. Mathematics, Greek.   2. Algebra—History.   I. Title.
QA22.K513   1992
512'.009—dc20
92–20992
CIP

Author's note

THIS STUDY was originally written and published in Germany during rather turbulent times. Were I writing it today, the vocabulary would be less "scholarly," and the change from the ancient to the modern mode of thinking would be viewed in a larger perspective.

Some of the references could have been brought up to date and made more accessible to English-speaking readers, but this would have entailed a labor out of proportion to its usefulness. However, a few additions to the references as well as some minor changes in the text have been made by the translator and by myself.

JACOB KLEIN

St. John's College
Annapolis, Maryland
November 1966

Translator's note

THE GREEK word *arithmos* (ἀριθμός) is rendered in the German text as *Anzahl*: "a number of [things]," to distinguish it from our modern *Zahl*: "number." Since English approximations to *Anzahl* are either obsolescent (e.g., "tale") or awkward (e.g., "counting-number," "numbered assemblage"), *Anzahl*, like *Zahl*, has been rendered simply as "number," *although it is a chief object of this study to show that Greek "arithmos" and modern "number" do not mean the same thing,* that they differ in their *intentionality,* for the former intends *things,* i.e., a number of them, while the latter intends a *concept,* i.e., that of quantity (cf. Pp. 206–208). *Intentionality* and *conceptualization* are both used to translate the German word *Begrifflichkeit.*

The following Greek terms occur so frequently that they have been incorporated into the text in transliterated form:

aísthesis (αἴσθησις) — sense perception
aisthetón (αἰσθητόν) — object of sense
analogía (ἀναλογία) — proportion
aóristos [*dyás*] *(ἀόριστος δυάς)* — indeterminate, infinite [dyad]
ápeiron (ἄπειρον) — sb.: the limitless, infinite

vii

apódeixis (ἀπόδειξις) — demonstration, [strict] proof

aporía (ἀπορία) — quandary

arché, pl. *archaí (ἀρχή, ἀρχαί)* — [governing] source, beginning

chorismós (χωρισμός) — separation

diánoia (διάνοια) — [the faculty and activity of] thinking

dýnamis (δύναμις) — power

eidetikós (εἰδητικός) — adjective from *eidos*

eídos, pl. *eíde (εἶδος, εἴδη)* — literally: "looks"; kind, form, species, "idea"; sometimes: "figure"

epistéme (ἐπιστήμη) — Lat. *scientia*, knowledge, science

génos, pl. *gene (γένος, γένη)* — genus, family, class; often: the higher *eide*

hen (ἕν) — one

hýle (ὕλη) — material

idéa (ἰδέα) = *eidos*

koinón (κοινόν) — sb.: common thing

koinonía ton eidón (κοινωνία τῶν εἰδῶν) — community of the *eide*

katá to pléthos (κατὰ τὸ πλῆθος) — according to multitude

kath' autó (καθ' αὐτό) — by itself

kathólou pragmateía (καθόλου πραγματεία) — general treatment or study

kat' eídos (κατ' εἶδος) — according to kind

kínesis (κίνησις) — change

logistiké (λογιστική) — logistical art

lógos, pl. *lógoi (λόγος, λόγοι)* — reasonable speech; also: ratio

máthema (μάθημα) — learning matter

mathematiká (μαθηματικά) — sb.: mathematical objects

máthesis (μάθησις) — study, discipline

méthexis (μέθεξις) — participation

monás (μονάς) — monad, unit

morphé (μορφή) — shape

noetón (νοητόν) — that which is for thought, object of thought *nóesis (νόησις)* — activity of thought, intellection

nous (νοῦς) — [direct and perfect activity of] thought, intellect

on; me on (ὄν, μὴ ὄν) — being; non-being

páthos, pl. *páthe (πάθος, πάθη)* — characteristic, property

posón (ποσόν) — Lat. *quantum*, sb.: the object of the question "how many?", that which has quantity; to avoid awkwardness, rendered as "quantity."

pros állo (πρὸς ἄλλο) — in relation to another

prós ti (πρὸς τι) — sb.: [that which is] in relation to something

stásis (στάσις) — rest

tautón (ταὐτόν) — the same

táxis (τάξις) — order

téchne (τέχνη) — art, skill

tháteron (θάτερον) — the other

Eva Brann

St. John's College
Annapolis, Maryland
Summer 1963

Short titles frequently used

Ad.-Tann.: C. Adam and P. Tannery, *Oeuvres de Descartes*, I–XIII; Paris, 1897–1913.

Alexander, *in metaph.*, Hayduck: Alexander of Aphrodisias, *In Aristotelis metaphysica commentaria*, ed. M. Hayduck, *Commentaria in Aristotelem Graeca*, I; Berlin, 1891.

Archimedes, Heiberg: Archimedes, *Opera omnia* (with the commentaries of Eutocius), ed. J. L. Heiberg; Leipzig, 1880–1881, 1910–1915.

Cantor I³, II³: M. Cantor, *Vorlesungen über die Geschichte der Mathematik*, I–IV; 3d ed., Leipzig, 1907.

Diels I³: H. Diels, *Die Fragmente der Vorsokratiker*, I–III; 3d. ed., Berlin, 1912.

Diophantus edition, Tannery: *Diophantus* (Greek and Latin), ed. P. Tannery, Leipzig, 1893–1895.

Eutocius, *in Archim.*: see Archimedes, Heiberg.

Heath, *Diophantus of Alexandria:* T. L. Heath, *Diophantus of Alexandria, A Study in the History of Greek Algebra*, 2d ed., Cambridge, 1910.

Heiberg-Menge: Euclid, *Opera omnia* (Greek and Latin), eds. J. L. Heiberg and H. Menge; Leipzig, 1883–1916.

Hermann: Plato, *Dialogi*, ed. C. F. Hermann, Leipzig, 1927.

Heron, Schmidt-Heiberg, or Schöne: Hero of Alexandria, *Opera* (Greek and German), eds. W. Schmidt, L. Nix, H. Schöne, and J. L. Heiberg, Leipzig, 1899–1914.

Iamblichus, Pistelli: Iamblichus, *In Nichomachi introductionem arithmeticam commentarius*, ed. H. Pistelli, Leipzig, 1894.

Iamblichus, *Theol. arith.*, de Falco: Iamblichus, *Theologoumena arithmeticae*, ed. V. de Falco, Leipzig, 1922.

Nicomachus, Hoche: Nicomachus of Gerasa, *Introductio arithmetica*, ed. R. Hoche, Leipzig, 1866.

Pappus, Hultsch: Pappus, *Collectio mathematica*, ed. F. Hultsch, Berlin, 1876–1878.

Pauly-Wissowa: *Real-Encyclopädie der klassischen Altertumswissenschaft*, eds. A. Pauly and G. Wissowa, Stuttgart, 1894–.

Proclus, *in Euclid.*, Friedlein: Proclus Diadochus, *In Euclidis elementorum librum primum commentaria*, ed. G. Friedlein, Leipzig, 1873.

Proclus, *in Tim.*, Diehl: Proclus Diadochus, *In Platonis Timaeum commentarii*, ed. E. Diehl, Leipzig, 1903.

Ptolemy, *Syntaxis*, Heiberg: Claudii Ptolemaei *Opera quae exstant omnia*, ed. J. L. Heiberg, I–II; Leipzig, 1898.

Simplicius, *in de cael.*: Simplicius, *In Aristotelis de caelo libros commentaria*, ed. J. L. Heiberg, *Commentaria in Aristotelem Graeca*, VII; Berlin, 1894.

Stallbaum: Plato, *Opera omnia* (with commentaries), ed. G. Stallbaum, Gotha and Erfurt, 1838; Leipzig, 1850.

Stevin, Girard: *Les Oeuvres Mathématiques de Simon Stevin de Bruges*, ed. A. Girard, Leyden, 1634.

Syranus, *in Arist. metaph.*, Kroll: Syranus, *In Aristotelis metaphysica commentaria*, ed. G. Kroll, *Commentaria in Aristotelis Graeca*, VI, 1; Berlin, 1902.

Tannery, *Mém. scient.*: P. Tannery's publications in *Mémoires scientifiques*, I–III; Paris, 1912–1915.

Theon, Hiller: Theon of Smyrna, *Expositio rerum mathematicarum ad legendum Platonem utilium*, ed. H. Hiller, Leipzig, 1878.

Vieta, van Schooten: Franciscus Vieta, *Opera mathematica*, ed. F. van Schooten, Leyden 1646.

Contents

APPENDIX

Introduction to the Analytical Art, by François Viète (Vieta).

PART I

I

Introduction:
Purpose and plan of the inquiry

THE CREATION of a formal mathematical language was of decisive significance for the constitution of modern mathematical physics. If the mathematical presentation is regarded as a mere device, preferred only because the insights of natural science can be expressed by "symbols" in the simplest and most exact manner possible, the meaning of the symbolism as well as of the special methods of the physical disciplines in general will be misunderstood. True, in the seventeenth and eighteenth century it was still possible to express and communicate discoveries concerning the "natural" relations of objects in nonmathematical terms, yet even then — or, rather, particularly then — it was precisely the mathematical form, the *mos geometricus*, which secured their dependability and trustworthiness. After three centuries of intensive development, it has finally become impossible to separate the content of mathematical physics from its form. The fact that elementary presentations of physical science which are to a certain degree nonmathematical and appear quite free of presuppositions in their derivations of fundamental concepts (having recourse, throughout, to immediate "intuition") are still in vogue should not deceive us about the fact that it is impossible, and has always been impossible, to grasp the meaning of what we nowadays call physics independently of its mathematical form. Thence arise the insurmountable difficulties in which discussions of modern physical theories become entangled as soon as physicist or nonphysicists attempt to disregard the mathematical apparatus and to present the results of scientific research in popular form. The intimate connection of the formal mathematical language with the content of mathematical physics stems from the special kind of conceptualization which is a concomitant of modern science and which was of fundamental importance in its formation.

Before entering upon a discussion of the problems which mathematical physics faces today, we must therefore set ourselves the task of inquiring into the origin and the conceptual structure of this formal language. For this reason the fundamental question concerning the inner relations between mathematics and physics, of "theory" and "experiment," of "systematic" and "empirical" procedure within mathematical physics, will be wholly bypassed in this study, which will confine itself to the limited task of recovering to some degree the sources, today almost completely hidden from view, of our modern symbolic *mathematics*. Nevertheless, the inquiry will never lose sight of the fundamental question, directly related as it is to the conceptual difficulties arising within mathematical *physics* today. However far afield it may run, its formulation will throughout be determined by this as its ultimate theme.

The creation of the formal language of mathematics is identical with the foundation of modern algebra. From the thirteenth until the middle of the sixteenth century, the West absorbed the Arabic science of "algebra" (*al-g'abr wa'l-muqābala*) in the form of a theory of equations, probably itself derived from Indian as well as from Greek sources.[1] As far as the Greek sources are concerned, the special influence of the *Arithmetic* of Diophantus on the content, but even more so on the form, of this Arabic science is un-mistakable[2] — if not in the *Liber Algorismi* of al-Khowarizmi himself, at any rate from the tenth century on.[3] Now con-currently with the elaboration, particularly in Italy, of the theory of equations which the Arabs had passed on to the West, the original text of Diophantus began, as early as the fifteenth century, to become well known and influential. But it was not until the last quarter of the sixteenth century that Vieta undertook to broaden and to modify Diophantus' technique in a really crucial way. He thereby became the true founder of modern mathematics.

The conventional presentations of the history of this development do not, indeed, fail to see the significance of the revival and assimilation of Greek mathematics in the sixteenth century. But they always take for granted, and far too much as a matter of course, the *fact* of symbolic mathematics. They are not sufficiently aware of the *character* of the conceptual transformation which occurred in the course of this assimila-tion and which constitutes the indispensable condition of modern mathematical symbolism. Moreover, most of the standard histories attempt to grasp Greek mathematics itself with the aid of modern symbolism, as if the latter were an altogether external "form" which may be tailored to any desirable "content." And even in the case of investigations intent upon a genuine understanding of Greek science, one finds that the inquiry starts out from a conceptual level which is, from the very beginning, and precisely with respect to the fundamental concepts, determined by modern modes of thought. To disengage ourselves as far as possible from these modes must be the first concern of our enterprise.

Hence our object is not to evaluate the revival of Greek mathematics in the sixteenth century in terms of its results retrospectively, but to rehearse the actual course of its genesis prospectively. Now in Vieta's assimilation and transforma-tion of the Diophantine technique, we have, as it were, a piece of the seam whereby the "new" science is attached to the old. But in order to be able to throw light on the essential features of this assimilation and transformation, we must first of all see the work of Diophantus *from the point of view of its own presuppositions*. Only then can we begin to distinguish Vieta's "Ars analytice" from its Greek foundations so as to reveal the conceptual transformation which is expressed in it.

The *Arithmetic* of Diophantus must, then, be given its proper place within the general framework of Greco-Hellenistic science, whatever one may imagine its prehistory to have been. This, however, immediately leads to a comparison of the foundations of the *Arithmetic* with those of the Neoplatonic "arithmetical" literature which forms its background, although the Neoplatonic categories were such as to prevent the integration of the *Arithmetic* into this literature. Sections 2–4 of Part I are devoted to the investigation of the classification of mathematical sciences in the Neoplatonic writers; these classifications go back to corresponding formulations in Plato, without, however, being identical with them. It will be shown that the Neoplatonic division of the science of numbers into "theoretical arithmetic" and "practical logistic" (the art of calculation) cannot assign an unambiguous position to the "theory of ratios and proportions." The latter does, on the other hand, seem identical with the "theoretical logistic" postulated by Plato. For Plato, this "theoretical logistic" bears a relation to "practical logistic" similar to that which "theoretical arithmetic" has to "practical arithmetic." "Theoretical logistic" and "theoretical arithmetic" both have as objects — in contrast to the corresponding practical arts — not things experienced through the senses but indivisible "pure" units which are completely uniform among themselves and which can be grasped as such only in thought. Both theoretical disciplines arise directly, on the one hand from actual *counting*, and on the other from *calculating*, i.e., from the act of relating numbers to one another; and the task of the theoretical disciplines is to reduce these "practical" activities to their true presuppositions. The Neoplatonic commentaries on the Platonic definitions of arithmetic and logistic in the *Charmides* and in the *Gorgias* show that in this "reduction" arithmetic is concerned with the "kinds" ($\epsilon \H{\iota} \delta \eta$) of numbers while logistic is concerned with their "material" ($\H{\upsilon} \lambda \eta$).

The Platonic postulation of a theoretical logistic as a noetic analogue for, and as the presupposition of, any art of calculation was ignored, as Section 5 will show, by the Neoplatonists, essentially because of the property of in-divisibility of the noetic monads; the use of fractional parts of the unit of calculation, which is unavoidable in calcula-tions, cannot be justified on the basis of such monads. An additional reason was the elaboration of the theory of ratios into a *general* theory of proportion, which depended on the discovery of incommensurable magnitudes and which led altogether beyond the realm of counted collections.

However, the difficulties which arise from the Platonic postulation of a theoretical logistic can be fully understood only if the ontological foundations which determine this conception are called to mind. And this requires, in turn, a thoroughgoing clarification of the *arithmos* concept which forms the basis of all Greek arithmetic and logistic. It can be shown (Section 6) that *arithmos never* means anything other than "a definite number of definite objects." Theoretical arithmetic grows initially out of the understanding that in the process of "counting off" any objects whatever we make use of a prior knowledge of "counting-numbers" which are already in our possession and which, as such, can only be collections of "undifferentiated" objects, namely assem-blages of "pure" units. The problem of the possibility of such assemblages, i.e., the question how it is possible that *many* "ones" should ever form *one* collection of "ones," leads to the search for *eide* with definite "specific properties" such as will give unity to, and permit a classification of, all counted collections. Greek arithmetic is therefore originally

nothing but the theory of the *eide* of numbers, while in the
art of "calculating," and therefore in theoretical logistic as
well, these counted collections are considered only with
reference to their "material," their *hyle*, that is, with
reference to the units as such. The possibility of theoretical
logistic is therefore totally dependent on the mode of being
which the pure units are conceived to have.

For this reason Pythagorean and Platonic philosophy in
their relation to the fundamental problems of Greek
mathematics are considered next (Section 7). In the first part
(7A), the general point of view of Pythagorean cosmological
"mathematics" and its connection with the *arithmos* concept
as such is presented. In the second part (7B), the significance
which is attached to "the ability to count and calculate" in
Platonic philosophy is discussed: In "pure" arithmetic and
"pure" logistic, human thinking *(διάνοια)* succeeds in
becoming conscious of the true object and the true presuppositions of its activity, an activity which always remains
tied to sense perception *(αἴσθησις)*. A third part (7C) follows
through the consequences which arise for Plato from the
privileged position he assigns to the theory of number: In
the structure of the *arithmos* concept he discovers the
possibility of a fundamental solution of the problem of
participation *(μέθεξις)* to which his "dialectic" necessarily
leads, without, however, being of itself able to provide a
solution. Thus the Pythagorean attempt at an "arithmo-
logical" ordering of all being is repeated by Plato within the
realm of the ideas themselves; this amounts to a decisive
correction of the Eleatic thesis of the "One."

This conception of numbers, eidetic as well as mathematical, as assemblages whose being is self-subsistent and
originally "separate" from sense perception, a conception
which is basic in Platonic teaching, is then criticized by
Aristotle (Section 8). He shows that the "pure" units are
merely the product of a "reduction" performed in thought,
which turns everything countable into "neutral" material.

The "pure" units have, therefore, no being of their own. Their indivisibility is only an expression of the fact that counting and calculating always presuppose a last, irreducible "unit," which is to be understood as the given "measure." It follows that there is nothing to prevent the introduction of a new and "smaller" measure; in other words, we may operate with fractional parts of the former unit. Only on the basis of this Aristotelian conception can the Platonic demand for a "scientific" logistic be realized.

In Part II of this study we turn to the relation of symbolic algebra to the *Arithmetic* of Diophantus. After a general consideration of the difference between ancient and modern concept formation, the work of Diophantus is, on the basis of the results of Part I, interpreted as a "theoretical logistic" (Section 9). In the formulation and solution of problems, this theoretical logistic always retains a dependence on the Greek *arithmos* concept, although it apparently incorporates a more general, pre-Greek "algebraic" tradition as well (Section 10).

Finally, in Sections 11 and 12 the transformation of the Diophantine technique at the hands of Vieta and Stevin is described. In these concluding sections we show that the revival and assimilation of Greek logistic in the sixteenth century are themselves prompted by an already current *symbolic* understanding of number, and we attempt to clarify the conceptual structure of the algebraic symbolism which is its product. At the same time we trace out the general transformation, closely connected with the symbolic understanding of number, of the "scientific" consciousness of later centuries. This transformation will be shown to appear characteristically in Stevin, Descartes, and Wallis.

2

The opposition of logistic and arithmetic in the Neoplatonists

NEOPLATONIC mathematics is governed by a fundamental distinction which is, indeed, inherent in Greek science in general, but is here most strongly formulated. According to this distinction, one branch of mathematics participates in the contemplation of that which is in no way subject to change, or to becoming and passing away. This branch contemplates that which is always such as it is and which alone is capable of being known: for that which is known in the act of knowing, being a communicable and teachable possession, must be something which is once and for all fixed. Within the realm of being, which has this character, a certain territory belongs to mathematics — namely whatever pertains to the questions: How large? How many? Insofar as the objects of mathematics fulfill the conditions set by the Greeks for objects of knowledge, they are not objects of the senses *(αἰσθητά)*, but only objects of thought *(νοητά)*. Such mathematical *noeta* fall into two classes. These are: (1) the "continuous" magnitudes — lines, areas, solids; (2) the "discrete" numbers — two, three, four, etc. Corresponding to these classes, two parts of the noetic branch of mathematics may be distinguished: geometry and *arithmetic.*

The other branch of mathematics, on the other hand, has for its object the treatment and manipulation of *aistheta* insofar as they are subject to determinations of size or to counting. To this branch belongs mensuration (geodetics), i.e., the art of measuring land, and the art of measuring in general; also *logistic* as the art of calculation; and furthermore, music (harmonics), optics, and mechanics. Astronomy occupies a special position insofar as it is assigned now to geometry, now to arithmetic. Like all of these distinctions, the opposition of a "pure" science of numbers and a "practical" art of calculation goes back to Plato. Yet in Plato this opposition is by no means fixed as unambiguously, either in terminology or, more important, in content, as it is for the Neoplatonists — a fact which has frequently been over-looked.

The chief Neoplatonic sources bearing on this point are: (1) a reference in the Euclid commentary of Proclus (Friedlein, 38–40), where Proclus cites Geminus' opinion concerning the classification of the mathematical sciences; (2) a frequently cited *scholium* to *Charmides* 165 E (Hermann, VI, 290), which is largely identical with Sections 1 and 5 of the so-called Geminus fragments in the *Definitions* of Heron (Schmidt-Heiberg, IV, 98 ff., Def. 135, 5–6);[4] (3) some lesser-known passages in Olympiodorus' *scholia* to Plato's *Gorgias*, 450 D and 451 A–C (*Neue Jahrbücher für Philologie und Pädagogik, Jahns Jahrbücher*, Suppl. 14 [Leipzig, 1848], Pp. 131 f.); (4) another anonymous *scholium*, rarely mentioned, to the same passage in the *Gorgias* (Hermann, VI, 301).

1. The Proclus passage, which follows immediately upon a discussion of the relations of mensuration to geometry, reads as follows: "Nor, again, does the man skilled in calculation study the properties of numbers as they are by themselves [which the arithmetician studies], but [he studies them] in the objects of sense, whence he gives them appellations taken from the measured [counted] objects, calling

some 'apples' and others 'bowls.'" *(οὐδ' αὖ ὁ λογιστικὸς αὐτὰ καθ' ἑαυτὰ θεωρεῖ τὰ πάθη τῶν ἀριθμῶν, ἀλλ' ἐπὶ τῶν αἰσθητῶν, ὅθεν καὶ τὴν ἐπωνυμίαν αὐτοῖς ἀπὸ τῶν μετρουμένων τίθεται, μηλίτας καλῶν τινας καὶ φιαλίτας.)* — The last is a direct reference to Plato, *Laws* VII, 819 B–C, where instructions are given for teaching children how to calculate.

2. The *Charmides scholium* is much more detailed: "Logistic is a science which concerns itself with *counted things*, but not with *numbers*, not handling that number which is truly a number, but positing that which is one [*namely one definite thing*] as the unit itself, and that which is being counted [namely the particular assemblage] as the number itself, so that this science takes, for example, three things as 'three' and ten things as 'ten';[5] to these things it applies the theorems of [pure] arithmetic. Thus it investigates both that problem which is called the 'cattle problem' by Archimedes[6] and also the 'sheep-numbers' and 'bowl-numbers,' the latter with reference to bowls, the former with reference to a flock of sheep;[7] and also with respect to other classes of bodies perceptible by the senses, it investigates their multitudes and comments on these as if they were perfect. All countable things are its material. Its parts are the methods of multiplication and division known as Hellenic and Egyptian[8] as well as the addition and division of fractions, by means of which it tracks down what is concealed with respect to its material in problems belonging to the study of triangular and polygonal numbers.[9] Its end is what is public in life and useful in contracts, even though it seems to make pronouncements about sensible things as if they were perfect."

(λογιστική ἐστι θεωρία τῶν ἀριθμητῶν, οὐχὶ δὲ τῶν ἀριθμῶν μεταχειριστική, οὐ τὸν ὄντως ἀριθμὸν λαμβάνουσα, ἀλλ' ὑποτιθεμένη τὸ μὲν ἓν ὡς μονάδα, τὸ δὲ ἀριθμητὸν ὡς ἀριθμόν, οἷον τὰ τρία τριάδα εἶναι καὶ τὰ δέκα δεκάδα· ἐφ' ὧν ἐπάγει τὰ κατὰ ἀριθμητικὴν θεωρήματα. θεωρεῖ οὖν τοῦτο μὲν τὸ κληθὲν ὑπ' Ἀρχιμήδους βοεικὸν πρόβλημα, τοῦτο δὲ μηλίτας καὶ φιαλίτας ἀριθμούς, τοὺς μὲν ἐπὶ φιαλῶν, τοὺς δὲ ἐπὶ ποίμνης· καὶ ἐπ' ἄλλων δὲ γενῶν τὰ

πλήθη τῶν αἰσθητῶν σωμάτων σκοποῦσα, ὡς περὶ τελείων ἀποφαίνεται. ὕλη δὲ αὐτῆς πάντα τὰ ἀριθμητά· μέρη δὲ αὐτῆς αἱ Ἑλληνικαὶ καὶ Αἰγυπτιακαὶ καλούμεναι μέθοδοι ἐν πολλαπλασιασμοῖς καὶ μερισμοῖς, καὶ αἱ τῶν μορίων συγκεφαλαιώσεις καὶ διαιρέσεις, αἷς ἰχνεύει τὰ κατὰ τὴν ὕλην ἐμφωλευόμενα τῶν προβλημάτων τῇ περὶ τοὺς τριγώνους καὶ πολυγώνους πραγματείᾳ. τέλος δὲ αὐτῆς τὸ κοινωνικὸν ἐν βίῳ καὶ χρήσιμον ἐν συμβολαίοις, εἰ καὶ δοκεῖ περὶ τῶν αἰσθητῶν ὡς τελείων ἀποφαίνεσθαι.)

3. Olympiodorus describes the contrast somewhat differently: "It must be understood that the following difference exists: arithmetic concerns itself with the kinds of numbers, logistic, on the other hand, with their material. There are two kinds of number: the even and the odd; of the even, in turn there are three kinds, etc." (δεῖ εἰδέναι ὅτι διαφέρουσι, τῷ τὴν μὲν ἀριθμητικὴν περὶ τὰ εἴδη τῶν ἀριθμῶν καταγίνεσθαι, τὴν δὲ λογιστικὴν περὶ τὴν ὕλην. εἴδη δὲ ἀριθμοῦ δύο· τό τε ἄρτιον καὶ τὸ περιττόν. καὶ πάλιν τοῦ μὲν ἀρτίου εἴδη τρία, κτλ.) (There follows the classification of the even and the odd according to Nicomachus, Hoche, 14 ff., 25 ff., to whom Olympiodorus explicitly refers.) He continues: "The material of numbers, on the other hand, is the multitude of the units [which are in each case to be counted or calculated]. For instance, multiplication, such as four times four, five times five, etc. [affects this material]. But this is not all. For in that case logistic would be accessible to all, since even little children know how to multiply. But it teaches also certain subtleties." (ὕλη δέ ἐστι τὸ πλῆθος τῶν μονάδων, οἷον ὁ πολλαπλασιασμός, ὅ ἐστι τετράκις δ καὶ πεντάκις ε καὶ τὰ τοιαῦτα. καὶ οὐ μόνον τοῦτο. οὕτω γὰρ ἂν [ἡ] εὐεπίβα[σ]τος (Jahn) ἦν πᾶσιν, εἴγε καὶ οἱ μικροὶ παῖδες ἴσα[σιν] ⟨J.K.⟩ τοὺς πολλαπλασιασμούς. ἀλλὰ καὶ γλαφυρά τινα διδάσκει.) There follow two verbal problems, one of them with the numerical values given, which correspond to the so-called arithmetic epigrams 51^{10} (cf. also 13) and 7 from Book XIV of the Anthologia Palatina (printed in the second volume of

Tannery's Diophantus edition; pp. 53 and 46 f.) In accordance with this, Olympiodorus, closely following the Platonic text (*Gorgias* 451 A–C), says further on: "It must be understood that that with which arithmetic deals is also the subject of logistic, namely the even and the odd. But arithmetic concerns itself with their kind, while logistic concerns itself with their material, investigating not only the even and the odd as they are by themselves but also their relation to one another in respect to *their multitude*. For multiplication is either among the same kind or is of different kinds with one another, the former, when I multiply even with even or odd with odd, the latter, when I multiply odd with even or even with odd." (ἰστέον γὰρ ὅτι περὶ ἃ ἡ ἀριθμητικὴ καταγίνεται, περὶ ταῦτα καὶ ἡ λογιστική, περὶ τὸ ἄρτιον καὶ περὶ τὸ περιττόν. ἀλλ᾽ ἡ μὲν ἀριθμητικὴ περὶ τὸ εἶδος αὐτῶν, ἡ δὲ λογιστικὴ περὶ τὴν ὕλην· τῷ μὴ μόνον καθ᾽ αὑτὰ ἀλλὰ καὶ πρὸς ἄλληλα πῶς ἔχει πλήθους ἐπισκοπεῖν. ὁ γὰρ πολλαπλασιασμὸς ἢ πρὸς ἑαυτὸν γίνεται, ἢ πρὸς ἄλληλα.[11] πρὸς ἑαυτόν, ὅταν πολλαπλασιάσω τὸν ἄρτιον ἐπὶ ἄρτιον ἢ περιττὸν ἐπὶ περιττόν· πρὸς ἄλληλα[11] δέ, ὅταν περιττὸν ἐπὶ ἄρτιον ἢ ἄρτιον ἐπὶ περιττόν.)

4. Olympiodorus uses multiplication to exemplify this state of affairs. The anonymous *scholium* to the same passage in the *Gorgias* speaks of this more clearly: "Logistic studies the multiplications and divisions of numbers, from which it is clear that it works with their quantity [determinate for each case] and the material in them. For a number is multiplied by a number not with reference to its kind but with reference to the units which are its *material*; and a number is divided according to its quantity, not according to the definition [of its kind] whereby numbers differ from one another. The words 'how they are related to one another in respect to multitude' [a paraphrase of the Platonic text] mean this: in what manner they multiply and divide each other according to the quantity which is in them." (τοὺς γὰρ πολλαπλασιασμοὺς καὶ τοὺς μερισμοὺς ἡ λογιστικὴ θεωρεῖ τοὺς τῶν

ἀριθμῶν, ᾧ καὶ δῆλον ὅτι τὸ ποσὸν τὸ ἐν αὐτοῖς καὶ τὴν ὕλην περιεργάζεται. πολλαπλασιάζεται γὰρ ἀριθμὸς ἐπὶ ἀριθμὸν οὐ κατὰ τὸ εἶδος ἀλλὰ κατὰ τὰς ὑλικὰς μονάδας, καὶ μερίζεται κατὰ τὸ ποσὸν ἀλλ' οὐ κατὰ τὸν λόγον ᾧ διαφέρουσιν ἀλλήλων οἱ ἀριθμοί. τὸ τοίνυν πῶς ἔχει πρὸς ἄλληλα κατὰ τὸ πλῆθος τοῦτο ἐνδείκνυται, πῶς πολλαπλασιάζονται καὶ μερίζονται παρ' ἀλλήλους κατὰ τὸ ἐν αὐτοῖς ποσόν.)

This much, at least, can be ascertained from these sources: The reference of Proclus and of the *Charmides scholium* to Plato's instructions for teaching children to calculate, as well as the examples given in the Olympiodorus and the *Gorgias scholium* (i.e., multiplication, division, fractions, verbal problems which lead to what we would today call equations in several unknowns), shows that logistic teaches how to *proceed* in order to solve problems relating to one or more multitudes of countable objects. It shows how to make the respective number of such objects definite, i.e., how to "calculate." In calculation the *result* alone, which *varies* as the given multitudes vary, matters. But the possibility of calculation is grounded in certain *immutable* characteristics of the numbers "themselves." With these arithmetic deals, which does not "calculate" with numbers but studies their properties and kinds as they are in themselves,[12] not as they may be read off the countable things. Calculation with numbers is nothing but the "application" of the facts of "pure" arithmetic; logistic is nothing but "applied" arithmetic, which serves, above all, practical ends. What, on the other hand, the Neoplatonic commentators and scholiasts have in mind when they speak of "pure" arithmetic are the classifications of numbers and the description of their mutual relations, as we know them from certain first indications in Plato's *Theaetetus* (147 D–148 B) and *Parmenides* (143 E–144 A); from several references of Aristotle (e.g., *Posterior Analytics* A 4, 76 b 7 f.; *Metaphysics* Δ 14, 1020 b 3 ff.); from Euclid VII, VIII, IX; and, finally, from the textbooks of

Theon of Smyrna, of Nicomachus of Gerasa (and his commentators), and of Domninus of Larissa.

But one matter remains unclear: while the two first *testimonia*, namely those which go back to Geminus, identify the area of logistic with that realm of the objects of sense in which the calculations take place, and presuppose over and above this a realm of pure numbers which are the objects of arithmetic, Olympiodorus (as also the last *Gorgias scholium*) seems to transfer the distinction between an immutable, i.e., noetic, and a changeable constituent to the numbers themselves. What Olympiodorus calls the "material" *(ὕλη)* of the numbers is clearly not attached to objects of sense but is identical with the *quantity* which each number designates. Only this quantity is affected by calculation, quite independently of the fact that the objects of the calculation might be objects of sense. In the number "six," for instance, the multitude "six," with its quantity, its *hyle*, must be distinguished from its *eidos*, namely the "even-times-odd" *(ἀρτιοπέριττον* — since six is composed of the even factor two and the odd factor three).[13] Arithmetic treats of the *eidos*, logistic of the multitude of "hylic" monads. What, now, is the relation of this conception of the objects of logistic to their inclusion within the realm of the *aistheta*? Is the multitude which is designated by any number after all always to be referred to objects of sense? These questions take us back directly to the Platonic passages themselves, since the precise interpretation of these was, in fact, the original concern prompting the Neoplatonic commentators. It may be that Olympiodorus was too little expert in mathematical matters to have done justice to this task; for instance, his opinion that one case refers to the multiplications of even (or odd) numbers with each other, while the second case involves the multiplication of odd with even or even with odd numbers, is clearly untenable.[14] But the deeper reason for this lack of clarity must, ultimately, lie in certain difficulties inherent in Plato's own conceptions.

3

Logistic and arithmetic in Plato

In the *Gorgias* (451 A–C), Socrates says that if he were asked "with what" *(περὶ τί)* arithmetic deals, he would answer: "It belongs to that knowledge which deals with the even and the odd, with reference to *how much* either happens to be." *(ὅτι τῶν περὶ τὸ ἄρτιόν τε καὶ περιττὸν [γνῶσις], ὅσα ἂν ἑκάτερα τυγχάνῃ ὄντα.)* And of logistic it is said further on: "It deals with the same thing, namely the even and the uneven; but logistic differs [from arithmetic] in so far as it studies the even and the odd with respect to the multitude which they [the single even and odd] make both with themselves and with each other." *(περὶ τὸ αὐτὸ γάρ ἐστι, τό τε ἄρτιον καὶ τὸ περιττόν· διαφέρει δὲ τοσοῦτον, ὅτι καὶ πρὸς αὐτὰ καὶ πρὸς ἄλληλα πῶς ἔχει πλήθους ἐπισκοπεῖ τὸ περιττὸν καὶ τὸ ἄρτιον ἡ λογιστική.)* This strangely involved definition — the word *arithmos* is carefully avoided[15] — coincides word for word with that in the *Charmides* (165 E–166 A–B): "Logistic, it seems, concerns even and odd, namely what multitude they make with themselves and with one another." *(ἡ λογιστική ἐστί που τοῦ ἀρτίου καὶ τοῦ περιττοῦ, πλήθους ὅπως ἔχει πρὸς αὐτὰ καὶ πρὸς ἄλληλα.)* This shows that we are here not dealing with an "accidental" assertion.[16] Of arithmetic it is said once again in the *Gorgias* (453 E) that it teaches "of the odd and

17

even, how much it is [in each case]" (περὶ τὸ ἄρτιόν τε καὶ τὸ περιττὸν ὅσον ἐστίν, cf. *Theaetetus* 198 A).

Usually these sentences, on which the *Charmides scholium* as well as the commentary of Olympiodorus are based, are said to affirm a direct opposition of arithmetic as a *theoretical* discipline to logistic as the *practical* art of calculation,[17] and this understanding is supported by an appeal precisely to the explications of Proclus and of the *Charmides scholium*. But it is hard to see how Plato's words can signify just this opposition. Is it really possible to understand the *arithmetike* mentioned there as "number theory" simply? Is the concept of *arithmos* really identical with our concept of number — vague as it generally is?

To do justice to the original Platonic definitions we must, above all, never lose sight of the context, which determines the manner of posing questions and the mode of conversation in Platonic dialogues, and which Neoplatonic systematizing tends to obscure only too easily. We must bear in mind that the Platonic opposition of arithmetic and logistic is, to begin with, not one between two scientific subjects which belong to different levels. Rather it concerns a "science" which we first acquire in our intercourse with objects of daily life and in which we may thereafter advance to special expert knowledge. The analysis of this science may, to be sure, take two directions: (1) In the face of definite multitudes of things we habitually *determine their exact number* — we "number," i.e., count, the things, and this presupposes a certain familiarity with numbers in general, especially in the case of larger multitudes. In order to be able to count we must know and distinguish the single numbers, we must "distinguish the one and the two and the three" (τὸ ἕν τε καὶ τὰ δύο καὶ τὰ τρία διαγιγνώσκειν — *Republic* VII, 522 C).[18] Plato calls the totality of this science of all possible numbers the "art of number" — "arithmetic." (2) But we are also in the habit of *multiplying or dividing these multitudes*. This means that we are no longer satisfied with the number by which we have enumerated the things in question, but that we bring to bear on this number new "numbers," whether we wish to separate off a "third" part of the respective quantity or wish to produce a multitude which amounts to "four" times the given one. In such multiplications and divisions, or, more generally, in all *calculations* which we impose on multitudes, we must *know beforehand* how the different numbers *are related to one another* and how they are constituted *in themselves*. This whole science, which thus concerns the behavior of numbers toward one another, i.e., their mutual relations, and which first enables us to *relate* numbers, i.e., to calculate with them, is called the "art of calculation" — "logistic."

"Arithmetic" is, accordingly, not "number theory," but first and foremost the art of correct counting. As Plato says explicitly in the *Theaetetus* (198 A–B): "Through this art, I think, one is oneself master of the sciences of number and is able as a teacher to pass them on to another." (ταύτῃ δή, οἶμαι, τῇ τέχνῃ αὐτός τε ὑποχειρίους τὰς ἐπιστήμας τῶν ἀριθμῶν ἔχει καὶ ἄλλῳ παραδίδωσιν ὁ παραδιδούς.) If "one becomes *perfect* in the arithmetical art" (ἀριθμητικὸς ὢν τελέως), then "he knows also *all* the numbers" (πάντας ἀριθμοὺς ἐπίσταται).[19] Only on the basis of this art of counting may further knowledge in the realm of numbers be gained. Logistic too is possible only on this basis. "Logistic" is not merely the art of calculation in the sense of "operating" with numbers, i.e., an art teaching the procedures to be applied in multiplication, division, taking of roots, and in the solution of verbal problems. All meaningful operations on number presuppose knowledge of the relations which connect the single numbers. This knowledge, which we acquire in childhood and which we use in every calculation, although it is not always present to us as a whole, is *logistike*. The fact that from it the possibility of numerical operations may be derived and that from these operations certain "mechanical" rules of reckoning may in turn arise has a significance which, though of importance for our practice, is yet subordinate. We can, it is true, acquire this knowledge best by making ourselves familiar with the elementary operations as applied to countable things. This is why Plato, in that passage in the *Laws* (819 B–C; see also *Republic* 536 D ff.) cited by the Neoplatonists, demands that calculation be taught through play (according to the Egyptian model), thus making it possible for children to acquire correctness in counting and in combining numbers painlessly. The main things to be learned in this way are exactly the multifarious relations which exist between different numbers: that the quintupling of a multitude of twenty wreaths yields a hundred wreaths, that a multitude of 221 apples can be divided into thirteen or seventeen equal parts and no more, etc.

From what has been said it follows that "arithmetical" and "logistical" science are hard to distinguish on this primary level, and, as we shall see, not only on this level. Thus the simplest relation between the parts of a multitude and the multitude itself (as their sum) is given as soon as they are counted or added up, i.e., by a counting process which extends over all members of all partial multitudes (cf. *Theae-tetus* 195 E–196 A and 198 A–C, where the task of adding five and seven is discussed and assigned to *arithmetike*; also 204 B–C). Addition and also subtraction are only an extension of counting. Furthermore, all the remaining relations between numbers, on which the complicated operations of reckoning are based, may also, in the last analysis, be reduced to that ordering of numbers which is ascertained by counting. On the other hand, counting itself already presupposes a continual relating and distinguishing of the numbered things as well as of the numbers. Thus it happens that in Plato's language "counting" and "calculating" frequently occur together: *Republic* VII, 522 E: λογίζεσθαί τε καὶ ἀριθμεῖν; 525 A: λογιστική τε καὶ ἀριθμητική; 522 C: ἀριθμόν τε καὶ λογισμόν, which is here intended as a summary of the realm to which the ability "to distinguish 'one' and 'two' and 'three'" *(τὸ ἕν τε καὶ τὰ δύο καὶ τά τρία διαγιγνώσκειν)* pertains, cf. P. 18; *Phaedrus* 274 C: ἀριθμόν τε καὶ λογισμόν; *Hippias Minor*: περὶ λογισμὸν καὶ ἀριθμόν. In the *Laws* (817 E), "calculations" *(λογισμοί)* and "what concerns numbers" *(τὰ περὶ ἀριθμούς)* are then expressly designated as "one learning matter" *(ἐν μάθημα)*.[20] And yet the fundamental significance of counting necessitates the emphasis on and the relative isolation of *arithmetike* as such. For to Plato that ability proper to man, *to be able to count* — an ability to which corresponds the *countableness* of things in the world — is a fact fundamental beyond all special problems, a fact which determines the systematic aspect of his teaching.[21]

The preceding remarks contribute only this much toward an understanding of the definitions given in the *Gorgias* (and *Charmides*): from the kind of knowledge which is conveyed in "arithmetic" and "logistic" we may infer the fact of their difference as well as of their unity. This unity ultimately has its roots in the object of which both treat. Both are concerned with "numbers" and with "number" in general, with the *arithmos* (cf. *Republic* 525 A). But what is the significance of the fact that the "even" and the "odd" occur in the definition, but not number? If it were to be the case that an attempt is here being made to bring to bear an alien technique of definition on perfectly familiar facts in order to formulate them precisely, and if it were in response to *this* that the "odd" and the "even" are regarded as characteristic of arithmetic and logistic,[22] then, clearly, we could elucidate the matter only through an analysis of the meaning of *arithmos*. But before we do this we must pursue further those of Plato's remarks which bear on the development of "arithmetical" and "logistical" knowledge.

In the form in which we have looked at arithmetic and logistic up till now, these represent a purely "practical" knowledge, which we must acquire and use with a view to the necessities of life (cf. especially *Laws* 819 C). Once practical aims are abandoned, these studies acquire an altogether different standing. At the same time our attention is freed to turn to certain peculiarities of that knowledge which is now raised to "knowledge" proper, to *episteme*. Thus in the *Republic* and in the *Philebus* Plato contrasts "practical" arithmetic *and* "practical" logistic with their respective "theoretical" counterparts. In the *Philebus* (56 D) Socrates asks "Must it not first be said that the arithmetic of the crowd is one thing, and that of lovers of wisdom another?" (Ἀριθμητικὴν πρῶτον ἆρ' οὐκ ἄλλην μέν τινα τὴν τῶν πολλῶν φατέον, ἄλλην δ'αὖ τὴν τῶν φιλοσοφούντων;) And to the question of his interlocutor Protarchus "what difference there might be?" he replies (56 D–E): "The distinction is not small, Protarchus: Of those who deal with number some count off units which are in some way unequal, such as two armed camps, two head of cattle and two of the smallest or two of the largest of all things, while the others would not follow their practice unless someone posited a unit such that not one single unit of those in a myriad differed from any other." (οὐ σμικρὸς ὅρος, ὦ Πρώταρχε· οἱ μὲν γάρ που μονάδας ἀνίσους καταριθμοῦνται τῶν περὶ ἀριθμόν, οἷον στρατόπεδα δύο καὶ βοῦς δύο καὶ δύο τὰ σμικρότατα ἢ καὶ τὰ πάντων μέγιστα. οἱ δ' οὐκ ἄν ποτε αὐτοῖς συνακολουθήσειαν, εἰ μὴ μονάδα μονάδος ἑκάστης τῶν μυρίων μηδεμίαν ἄλλην ἄλλης διαφέρουσάν τις θήσει.) And Protarchus makes the point once more with emphasis: "You are quite right in saying that there is no mean difference among those who are concerned with number, so that it is reasonable to say that there are two [kinds of arithmetic]." (καὶ μάλα γ'εὖ λέγεις οὐ σμικρὰν διαφορὰν τῶν περὶ ἀριθμὸν τευταζόντων, ὥστε λόγον ἔχειν δύο αὐτὰς εἶναι.) This distinction is immediately expanded to include, among other things, *logistic*, so that Plato here explicitly postulates a "theoretical" logistic. What differentiates this theoretical logistic from practical logistic is the kind of multitude with which each deals; in the one case we are concerned with multitudes of "unequal" objects — and obviously *all* objects of sense are such — in the other with multitudes of wholly similar units, namely precisely those which cannot occur in the realm of objects of sense. We shall have to investigate later how far the consequences which arise from this distinction are compatible with the distinction itself. But this much is already clear — theoretical logistic raises to an explicit science that knowledge of relations among numbers which, albeit implicitly, precedes, and indeed must precede, all calculation. Only in doing this does logistic make the right use of this knowledge (*Republic* 523 A). As soon as that aspect which binds this science to objects of sense is relinquished, we see that it cannot pertain to a special realm of objects in the world of sense, but can only refer to a "neutral" material, namely the homogeneous monads. Thus theoretical logistic arises from practical logistic when its practical applications are neglected and its presuppositions are pursued for their own sake.

Logistic is spoken of in the *Republic* (525 C–D) in just this sense as not to be pursued "privately, for the sake of buying and selling as merchants or shopkeepers pursue it" (μὴ ἰδιωτικῶς, . . . οὐκ ὠνῆς οὐδὲ πράσεως χάριν ὡς ἐμπόρους ἢ καπήλους μελετῶντας), but for the public interest,[23] and especially "for the sake of knowing" (τοῦ γνωρίζειν ἕνεκα). Through it the soul is said to be brought to the point where it grasps the nature of the numbers as they are in themselves, i.e., not as they are grasped when turned toward objects of sense, but by "thought alone" (τῇ νοήσει αὐτῇ). It does not treat of numbers "which have visible or tangible bodies" (ὁρατὰ ἢ ἁπτὰ σώματα ἔχοντας, cf. *Epinomis* 990 C), but only "of numbers themselves" (περὶ αὐτῶν τῶν ἀριθμῶν). The same opposition arises in the *Theaetetus* (195 D–196 B) in a different context: It is easier for someone to make a mistake in the addition of a number of things which are seen or tasted or otherwise perceived by the senses, e.g., a multitude of five or a multitude of seven people, than when he considers the respective number "itself," in this case "five itself and seven itself" (αὐτὰ πέντε καὶ ἑπτά), which he has only "in his thinking" *(ἐν τῇ διανοίᾳ ἔχει)*, although error, especially when larger numbers are involved, is by no means excluded even then. These numbers, as also "the eleven which one cannot grasp except by thinking" *(τὰ ἕνδεκα, ἃ μηδὲν ἄλλο ἢ διανοεῖταί τις)* and *every* other number of similar kind, are cases of numbers of "pure" units — "for I think you are speaking of all number" *(οἶμαι γάρ σε περὶ παντὸς [μᾶλλον] ἀριθμοῦ λέγειν)*, says Theaetetus to Socrates, who assents to this supposition. That is, they are units accessible only to thinking, which one can, therefore, apprehend and count only "by himself" *(αὐτὸς πρὸς αὑτόν)* in distinction from "whatever outside has number" *(τῶν ἔξω ὅσα ἔχει ἀριθμόν* — 198 C). The fact that these numbers "admit only of being thought" *(ὧν διανοηθῆναι μόνον ἐγχωρεῖ* — *Republic* 526 A) has its basis in the "purity" of the units, of which "each and every one is equal and not in the least different and has no parts within itself." *(ἴσον τε ἕκαστον πᾶν παντὶ καὶ οὐδὲ σμικρὸν διαφέρον, μόριόν τε ἔχον ἐν ἑαυτῷ οὐδέν* — *ibid.*) More will have to be said of this last property of pure units, their indivisibility. First, however, we shall try to imagine what may have been Plato's notion of theoretical logistic.

For this purpose, the definitions of arithmetic and logistic given in the *Gorgias* and the *Charmides* must always be kept in mind. For their meaning, although independent of the practical or theoretical character of the two disciplines, can be *understood* only from the *theoretical* side. It is precisely because a "scientific" definition properly so called is here applied to a subject which ordinarily serves only practical purposes that we find such a strangely elaborate formulation. According to this definition, theoretical logistic would have to include primarily knowledge concerning all those *relations*, i.e., ratios *(λόγοι)* among "pure" units, on which the success of any calculation depends, while knowledge of these "pure" *numbers themselves* would be reserved for theoretical arithmetic. Perhaps it will be possible to find some indications of this in the later mathematical textbooks of the Neoplatonists and the Neopythagoreans, for in these textbooks scientific material going back to Platonic and pre-Platonic times was undoubtedly absorbed, although its arrangement, its nomenclature, and the general manner of presentation may well have undergone great changes.

4

The role of the theory of proportions in Nicomachus, Theon, and Domninus

In the *Introductio Arithmetica* of Nicomachus the field of quantity *(τὸ ποσὸν)* which is to be investigated is divided into two parts (I, 3 — Hoche, 5, 13 ff.): "Of quantity one part is studied *by itself,* namely that which has no sort of relation to another" *(τοῦ ποσοῦ τὸ μὲν ὁρᾶται καθ᾽ ἑαυτό, μηδεμίαν πρὸς ἄλλο σχέσιν ἔχον),* "and the other as having *some sort of relation to another* and capable of being thought of only in its relations to another" *(τὸ δὲ πρὸς ἄλλο πως ἤδη ἔχον καὶ σὺν τῇ πρὸς ἕτερον σχέσει ἐπινοούμενον).* Taking his departure from the traditional "Pythagorean" division of the mathematical sciences into arithmetic, geometry, music, and astronomy,[24] Nicomachus now assigns arithmetic to the realm of that which is "by itself" *(καθ᾽ αὐτό),* music to the realm of that which is related "to another" *(πρὸς ἄλλο* — 6, 1–3). Nevertheless a large part of both of his "arithmetical" books is devoted precisely to this second realm of the *poson.* In Chapters 7–13 of the first book he first of all investigates the two "primary kinds" *(πρώτιστα εἴδη)* of "number as it is scientifically studied" *(ἐπιστημονικὸς ἀριθμός),* namely the odd and the even, of which he says that "they have the being of quantity" itself *(οὐσίαν ἔχοντα τὴν τῆς ποσότητος* — cf.

12, 20–13, 2). Next he presents their subspecies, such as the even-times-even, the odd-times-even, etc. In contrast to Euclid, Theon, and Domninus, Nicomachus proceeds in a completely schematic manner and patently exaggerates the differentiation. It must be emphasized that in this procedure even those numbers which are "prime *to one another* and not composed [of others]" *(τὸ πρὸς ἄλλο πρῶτον καὶ ἀσύνθετον — 28, 20 ff.)* are subsumed under a separate species (namely the third subspecies of the odd), although according to Nicomachus himself only the realm of the *kath' auto* is here to be considered.[25] Next (Chapters 14–16) are presented the "perfect," the "superabundant," and the "deficient" numbers, which are determined by the sum and product of their aliquot parts, or more generally, those numbers for which interest is centered on the manner in which they are themselves composed of others. Then he says (44, 8 f.); "Having given a preliminary treatment of quantity by itself we now pursue it also in its relations." *(προτετεχνολογημένου δὲ ἡμῖν περὶ τοῦ καθ' αὐτο ποσοῦ νῦν μετερχόμεθα καὶ ἐπὶ τὸ πρός τι.)* Accordingly, in Chapters 17–23 the "ten arithmetical relations" or ratios *(αἱ δέκα ἀριθμητικαὶ σχέσεις — 64, 21)* are presented. They appear as "kinds" *(εἴδη)* of "inequality" *(ἀνισότης)* and are further pursued with respect to their mutual dependence and their relationship to "equality" (up to Chapter 5 of the second book). They are comprehended under two groups of five kinds each, to which correspond the following examples: four to three, five to three, six to three, seven to three, eight to three, and their inverses. These kinds of ratios do, indeed, have their most immediate application in the field of music; yet we should not forget that they belong to the theoretical foundation of all calculations, especially those dealing with fractions, as their very names and the names of the numbers they involve show: for example, "superparticular number" *(ἀριθμὸς ἐπιμόριος)*, "superpartient relation" *(σχέσις ἐπιμερής)*, etc., and especially the fractional terms "sesquitertian" (4:3), "superquintipartient" (11:6), "double sesquiquartan" (9:4) (ἐπίτριτος, ἐπιπενταμερής, διπλασιεπιτέταρτος), etc. With the fifth chapter of the second book the discussion of "relational quantity" (τοῦ πρὸς ἕτερόν πως ἔχοντος ποσοῦ — 82, 10 f.) is broken off for the time being and in Chapters 6–20 the theory of figurate numbers, plane as well as solid, is presented,[26] leading at last to the interrelation of the sequences of triangular, quadratic, and oblong (heteromecic) numbers. Aside from its significance in other respects, this theory too is concerned with the *composition* of numbers either by summation or by multiplication. From the twenty-first chapter to the end, finally, the theory of proportions and means is discussed, and with these we return to the realm of the *pros ti*.[27] But this return is accompanied by an express emphasis on the usefulness of this theory for natural science (by which the cosmology of the Platonic *Timaeus* is meant); for music, astronomy, geometry; and, last but not least, for understanding the work of the ancients. Nicomachus here seems to be *justifying* the fact that the study of proportions and means, which belongs to the realm of the *pros ti*, forms the conclusion of an "Introduction to *Arithmetic*" (cf. 199, 19–120, 2 and also 64, 21 ff.).

Theon of Smyrna, for all his arbitrariness in particulars, is more consistent insofar as he is careful to keep the realms of the *poson* strictly separate. He does not, it is true, mention the distinction between the *kath' auto* and the *pros allo*. But the structure of his book is determined by the distinction between "the theory of *mere* numbers" (ἡ περὶ ψίλους ἀριθμοὺς θεωρία — cf. Plato, *Statesman* 299 E), i.e., arithmetic, and "arithmetical music" (ἡ ἐν ἀριθμοῖς μουσική, — Hiller, 17, 12 f.), a distinction which is, as we saw, identified by Nicomachus with the distinction between *kath' auto* and *pros allo* — with which it is indeed in theory identical. For the first, the "arithmetical" part in the narrower sense, treats only of the different kinds of numbers *taken by themselves*, that is, of the even and the odd together with their subspecies, of primes and "composite" numbers, of figurate numbers (plane and solid), and finally of "perfect," "super-abundant," and "deficient" numbers. To be sure, even in this part some incidental reference to the relations between the single numbers and between their kinds cannot be avoided, e.g., in the descriptions of the series of number (22, 19 ff.) and in the relations between the squares (τετράγωνοι) and the oblongs (ἑτερομήκεις — 28 ff.). But the detailed discussion of these relations, i.e., the ten ratios (λόγοι) as well as the proportions (ἀναλογίαι) and means (μεσότητες), is reserved entirely for the second, "musical" part, where it is closely connected with the presentation of the theory of musical intervals, but at the same time also *clearly contrasted with it.* The theory of musical intervals is divided into two parts (46 f.): the first is concerned with the "harmony perceptible by sense and made by instruments" (ἡ ἐν ὀργάνοις αἰσθητὴ ἁρμονία), the second, which really first provides the theoretical foundation of the first, with "the harmony apprehended by thought and made by numbers" (ἡ ἐν ἀριθμοῖς νοητὴ ἁρμονία). The proportions treated by *arithmetic as such* are, in turn, distinguished from the proportions of "arithmetical" harmonics, because the latter are only special cases of the former (cf. 75, 17 ff.). Thus Theon (76–81), citing Adrastus, discusses the ten kinds of ratios "*according to the arithmetical tradition,*" (κατὰ τὴν ἀριθμητικὴν παράδοσιν — 76, 1 ff.) which means precisely: *not only* as the subject of arithmetical harmonics. This discussion, which is prefaced by some general considerations (72–74), occurs after he has only just touched on arithmetical harmonics (74 f.) while, on the other hand, he treats "aisthetic" harmonics in detail. In the same way the proportions and means which may be studied "in numbers as well as otherwise" (ἐν ἀριθμοῖς καὶ ἄλλως — 85, 16) are first (85–93) discussed in connection with the "cutting of the canon" (following Thrasyllus). This discussion of proportions and means is connected quite naturally with the presentation of the "Pythagorean" doctrine of the *tetraktys* or decad (i.e., the sum 1+2+3+4, 95–106, cf. 87, 4 ff.),[28] and recurs finally in a very generalized presentation,[29] again following Adrastus (and Eratosthenes[30]), as a purely mathematical subject (106–119). There is thus in Theon, as well as in Nicomachus, a certain tension between the area of *application* of arithmetical knowledge and the *conception* of arithmetic as a science which should be confined to the study of numbers taken by themselves and not in relation to one another. In Theon's work this tension becomes especially clear in the introductory part, where he speaks about the arrangement and the sequence of the mathematical sciences (16 f.). "Within the natural order" (πρὸς τὴν φυσικὴν τάξιν), he says, "music would come only in the fifth place, namely after arithmetic, geometry, stereometry, and astronomy." (This is the order of sciences sketched out by Plato in the seventh book of the *Republic* with reference to the education of the future rulers of the state.) "Within our study, let arithmetical music follow on arithmetic" (πρὸς τὴν ἡμετέραν θεωρίαν μετ' ἀριθμητικὴν τετάχθω ἡ ἐν ἀριθμοῖς μουσική), since "without music made of number and grasped by thought" (ἄνευ τῆς ἐξαριθμουμένης καὶ νοουμένης μουσικῆς) "cosmic harmony and cosmic music" (ἡ ἐν κόσμῳ ἁρμονία καὶ ἐν τούτῳ μουσική), which alone is of importance, cannot be grasped, while "we do not absolutely need instrumental music" (τῆς . . . ἐν ὀργάνοις [sc. μουσικῆς] οὐ παντάπασι προσδεόμεθα). Thus Theon, basing himself on Plato, *Republic* VII, 530D–531 C, but also "for the sake of the convenience of our study" (πρὸς τὴν τῆς ἡμετέρας θεωρίας εὐμάρειαν[31]), "yokes together" (συνέξευκται) "arithmetical music" and arithmetic, while "cosmic" music is relegated to the fifth place. In fact, Theon returns to cosmic music only at the end of the third, the astronomical part of his work. But, as we have seen, he is far from neglecting the theory of "aisthetic" harmony, for, as he himself observes, "noetic harmony is more easily understood [starting] from sensible harmony" (. . . ἡ δε νοητὴ ῥᾷον ἀπὸ τῆς αἰσθητῆς κατανοεῖται

— 47, 5 f.). On the other hand, arithmetical harmonics is, as we have seen, entirely eclipsed by the purely arithmetical theory of relations. In view of the rigour of Greek scientific methodology, we cannot ascribe all these incongruities simply to the inadequacies of Theon or Nicomachus. In this context it is certainly of crucial importance that the theory of proportions owes its first formulation to this very investigation of musical intervals. When Nicomachus and Theon assign the theory of proportions to musical theory they are following, as Tannery notes at one point,[32] a major tradition, which probably goes back mainly to Archytas. The fact that neither succeeds in maintaining this arrangement of subject matters shows, again, that both have yielded, more or less inconsistently, to the influence of another, classical, tradition which goes back to Euclid and Aristoxenus and which is, as far as Theon is concerned, essentially represented by the Peripatetic Adrastus. Within this tradition the theory of proportions, insofar as it is concerned only with the relations between numbers, has, as the "arithmetical" books of Euclid immediately show, no separate existence, but is passed on "according to the arithmetical tradition" (κατὰ τὴν ἀριθμητικὴν παράδοσιν; cf. Theon 76, 1–3);[33] thus it cannot be naturally isolated from the arithmetical theory of proportions whose very foundation it provides. Therefore the incoherence of Theon's and Nicomachus' presentations has its origin in the fact that in their programmatic scheme they do not *want* to follow the classical tradition, while in its execution they *cannot* follow the "Pythagorean" tradition, though they nevertheless consider it necessary to retain the division of the *poson* into the realms of the *kath' auto* and the *pros allo*. This results in a certain displacement of the theory of proportions. The obvious solution is to assign to it that place which would correspond to *theoretical logistic* in Plato. This much, at least, is clear: *precisely that realm of knowledge is proper to it which the Platonic definition of logistic seems to mark out.* We shall have to discuss the reasons which make it impossible for Theon and Nicomachus to identify the theory of relations with Plato's theoretical logistic. But these reasons notwithstanding, the fact remains that the strange displacement of the theory of relations points to the very quarter whence the Platonic distinction between arithmetic and logistic first derives its meaning.

Finally, we must consider the arithmetical handbook of Domninus[34] (fifth century A.D.) who, in conscious contrast to Nicomachus and Theon, prefers the Euclidean tradition,[35] although he follows, as they do, the Platonic definitions (cf. 427, 17 f., and 429, 2 f.). Its whole internal structure is determined exclusively by two sets of distinctions. One of these is again that of *kath' auto* and *pros allo*, while the other, already known to us from Olympiodorus and the *Gorgias scholium* (see P. 13) distinguishes "according to kind" (κατ' εἶδος) and "according to multitude" (κατὰ τὸ πλῆθος). "Any number can first of all be studied with a view to the *kind* to which it belongs when taken by itself, either the even, or the odd," the even-times-odd or the odd-times-even, etc. — 416, 10 ff. (Τῶν γὰρ ἀριθμῶν ὁστισοῦν, αὐτὸς καθ' ἑαυτὸν θεωρούμενος, κατὰ μὲν τὸ εἶδος ἢ ἄρτιος ἐστὶν ἢ περιττός, κτλ. — 413–415.) But it can also be judged "with a view to the underlying *multitude* of units in it," (κατὰ δὲ τὸ ὑποκείμενον πλῆθος τῶν ἐν αὐτῷ μονάδων . . . — 416, 15 f.) and in this way also it can be "studied by itself" (αὐτὸς καθ' ἑαυτὸν θεωρούμενος). In another passage (415, 7) this multitude of units is said to be "that which underlies and is, as it were, the material of numbers" (ὑποκείμενον καὶ οἷον ὕλη . . . ἐν αὐτοῖς). Here the subject of discussion (415–416) is the decadic system of counting by ones, tens, hundreds, etc., which Domninus, in accordance with Apollonius' procedure, understands in terms of tetrads (the four orders — τέσσαρες τάξεις, i.e., the tenths, hundreds, thousands, ten-thousands of units, myriads, etc.) which are raised to infinity by raising the powers of the myriads. "But it is the part of logistical theory to speak more fully of these." (Ἀλλὰ περὶ μὲν τούτων ἐπιπλέον εἰπεῖν τῆς λογικῆς ἔχεται θεωρίας — 416, 6 f.)[36] After the numbers have been studied *by themselves* both according to their *eidos* and in terms of their *hyle*, "it is necessary to investigate their association with one another" (δεῖ δὲ καὶ τὴν πρὸς ἀλλήλους αὐτῶν ἐπισκέψασθαι κοινωνίαν — 416, 21 ff.). Thus numbers which are prime to one another and numbers which have a common measure, i.e., one or more common divisors (416–417) may be distinguished "according to kind" (κατ' εἶδος). According to material, namely "according to what underlies, that is, according to the multitude of units in them" (κατὰ τὸ ὑποκείμενον, τουτέστι κατὰ τὸ πλῆθος τῶν ἐν αὐτοῖς μονάδων — 417, 10-12), they are either equal, that is to say, they contain the same multitude of units, or they are related by the "ten relations" (δέκα σχέσεις). Here, too, belongs that other approach discovered by the ancients, which divides the numbers into "perfect," "superabundant," and "deficient" (417–422). Furthermore, it is possible to study numbers "by themselves and in relations to one another at the same time" (καθ' ἑαυτούς τε ἅμα καὶ πρὸς ἀλλήλους —422, 6). Two numbers are either prime in themselves and to one another, or composite in themselves but prime to one another, etc. (422). The theory of proportions and means represents then a more comprehensive study of their mutual relations according to their material, a study "concerned with the multitude underlying them" (περὶ τό τε ὑποκείμενον αὐτοῖς πλῆθος — 423, 11), more comprehensive insofar as in proportions always more than two numbers are related to one another (423–425). There is, finally, yet another theory which was introduced by the ancients, namely that which assimilates numbers to geometric figures; here Domninus, citing Euclid and Plato, demands that the (plane and solid) figurate numbers be understood only in terms of multiplication (and not addition as well). This theory, too, admits both the point of view of the *kath' auto* and that of the *pros allo*, the first as long as the different kinds of figurate numbers are treated alone, the second, when their "similar-ity" is considered (425–428). In summarizing, Domninus once again emphasizes the set of distinctions which guides the arrangement of this compressed survey: "What then the kinds of numbers are when they are studied separately, first by themselves and again in relation to one another, and also when they are studied at the same time by themselves and in their relation to one another, as well as when they are studied in the image of their geometric figures, we have sufficiently said." (Τίνα μὲν οὖν ἐστὶ τὰ τῶν ἀριθμῶν εἴδη ἔν τε τῇ καθ᾽ αὐτοὺς καὶ τῇ πρὸς ἀλλήλους θεωρίᾳ, κἂν τῇ καθ᾽ αὐτούς τε ἅμα καὶ πρὸς ἀλλήλους, καὶ δὴ καὶ ἐν τῇ πρὸς τὰ γεωμετρικὰ [Tannery, instead of μέγιστα] σχήματα αὐτῶν ἀπεικασίᾳ, εἴρηται ἀρκούντως ἡμῖν — 428,9 ff.). One cannot fail to see that Domninus is successful in avoiding all the inconsistencies arising from systematiz-ation by assigning the study both of the *kinds* of numbers and of their *relations* to arithmetic from the very beginning. Thus we do indeed have here, as in Euclid (cf. P. 31 and Pp. 43–44), one single subject matter (cf. the expression "one learning matter" — ἓν μάθημα — in Plato, *Laws* 817 E), which can be ordered from two main points of view, namely according to the distinction between *kath' auto* and *pros allo*. At the same time Domninus' second and related set of distinctions affords a better understanding of the com-mentaries of Olympiodorus and the scholiast of the *Gorgias* quoted in Section 2. As we have seen, both connect Platonic "logistic" with the "material" of numbers. Although the very same reasons (see Section 5) which prevent Nico-machus and Theon from identifying the "pure" theory of proportions with theoretical logistic lead both Olympio-dorus and the *Gorgias* scholiast to take it for granted that logistic is to be understood only as a *practical art of calculation* (i.e., as a set of instructions for the computation of the multitudes in question), yet both understand these quantities, independently of whether they are objects of sense or "pure" units, as the "material" of numbers. For only this *material* can in Domninus' sense be subject to those relations which make calculation possible to begin with. Now closer inspection shows that in Domninus the distinction of *kat' eidos* and *kata to plethos* coincides, on the whole, with that of *kath' auto* and *pros allo* as far as their area of application is concerned. For as opposed to the traditional theory of the *kinds* of number, which corresponds to the study "*kath' auto* and *kat' eidos*," the introduction of the theory of decadic counting as a study "*kath' auto* and *kata to plethos*" clearly represents Domninus' own innovation, an innovation which can be understood only as arising from his concern with "systematizing," as the reference to the *logike*, or *logistike*, *theoria* shows (cf. Note 36). On the other hand, the subject matter of the study "*pros allo* and *kat' eidos*," namely the numbers which are or are not prime to one another, is very negligible in comparison with the subject matter of the study "*pros allo and kata to plethos*," which contains all of the traditional theory of relations as well as to that of proportions and means. Furthermore, the third section, where the distinction *kat' eidos* — *kata to plethos* is not even used (as it is neither, incidentally, in the two last sections), contains the same subject matter, strictly speaking, as the study "*pros allo* and *kat' eidos*." The systematization undertaken by Domninus should therefore not obscure the fact that (disregarding the "perfect," "suberabundant," and "deficient" numbers) actually *only* the theory of proportions is subject to the approach *pros allo* and *at the same time* to that *kata to plethos*. If we compare this with the interpretation of Olympiodorus and of the *Gorgias* scholiast, according to which arithmetic treats of the kinds, while logistic considers the material of numbers (although on the whole Olympiodorus makes very meager use of this insight), a striking connection between the traditional theory of relations and the logistic of which Plato speaks appears. With respect to Olympiodorus and the scholiast, the question concerning the connection between the "material" of numbers and the realm of the *aistheta* (cf. P. 16) still remains unanswered. Another question which arises, namely what might be the real point of the distinction *kat' eidos* — *kata to plethos*, is directly connected with the former. Does not this distinction ultimately refer back to the old Platonic opposition of "theoretical" to "practical" studies? We shall have to return to this question in Section 6.

Although we may, therefore, in accordance with our conjecture, assign to theoretical logistic as understood by Plato primarily the study of the relations of numbers,[37] we must nevertheless remember that for Plato, too, logistic is originally related to the possibility of calculating. Thus, in analogy to geometrical usage, it may have contained, besides its own "theorems," "problems" and "porisms" (cf. Proclus, *in Euclid.*, Friedlein, Pp. 77; 178 f.; 201; 212), intended to determine special relations between numbers and serving as the counterparts in the realm of "pure" units of the computational problems proper to practical logistic. (The famous "bloom" — ἐπάνθημα — of Thymaridas mentioned by Iamblichus, Pistelli, Pp. 62 ff., might be cited as an example; according to Tannery and Cantor, Thymaridas lived at least not after Plato.[38]) These considerations, however, lead us into a discussion of the difficulties inherent in the whole conception of theoretical logistic.

5

Theoretical logistic and the problem of fractions

THE QUESTION before us is: What prevents later writers from interpreting the arithmetical theory of relations, i.e., proportions, as theoretical logistic? Or, in other words: How did it happen that the Platonic double distinction of theoretical and practical arithmetic, on the one hand, and theoretical and practical logistic, on the other, was reduced to the single distinction between theoretical arithmetic and practical logistic?

First of all, even supposing that some memory of a possible separate treatment of the arithmetical theory of proportions had remained alive within the Platonic tradition, Plato's *demand* for a logistic to be co-ordinated with, but distinct from, theoretical arithmetic need certainly not have corresponded to classifications current among the mathematicians of his own or of immediately subsequent times. For instance, Plato, as is well known, also demands a "pure" astronomy which ascends from the observed processes of the visible heavens to an understanding of the invisible spheres (*Republic* 529 A–530 C; cf. *Philebus* 62 A); this causes so late an author as precisely Olympiodorus to contrast the "pure" spheric astronomy of Theodosius of Tripolis (or Bithynia)[39] with Ptolemaic astronomy as a science of the visible world (in the *scholia* to the *Philebus*, ed. Stallbaum [1820], P. 280). The same holds also for music, and so in general: Plato postulates, in contrast to the tetrad of "Pythagorean" disciplines, a science similarly organized but completely freed of sense perception, a science whose ultimate object is the one invisible and inaudible "cosmic" order on which our world of sense is founded.[40] To this one order corresponds the close inner relationship of all sciences based on measuring and counting (cf. *Philebus* 55 D ff.; *Republic* 530 D, 531 D, 537 C; and Archytas, Diels I³, 331, 7–9). Thus in the *Theaetetus* (145 A) Theodorus is called *geometrikos* as well as *astronomikos*, *logistikos* and *mousikos*. Thus, criticizing the theory of harmony of the "Pythagoreans," Plato says (*Republic* 531 C) in reproof that though they search for the numbers which are responsible for audible consonances, they fail to attempt to determine without reference to anything audible "which numbers are consonant, which are not, and why" (τίνες ξύμφωνοι ἀριθμοὶ καὶ τίνες οὔ, καὶ διὰ τί ἑκάτεροι). Evidently Plato holds that "consonance" and "dissonance" exist also in the realm of the purely noetic number relations, in other words, in the realm of theoretical logistic — a point of view which later turns out to be the basis of Nicomachus' and Theon's understanding.[41] (Cf. also the terminology of the difficult passage in the *Republic*, 546 C–D, where the number which controls good births is described.) Plato's special demand for a theoretical logistic corresponds to the understanding that within the *unified* framework of the purely noetic sciences there should also be a science addressed to the pure relations of numbers as such, which would correspond to the common art of calculation and provide its foundation. This science, which is "in the service of the search for the noble and the good" (πρὸς τὴν τοῦ καλοῦ τε καὶ ἀγαθοῦ ζήτησιν), inquires into the *presuppositions* of common calculation and also of harmony, and ignores the manner in which these sciences might be pursued in other contexts.

But even if this demand is taken seriously *within* the sciences and if one sets about the construction of a theory of relations of numbers which is intended to stand beside the theory of numbers as such, i.e., of their different kinds, it soon becomes clear that such a division in the presentation of these subject matters is hard to maintain. We were already able to see this on the original "practical" level (P. 20), and we then had an ample view of the difficulties with which the Neoplatonic presentations have to struggle in this connection. A further indication of these difficulties may be seen in the fact that Plato (*Statesman* 259 E) refers the knowledge of the "difference among numbers" (τὴν ἐν τοῖς ἀριθμοῖς διαφοράν) to logistic, although this might as well be said to be the business of arithmetic (cf. 258 D, also *Republic* 587 D).

But the crucial obstacle to theoretical logistic — keeping in mind its connection with calculation — arises from *fractions*, or more exactly, from the fractionalization of the unit of calculation. Logistic is, as we have seen (P. 22), concerned with units which can be grasped only in thought, which are all equal to one another and which, above all, *defy all partition*. Should someone attempt to partition such a unit, so Plato himself says (*Republic* 525 E), all the expert mathematicians "would laugh at him and would not allow it, but whenever you were turning it into small change, they would multiply it, taking care lest the one should ever appear not as one, but as many parts." (καταγελῶσί τε καὶ οὐκ ἀποδέχονται, ἀλλ' ἐὰν σὺ κερματίζῃς αὐτό, ἐκεῖνοι πολλαπλασιασοῦσιν, εὐλαβούμενοι μή ποτε φανῇ τὸ ἓν μὴ ἓν ἀλλὰ πολλὰ μόρια — cf. also *Parmenides* 143 A and *Sophist* 245 A.) The fundamental significance of this fact is sufficiently emphasized by later writers as well. Theon renders the paradox inherent in the partition of the unit as follows: Every number can be diminished by partition and decomposed into smaller numbers, "but the one, when it is divided within the realm of sense, is on the one hand diminished *as a body* and is, once the cutting-up has taken place, divided into parts which are smaller than it, while, on the other hand, *as a number*, it is augmented." (τὸ δὲ ἕν ἂν μὲν ἐν αἰσθητοῖς διαιρῆται, ὡς μὲν σῶμα ἐλαττοῦται καὶ διαιρεῖται εἰς ἐλάττονα αὐτοῦ μόρια τῆς τομῆς γινομένης, ὡς δὲ ἀριθμὸς αὔξεται — Hiller, 18, 18–21.) That is, when an object of sense is partitioned into several pieces, the object is diminished in respect to its bodily extent but is multiplied in the process with respect to its former unity; "For instead of one thing there arise many things." (ἀντὶ γὰρ ἑνὸς γίνεται πολλά.) In other words, the bodily *thing* is partitioned, but not that which we call "one" in it, namely the *unit* itself, whose being consists in nothing but *being one*, in being the "unit in respect to which each of the things that are is called one" (... μονάς, καθ᾽ ἣν ἕκαστον τῶν ὄντων ἓν λέγεται — Euclid VII, Def. 1; cf. Theon 20, 17 f.). After partitioning, every part assumes the property of being one, i.e., there now exist several "units" — the one has multiplied itself! In every calculation or counting process we work from the outset with many units, and if in the course of the calculation we are forced to partition one of these units, then what we do is precisely to substitute something else for the indivisible unit which is subject to partitioning, while the unit itself *is not partitioned but only further multiplied*. This process can be repeated over and over (Theon 18, 9–14) without touching the impartible and indivisible nature of the unit or units: "The one as one is impartible and indivisible" (ἀμέριστον καὶ ἀδιαίρετον τὸ ἓν ὡς ἕν — 18, 15), "and even if we multiply the unit to infinity it [and therefore every single unit] yet remains — a unit"[42] (καὶ μέχρις ἀπείρου ἐὰν πολλαπλασιάζωμεν τὴν μονάδα, μένει μονάς — 19, 10 f.). Hence arises the necessity of making a strict distinction between the *one object of sense* which is subject to counting and calculation and the *one as such*, i.e., of keeping each "one" thing strictly separate from all "ones." Each single thing can be infinitely partitioned because of its bodily nature as an object of sense. The unit which can only be grasped in thought is, on the other hand, indivisible simply, precisely in virtue of its purely noetic character: "So that the unit, since it is noetic, is indivisible while the one, inasmuch as it is sensible, can be cut infinitely." (ὥστε ἡ μὲν μονὰς νοητὴ οὖσα ἀδιαίρετος, τὸ δὲ ἐν ὡς αἰσθητὸν εἰς ἄπειρον τμητόν — 20, 2–4.) This property of the noetic unit clearly precludes calculation with it. Thus Proclus says about the man who calculates (immediately following quotation no. 1 in Section 2, P. 11; Friedlein, 40, 5 ff.): "The logistician does not admit that there is a least [simply] as the arithmetician does; however, he takes the least in relation to some class [of objects of sense]. For the *one man* becomes for him the measure of the multitude as the unit is for numbers [in arithmetic].[43] (καὶ ἐλάχιστον μὲν οὐδὲν εἶναι συγχωρεῖ [sc. ὁ λογιστικός] καθάπερ ὁ ἀριθμητικός, ὡς μέντοι πρός τι γένος λαμβάνει τὸ ἐλάχιστον. ὁ γὰρ εἷς ἄνθρωπος μέτρον αὐτῷ γίνεται τοῦ πλήθους ὡς μονάς.) And the sixth paragraph of the fragment ascribed to Geminus or Anatolius (Heron IV, 98, 24–100, 3 — this fragment is a direct continuation of the discussion of logistic in the *Charmides scholium*, cf. Pp. 12 f.) contains an even more precise statement: "Since in the *material* realm the one [single thing] is the least [element for each particular case — not in general], just as the unit is in arithmetic, [logistic] uses the one [single definite thing] as the least [element] of things of the same genus comprehended under one and the same multitude. Thus it posits the [single] man as indivisible in a multitude of men, though not once and for all [for only the unit itself is thus indivisible]; and in [a multitude of] drachmae it posits the [single] drachma as not to be cut up [on the level of drachmae], although as a coin it *is* divisible." (ἐπεὶ δὲ τὸ ἕν [instead of μέν Heiberg] ἐστιν ἐν τῇ ὕλῃ ἐλάχιστον, ὁποῖον ἐν τῇ ἀριθμητικῇ ἡ μονάς, προσχρῆται τῷ ἑνὶ ὡς ἐλαχίστῳ τῶν ὑπὸ τὸ αὐτὸ πλῆθος ὁμογενῶν. ἕνα γοῦν τίθεται ἄνθρωπον ἐν πλήθει ἀνθρώπων ἀδιαίρετον, ἀλλ' οὐχ ἅπαξ, καὶ μίαν δραχμὴν ἐν δραχμαῖς ἄτομον, εἰ καὶ ὡς νόμισμα διαιρεῖται.) The unit of calculation is indivisible only in its function as "substrate." If some calculation is performed on a multitude of apples, one apple is that "least" element which is not

further reducible since all computational operations are
referred to it as to their ultimate basis. The series of *whole*
apples *one* by *one* here furnishes that "homogeneous"
medium in which counting and calculating takes place. Now
if it proves to be necessary to operate with fractional parts of
apples, the basis of the calculation, namely the one whole
apple, still remains untouched. For once an apple has been
divided into fractional parts, it loses its significance as that
which underlies the counting and is the fundamental element
of all calculation, and is instead understood as a *mere thing*
such as has the property of unlimited divisibility. As the
fundamental element of calculation the function of *one*
apple is precisely to represent the unit "within the material"
(ἐν τῇ ὕλῃ). "Fractions" are never anything but fractional
parts of the thing as such which underlies the counting and
which can, by reason of its *bodily nature*, be infinitely divided.
In the realm of "pure" numbers, on the other hand, the *unit
itself* provides the last limit of all possible partitions: all
partitioning "will stop at one" (καταλήξει εἰς ἕν — Theon 18,
11 and 13 f.).[44] Thus Proclus (*in Timaeum*, Diehl, II, p. 138,
23) says: "Since each number is with respect to its own *kind*
one and without parts but with respect to its own *material*, as
it were, [cf. Domninus 415, 7; see also P. 32] divisible into
parts, though not with respect to all of the material either;
but rather what is ultimate [in it, i.e., the unit] is without
parts even in the material, and in this ultimate thing [count-
ing or calculation and, above all, partitioning] comes to a
stop." (... ἐπεὶ καὶ ἕκαστος ἀριθμὸς κατὰ τὸ οἰκεῖον εἶδος εἷς
ἐστι καὶ ἀμερής, κατὰ δὲ τὴν οἷον ὕλην μεριστός, καὶ οὐδὲ ταύτην
πᾶσαν, ἀλλὰ τὸ ἔσχατον καὶ ἐν ταύτῃ ἀμερές, εἰς ὃ καὶ ἡ κατάληξις
— cf. also Aristotle, *Physics* Γ 7, 207 b 1 f. and Γ 6, 206 b
30–32.)

Now if one considers that in the majority of cases the
necessity for introducing fractional parts of the unit of
calculation does arise in the course of calculation, there
emerges a remarkable maladjustment between the material

within which such calculations are performed and that other "material" of "pure" numbers whose noetic character is expressed precisely in the indivisibility of the units.[45] Thus a calculation which is intended to be as exact as possible simply cannot be effected within the realm of these pure numbers. The immediate consequence of this insight, at least within the Platonic tradition, is the exclusion of all computational problems from the realm of the "pure" sciences.[46] But beyond this difficulty arises the question — and this is the crucial question to be asked about theoretical logistic — whether that which distinguishes *exact* calculation, namely operation with fractional parts of the unit of calculation, can really be sufficiently grounded in the science of the possible relations of numbers, i.e., in the "pure" theory of relations, alone.

We must first of all examine whether this is accomplished in the "arithmetical" books of Euclid. Keeping in mind that the tenth book, whose subject is "incommensurable" magnitudes, goes back to Theaetetus (and Theodorus) and that this book, in turn, presupposes Books IX, VIII and VII, we may conjecture with Zeuthen[47] that these latter books also essentially depend on the works of Theaetetus. Theaetetus, however, belongs to the Platonic circle. The thirteenth book of Euclid, which presents the construction of the so-called "Platonic solids" (i.e., of the five regular polyhedra) and thus furnishes a "Platonic" conclusion to the *Elements*,[48] is in all essentials *his* work.[49] From this point of view it is understandable that Books VII, VIII, and IX consistently avoid the introduction of fractional parts of the *unit* of calculation, while they certainly use the notion of the part or parts of a *number*, as previously defined (VII, Defs. 3 and 4; cf. especially VII, 37 and 38). It is true that in these books "arithmetic" and "logistic" matter can hardly be separated, but the "logistic" constituent undoubtedly predominates and is here understood precisely as "arithmetic"; it is obviously this fact which permits the later "arithmetical

tradition" (ἀριθμητικὴ παράδοσις) to include the theory of
relations as well. From the point of view of the "systematic"
context of the *Elements*, Books VII–IX attain their real
significance only with reference to the theory of incommensurables of the tenth book;[50] their content itself, however, amounts to a first presentation of the foundations for
any calculation which goes beyond simple addition and
subtraction, insofar as in such operations it becomes necessary
to decompose single numbers into their components
(factors), to find the greatest common measure (i.e., divisor)
of several numbers, to express their ratios in the "least"
terms, etc. Yet it is *not* possible in this way to find a foundation (consonant with Platonic presuppositions) for the
partitioning of the units themselves, although this is in most
cases quite unavoidable in the course of an exact calculation.
Hence the theory of relations of the "arithmetical" books of
Euclid cannot be understood as the noetic analogue of
practical logistic. "Calculation," with all its presuppositions,
must therefore be referred entirely to the realm of the practical arts and sciences, while the "pure" theory of relations
loses its fixed place and comes to be assigned now to arithmetic as the theory of the kinds of numbers, now to
harmonics as the theory of musical intervals based on ratios
of numbers.[51] Furthermore its distinctness is called in
question from the point of view of a totally different context
of problems: the discovery of incommensurable "irrational"
magnitudes causes the "pure" theory of number relations to
appear as nothing but a special case of the general theory of
relations and proportions of the fifth book of Euclid.[52] Thus
it loses all connection with the art of calculation, which is, in
turn, forced to give up any claim to the title of a strict
apodeictic discipline. Logistic is reduced to giving instructions for speedy and convenient addition, subtraction,
multiplication, and division not only as performed "in
writing" but particularly as accomplished "mechanically"
with the aid of the fingers and the abacus; furthermore it

teaches calculation with fractions and probably also extraction of (square) roots with the aid of geometry, develops the system of counting for large numbers, especially with a view to astronomical computations, and, finally, solves problems presented in words such as have come down to us in the so-called arithmetical epigrams of the *Anthologia Palatina*.[53] Thus logistic comprises approximately the subject matter of present-day elementary arithmetic. But we should not forget that this position of logistic is founded on that special conception of the "pure" numbers and their material which governed the Platonic tradition throughout. In order to understand this conception in its full significance and from its foundations we must at last begin to clarify the concept of the *arithmos* itself. Only then can we hope to make the meaning of those Platonic definitions of arithmetic and logistic which have formed the basis of the preceding discussion more clearly comprehensible.

6

The concept of arithmos

THE FUNDAMENTAL phenomenon which we should never lose sight of in determining the meaning of *arithmos* (ἀριθμός) is counting, or more exactly, the *counting-off*, of some number of things. These things, however different they may be, are taken as uniform when counted; they are, for example, either *apples*, or apples and pears which are counted as *fruit*, or apples, pears, and plates which are counted as "*objects*." *Insofar* as these things underlie the counting process they are understood as *of the same kind*. That word which is pronounced last in counting off or numbering, gives the "*counting-number*," the *arithmos* of the things involved. Thus Plato says (*Theaetetus* 198 C): "Socr.: We will then posit counting as nothing else than looking things over to see how great a number happens to be [in a given case]. Theaetet.: Just so." (*ΣΩ. Τὸ δὲ ἀριθμεῖν γε οὐκ ἄλλο τι θήσομεν τοῦ σκοπεῖσθαι πόσος τις ἀριθμὸς τυγχάνει ὤν. ΘΕΑΙ. Οὕτως.*) Thus the *arithmos* indicates in each case a *definite number of definite things*. It proclaims that there are precisely so and so many of these things. It intends the *things* insofar as they are present in this number, and cannot, at least at first, be separated from the things at all. When Plato speaks of numbers (*Republic* 525 D) which have

46

"visible and tangible bodies" (cf. P. 23), this expression must be taken quite literally. For instance, in counting dogs, horses, and sheep, these processes of counting yield as results a definite horse-, dog,- or sheep-number (cf. the expression "apple"- and "bowl"-number in Proclus and the *Charmides* scholium Pp. 11 f.).[54] Aristotle, who here, as always, refers to that which is really meant in speech, makes himself as plain as could be desired. Speaking of "equality of number," he says: "It is also quite rightly said that the *number* of sheep and dogs is the *same* if each is *equal* to the other, but the '*decad*' is *not the same* [in these cases], nor are the ten [sheep and dogs] the same [ten things], just as the equilateral and the scalene triangle are not the same triangles, although their *figure* is indeed the same, since both are '*triangles.*' For that is called *the same* [as some particular thing] which does not differ [from this thing] by a difference [of kind] but not as that from which it does differ [in this way], as for instance one of the aforementioned triangles differs from the other by a difference [of kind]; for they are indeed different triangles, though they do not differ in *figure*, but [in this respect] they belong to one and the same division [namely to one and the same subspecies of the genus '*figure*']. For *such and such* a figure is a circle and *such and such a one* a triangle, while of the latter *such a one* is equilateral and *such a one* scalene. The latter of these [the scalene one] is then indeed *the same figure* [as the equilateral triangle] — for it is a triangle — but *not the same triangle*. And in *just this way number is the same*: for the [horse-]number and the [dog-]number do not differ by a difference [in number, for that is in both cases 'ten']; but the '*decad*' *is not the same*, for it differs in that *of which it is said to be* [a decad], for this is in the former case *horses* and in the latter *dogs*." (λέγεται δ'ὀρθῶς καὶ ὅτι ἀριθμὸς μὲν ὁ αὐτὸς ὁ τῶν προβάτων καὶ τῶν κυνῶν, εἰ ἴσος ἑκάτερος, δεκὰς δὲ οὐχ ἡ αὐτὴ οὐδὲ δέκα ταὐτά, ὥσπερ οὐδὲ τρίγωνα τὰ αὐτὰ τὸ ἰσόπλευρον καὶ τὸ σκαληνές. καίτοι σχῆμά γε ταὐτό, ὅτι τρίγωνα ἄμφω· ταὐτὸ γὰρ λέγεται οὗ μὴ διαφέρει διαφορᾷ, ἀλλ' οὐχὶ οὗ διαφέρει, οἷον τρίγωνον τριγώνου διαφορᾷ διαφέρει· τοιγαροῦν ἕτερα τρίγωνα· σχήματος δὲ οὔ, ἀλλ' ἐν τῇ αὐτῇ διαιρέσει καὶ μιᾷ. σχῆμα γὰρ τὸ μὲν τοιόνδε κύκλος, τὸ δὲ τοιόνδε τρίγωνον, τούτου δὲ τὸ μὲν τοιόνδε ἰσόπλευρον, τὸ δὲ τοιόνδε σκαληνές. σχῆμα μὲν οὖν τὸ αὐτὸ καὶ τοῦτο (τρίγωνον γάρ), τρίγωνον δ'οὐ τὸ αὐτό. καὶ ὁ ἀριθμὸς δὴ ὁ αὐτός· οὐ γὰρ διαφέρει ἀριθμοῦ διαφορᾷ ὁ ἀριθμὸς αὐτῶν· δεκὰς δ'οὐχ ἡ αὐτή· ἐφ' ὧν γὰρ λέγεται, διαφέρει· τὰ μὲν γὰρ κύνες, τὰ δ'ἵπποι. — *Physics* Δ 14, 224 a 2 ff.) Here, then, "decad" is taken as a parallel to "triangle": just as there exists no triangle which is *neither* equilateral *nor* scalene, *so there can be no decad which is not this or that ten definite things.* A "triangle" is always a *definite* triangle, namely an isosceles or an equilateral or a scalene triangle. A "decad" is always a *definite number of definite things*, of apples, of dogs, of cattle and — in the limiting case — of pure units, accessible only to thought; great as the difference between this limiting case and all the others is, the character of the *arithmos* as a "definite number of . . ." *is preserved even there.* "For every number is [*a counting-number*] of something,"[55] *(πᾶς γὰρ ἀριθμός τινός ἐστι)*, says Alexander (in his commentary on the *Metaphysics*, Hayduck, 86, 5); and *this* is why the *arithmos* can be included in the category of the *pros ti.* But this means that a number is always and indissolubly related to that of which it is the number.

This fact cannot be adequately described by speaking of "concrete" or "specified" as opposed to "abstract" or "unspecified" numbers. For even a "pure" number, i.e., a number of "pure" units, is no less "concrete" or "specified" than a number of apples.What distinguishes such a number is in both cases its twofold determinateness: it is, first of all, a number of objects *determined in such and such a way*, and it is, secondly, an indication that there are *just so and so many* of these objects. Nor can it be said of such a number that it is "the comprehensive unity of a multiplicity." This description may hold for the *concept* of such numbers in general, but it is not what is meant by the numbers falling under that concept. In the process of counting, in the *actus exercitus* (to use scholastic terminology), it is only *the multiplicity of the counted things* which is the object of attention. Only that can be "counted" which is *not one*, which is before us in a certain number: neither an object of sense nor *one* "pure" unit is a *number* of things or units. The "unit" as such is no *arithmos*, a fact which seems strange only if we presuppose the notion of the "natural number series." The smallest number of things or units is: *two* things or units (cf. Aristotle, *Physics* Δ 12, 220 a 27; *Metaphysics* I 6, 1056 b 25 ff; M 9, 1085 b 10). The unit itself is, of course, still smaller than the smallest number. It is just because this is the case that it has the character of a "beginning" or "source" (ἀρχή) such as makes something of the nature of "counting" originally possible. (Cf. Pp. 40 f. and 53 f.). But this very fact — that a multitude, a *number* of objects can be grasped as *one* number, that the many can be "*one*" — was first treated explicitly as a fundamental *problem* within Pythagorean and Platonic philosophy. This problem carries us far beyond the theory of numbers, although it always remains tied in with it.

Before we go on, we must first attempt to understand how the conception of "pure" numbers, as opposed to the "visible or tangible" numbers, arises out of the natural phenomenon of counting. Plato's own indications can here serve as our guideposts. The continual practice of counting and calculation gradually fosters within us that familiarity with numbers and their relations which Plato terms "arithmetic and logistic art" (ἀριθμητική and λογιστική [τέχνη]) and which enables us to execute any operation of counting or calculating we wish. But those numbers which we have at our disposal *before* we begin counting or calculating and which must clearly be independent of the particular things which happen to undergo counting — *of what* are these the numbers? To pose this question means to begin to raise the problem of "scientific" arithmetic or logistic. In this context we are no longer interested in the requirements of daily life, namely those operations of counting and calculation whose objects are various and impermanent and yield ever-changing results; now our concern is rather with understanding the very possibility of this activity, with understanding the meaning of the fact that *knowing* is involved and that there must therefore be a corresponding *being* which possesses that *permanence* of condition which first makes it capable of being "known". But the soul's turning away from the things of daily life, the change in the direction of its sight, the "conversion" (περιαγωγή) and "turning-about" (μεταστροφή — *Republic* 518 D) which is implicit in this new way of posing questions leads to a further question concerning the special nature of the object of arithmetic and logistic as that which alone of all things is in the strict sense knowable, being in fact always to some degree already known. What is required is an object which has a purely noetic character and which exhibits at the same time all the essential characteristics of the *countable* as such. This requirement is exactly fulfilled by the "pure" units, which are "nonsensual," accessible only to the understanding, indistinguishable from one another, and resistant to all partition (cf. Pp. 23 ff. and 39 ff., also P. 53). The "scientific" arithmetician and logistician deals with *numbers of pure monads*. And, as we have seen, Plato stresses emphatically that there is "no mean difference" between these and the ordinary numbers. It would indeed seem strange if there should be numbers of — nothing. How will he who is involved in "natural" counting and calculating become aware of these monads which are in no way to be touched or seen? Only a careful consideration of the fact that it is really necessary to suppose that there are numbers different from the ordinary kind, if the possibility and the successful execution of counts and calculations are to be understood, forces us into the further supposition that there must indeed by a special "nonsensual" material to which these numbers refer. The immense propaedeutic importance which this understanding must have within Platonic doctrine is immediately clear, for is not a continual effort made in this doctrine to exhibit as the true object of knowing that which is *not* accessible to the senses? Here we have indeed a "learning matter" *(μάθημα)* which can be described as "capable of hauling [us] toward being" *(ὁλκὸν ἐπὶ τὴν οὐσίαν — Republic* 524 E; 523 A). It forces the soul to study, by thought alone, the truth as it shows itself by itself (526 A–B). Everyone is able to see — if only it has been emphatically enough pointed out to him — that his ability to count and to calculate *pre-supposes* the existence of "nonsensual" units.

Thus an *unlimited* field of "pure" units presents itself to the view of the "scientific" arithmetician and logistician (cf. Aristotle, *Posterior Analytics* A 10, 76 b 4 f.). The particular multitudes which may be chosen from this field are precisely those "pure" numbers (of units) with which he deals. This is how the traditional "classical" definitions of *arithmos* are to be understood; Eudoxus (Iamblichus, *in Nicom.* 10, 17 f.): "A number is a *finite* multitude [of units]" *(ἀριθμός ἐστιν πλῆθος ὡρισμένον)* — cf. Aristotle, *Metaphysics* Δ 13, 1020 a 13: "limited multitude" *(πλῆθος πεπερασμένον)*; Euclid (VII, Def. 2): "the multitude composed *of units*" *(τὸ ἐκ μονάδων συγκείμενον πλῆθος)* — cf. Aristotle, *Metaphysics* I 1, 1053 a 30: *πλῆθος μονάδων*; furthermore: "a set [composed] of units" *(μονάδων σύστημα* — Theon 18, 3; Nicomachus 13, 7 f.; Iamblichus 10, 9; Domninus 413, 5; *σύνθεσις μονάδων* — Aristotle, *Metaphysics* Z 13, 1039 a 12); "an aggregate in the realm of quantity composed of monads" *(ποσότητος χύμα ἐκ μονάδων συγκείμενον* — Nicomachus 13, 8). Numbers are, in short, many units: "for each number is many because it[consists of many] ones"[56] *(πολλὰ γὰρ ἕκαστος ὁ ἀριθμὸς ὅτι ἕνα . . .)*, i.e., because it represents nothing but several or many units (Aristotle, *Metaphysics* I 6, 1056 b 23 — cf. *Physics* Γ 7, 207 b 7).[57] "Multiplicity [manyness] is, as it were, the genus of number" *(τὸ δὲ πλῆθος οἷον γένος ἐστὶ τοῦ ἀριθμοῦ* — ibid., 1057 a 2 f.; cf. Iamblichus 10, 18 f.). Because of this, "number" and "one" are opposites (Aristotle, *Metaphysics* I 6, 1056 b 19 f.), although it is possible to speak of the one metaphorically as being "a certain, although a small, multitude" (πλῆθός τι, εἴπερ καὶ ὀλίγον — *ibid.*, 1056 b 13 f.), namely the multitude "one." Cf. the definition of Chrysippus in Iamblichus, *in Nicom.*, 11, 8 f. and Syrianus, *in Arist. metaph.*, Kroll, 140, 9 f.: "the unit is the multitude 'one'" (μονάς ἐστι πλῆθος ἕν); also *Theaetetus* 185 C–D. On this possibility are founded the definitions of the series of numbers: "a progression of multitude beginning from the unit and a recession ceasing with the unit" (προποδισμὸς πλήθους ἀπὸ μονάδος ἀρχόμενος καὶ ἀναποδισμὸς εἰς μονάδα καταλήγων — Theon 18, 3 ff.; cf. Iamblichus, *in Nicom.* 10, 16 f.; this definition may go back to Moderatus, first century A.D., cf. Theon 18, note). Thus also Domninus (413, 5 ff.): "The whole realm of number is a progress from the unit to the infinite by means of the excess of one unit [of each successive number over the preceding]." (ὁ δὲ σύμπας ἀριθμός ἐστι προκοπὴ ἀπὸ μονάδος κατὰ μονάδος ὑπεροχὴν ἄχρις ἀπείρου.) The series of numbers may be understood as the result of a successive reproduction of the unit: "It is in consequence of the monad that all the successive numbers beginning with the dyad form aggregates and produce the well-ordered kinds of that which is multiple in accordance with the proper sequence [of the numbers]." (πρὸς τὴν μονάδα πάντες οἱ ἐφεξῆς ἀριθμοὶ ἀπὸ δυάδος ἀρξάμενοι συγκρινόμενοι τὰ τοῦ πολλαπλασίου εὔτακτα εἴδη ἀπογεννῶσι τῇ οἰκείᾳ ἀκολουθίᾳ — Nicomachus 46, 13 ff.) The truth is that the unit can be spoken of as a "multitude" only improperly, "confusedly" (συγκεχυμένως — Iamblichus 11, 7). The unit is rather that permanently same and irreducible basic element which is met with in all counting — and thus in every number. To determine a number means to count off in sequence the given single units, be they single objects of sense, single events within the soul, or single "pure" units. What is countable must, *insofar* as it is countable, be articulated in such a way that the units in question are similar to one another (cf. P. 46) and yet separated and clearly "determined" *(διωρισμένα)*. This means that the single units possess similarity and perfect wholeness insofar as they are units of *counting*. These characteristics guarantee their internal indivisibility and their external *discreteness* — the essential marks of the field of "pure" units. The "discreteness" of numbers is based solely on the discreteness of the units, namely on the fact that the single units which are "parts" of numbers do not, in contrast to the sectional parts of continuous magnitudes, have a "common terminus" *(κοινὸν ὅρον* — cf. Aristotle, *Categories* 4, 4 b 25; also *Physics* E 3 and Z 1). It is precisely this discreteness which makes something like a "count" and a "number" possible: as a "number of..." every number presupposes definite discrete units. Such discrete units, however, form the "homogeneous" medium of counting only if each unit, whatever its nature, is viewed as an *indivisible whole* (cf. P. 42). In this sense a number is *always* "a multitude of indivisibles" *(πλῆθος ἀδιαιρέτων* — Aristotle, *Metaphysics* M 9, 1085 b 22). And the indivisible unit itself is always the last, the basic, element of all counting and all number (cf. P. 39). That is why Aristotle can say (*Metaphysics* I 1, 1052 b 22 ff.): "Every quantity is recognized as quantity through the one, and that by which quantities are primarily known [as quantities] is the one itself; therefore the one is the source of number as number." *(πᾶν τὸ ποσὸν γιγνώσκεται ᾗ ποσὸν τῷ ἑνί, καὶ ᾧ πρώτῳ ποσὰ γιγνώσκεται, τοῦτο αὐτὸ ἕν· διὸ τὸ ἓν ἀριθμοῦ ἀρχὴ ᾗ ἀριθμός.)* The one is the source of number as that which gives each number its character as a "number of...," thereby rendering it a "number." Now the possibility of recognizing a number of units as such presupposes a knowledge of the unit. In this sense the unit has priority of "intelligibility."[58] But this priority is only an expression of the fact that the possibility of the being of number is based on the being of the unit: "More knowable than the number is the unit; for it is prior and the source of every number." *(γνωριμώτερον . . . μονὰς*

ἀριθμοῦ· πρότερον γὰρ καὶ ἀρχὴ παντὸς ἀριθμοῦ — *Topics* Z 4, 141
b 5 ff.) Iamblichus speaks in the same vein (11, 1 f.): "The
monad is the least [element] of quantity; or the primary and
common part of quantity; or the source of quantity." (μονὰς
δέ ἐστι ποσοῦ τὸ ἐλάχιστον ἢ ποσοῦ τὸ πρῶτον καὶ κοινὸν μέρος ἢ
ἀρχὴ ποσοῦ — similarly Syrianus, *in Arist. metaph.*, Kroll,
140, 6 f.) The possibility of extending the count over ever
more units is unlimited, but in the decreasing direction there
is at last a barrier; here "it is necessary to stop on reaching
the indivisible" (ἀνάγκη στῆναι ἐπὶ τὸ ἀδιαίρετον — Aristotle,
Physics Γ 7, 207 b 8), i.e., the indivisible unit (cf. P. 42). All
these characterizations stem from one and the same point of
view, one grounded in the phenomenon of counting; they
precede all the possible differences of opinion regarding the
mode of being of the "pure" number units themselves or the
nature of the priority of the unit over number. These matters
are now to be taken up.

From what point of view does the man who knows numbers study the unlimited field of "pure" units? This question
brings us back to the definitions of arithmetic and logistic
with which we began (see P. 17). We are now better able to
understand why these definitions *fail* to name number as the
object of either of these sciences. The unlimitedness and the
homogeneity of the field of monads permit us to combine
units into assemblages of monads, i.e., into numbers of units,
in whatever way and as often as we please. Accordingly, from
the "*mathematical*" point of view there are *infinitely many*
such numbers — as many as there are units. Thus the first
task of "scientific" arithmetic — as contrasted with that
"practical" knowledge which is satisfied with "knowing"
these numbers without understanding what this "knowing"
implies — consists in finding such arrangements and orders
of the assemblages of monads as will completely comprehend their variety under well-defined properties, so that
their unlimited multiplicity may at last be brought within
bounds (cf. Nicomachus I, 2). We must find within the field

of "pure" units itself those properties which will permit us to collect the different assemblages of monads under some few aspects, so as to obtain a complete synopsis of all possible multitudes, all possible numbers, all possible multiplicity. When we recall how Plato (*Theaetetus* 147 C ff.) makes Theaetetus, speaking from a very advanced stage of scientific geometry *and* arithmetic, describe his procedure in studying lines and numbers, we are led to wonder what it is in this procedure that appears to Plato so exemplary for Socrates' present inquiry concerning "knowledge" (ἐπιστήμη), and indeed for every Socratic inquiry of this kind.[59] Theaetetus begins by showing that some square magnitudes to which certain numbers (of units of measurement) correspond have sides that, by themselves, cannot be measured by numbers (of those units) and are therefore called *dynameis*.[60] Since "such roots [and also the corresponding square numbers] appear to be unlimited in multitude" (ἐπειδὴ ἄπειροι τὸ πλῆθος αἱ δυνάμεις ἐφαίνοντο), he tries to "gather them into one" (λαβεῖν εἰς ἕν), so that he may designate them "all" (πάσας) properly. For this purpose he divides "the whole realm of number" (τὸν ἀριθμὸν πάντα) into two domains: to one of these belong all those numbers which may arise from a number when it is multiplied by itself (i.e., when it is taken as many times as the number of its monads indicates), to the other, all those which may arise from the multiplication of one number with another. The first number domain he calls "square," the second "promecic" or "heteromecic" (oblong), designations which recur in all later arithmetical presentations (cf. Diogenes Laertius III, 24). Thus two *eide* are indeed given which allow us to articulate and delimit a realm of numbers previously incomprehensible because unlimited, especially if we substitute the various *eide* of polygonal numbers for the one *eidos* of oblong numbers (προμήκεις). With the aid of the *gnomon* it can be shown immediately that the *eidos* for all similar numbers is in each case one and the same. Indeed operations with a gnomon, that is, with a figure or point configuration which, when fitted about a formation of a certain kind, produces a formation "similar" in form, i.e., *like in "looks"* or *kind* (cf. Heron IV, 44, Def. 58; Theon 37, 11–13; Iamblichus 58, 19 ff.), do generally make sense only when the aim of the investigation is the discovery of *kinds* of figures and numbers. Greek theoretical arithmetic does, in fact, deal first and last with different *kinds* of numbers. Were it otherwise, how would it be possible to come to terms with the limitlessness of the material with which arithmetic is confronted? Therefore in theoretical arithmetic the numbers themselves are *not studied directly* — the monads themselves as they occur in some given number are not its object, but rather it attempts to comprehend all possible groupings of monads in general under arrangements which are determinate, i.e., which possess unambiguous characteristics and which may, in turn, be reduced to their own ultimate elements, such as the "same," and the "other," the "equal" and the "unequal," the "limit" and the "limitless" (ταὐτόν — ἔτερον, ἴσον — ἄνισον, πέρας — ἄπειρον, cf. *Philebus* 25 A–B). Only through membership in an *eidos* "derivable" from such "sources" (ἀρχαί) does the being of a number become intelligible as determinate, i.e., as *delimited*, number, as *one* assemblage of just so and so many monads — whatever the mode of being of the *eide* themselves may be. While the determination of each number as a "number of something" is given by the pure units or the given objects, the other aspect of the determinateness of a number, namely the fact that it is always a *definite* number (of pure units or of things of some sort), can be understood only as the consequence of the special *kind* to which it belongs, i.e., by means of something *which is in itself one* and is thus capable of unifying, of making wholes — of delimiting. Precisely because *the arithmos as such is not one but many, its delimitation in particular cases can be understood only by finding the eidos which delimits its multiplicity, in other words, by means of* arithmetike *as a theoretical discipline.*

Now the most comprehensive *eide*, those which come closest to the rank of an *arche* and are therefore termed "the very first" (πρώτιστα — cf. P. 26), are the *odd* and the *even*. This "first cut" (πρώτη τομή — Theon 21, 20; Nicomachus 13, 9) in the realm of numbers in terms of "even" and "odd" (cf. Plato, *Statesman* 262 E: "...if someone should divide number in terms of *even and odd*" ... — εἰ τὸν ... ἀριθμὸν ἀρτίῳ καὶ περιττῷ τις τέμνοι ...) affects all numbers in such a way that "one whole half of the realm of number" (ὁ ἥμισυς τοῦ ἀριθμοῦ ἅπας — Plato, *Phaedo* 104 A) falls under the "odd" and the other under the "even," each of these halves nevertheless comprising an *unlimited* multitude of numbers. But each of these unlimited multitudes is now in turn gathered "into one" (εἰς ἕν) by means of certain unambiguous characteristics: all numbers which can be divided into two equal parts, i.e., which can be divided without remainder, are "even" (cf. Plato, *Laws* X, 895 E), while all numbers which cannot be thus divided, i.e., whose division yields a remainder of one indivisible unit, are "odd." This latter property of "being odd" can obviously occur only in a field of discrete and indivisible units, since it always depends on a single, "supernumerary" unit, indivisible by "nature."[61] The property of being divisible into two equal parts without "remainder" is, on the contrary, a property common to numbers and the continuous, i.e., infinitely divisible, magnitudes which are always capable of being further bisected[62] (cf. Plato, *Laws* 895 E: "number also can, I suppose, like other things be divided into two parts" — ἔστιν που δίχα διαιρούμενον ἐν ἄλλοις τε καὶ ἐν ἀριθμῷ ...). Consequently *only* "oddness" is characteristic of that which is countable as such, while "evenness" represents something within the realm of numbers, and of everything countable, which goes, as it were, beyond it — something "other," namely the possibility of unlimited divisibility and thus, in a way, the "unlimited" itself.[63] From this point of view it might even seem necessary to see the "odd" and "even" as *eide* which are mutually completely exclusive. Thus Philolaus (fragment no. 5, Diels, I³, 310, 11–14) speaks of the "even" in the sense of "even-times-even" (ἀρτιάκις ἄρτιον, i.e., in modern terminology, in the sense of powers of two — similarly Theon 25, 6 ff., Nicomachus 15, 4 ff., and Domninus 414, 4 ff., but *not* Euclid VII, Def. 8, cf. Euclid IX, 32–34), and of the "odd" in the sense of "odd-times-odd" (περισσάκις περισσόν, which includes the prime numbers with the exception of two). These two *eide* are accompanied by the *eidos* "mixed" from both, namely the "even-times-odd" (ἀρτιοπέριττον). Later on we shall have to examine the relation of this classification to the cosmological doctrine of the *Timaeus* and the *Philebus*. For the present we need emphasize only the designation of the *perisson* and the *artion* as the "two kinds proper" (δύο ἴδια εἴδη) of *arithmos*. In spite of the priority of the odd over the even (cf. *Theol. arithm.*, De Falco 83, 13 = Diels, I³, 304, 3 : "the odd number is always prior to the even" — . . . πρότερος ἀεί ἐστιν ὁ περισσὸς [sc. ἀριθμός] τοῦ ἀρτίου), which is based on the fact that the odd imposes a limit on unlimited divisibility in the form of an indivisible unit, the even still appears alongside the odd as a second essential characteristic of the discrete realm of numbers.[64] If the odd and the even are understood in the usual way, then the corresponding division of the realm of numbers is complete and unambiguous insofar as every number is either odd or even. In other words, "with every number one or the other belongs necessarily, either the odd or the even" (ἀναγκαῖόν γε θάτερον τῷ ἀριθμῷ ὑπάρχειν, ἢ περιττὸν ἢ ἄρτιον — Aristotle, *Categories* 10, 12 a 7 f.) ; "at least, there is nothing midway between these" (οὐκ ἔστι γε τούτων οὐδὲν ἀνὰ μέσον), as for instance between white and black (cf. *Metaphysics* I 4, 1055 b 24 f.). But this means that these properties belong to the *being* of number, of that which is countable as such (cf. P. 26), namely to that being (οὐσία) which an *arithmos* has as a counted collection always determinate in respect to its quantity. Externally this

fact is expressed by the general familiarity of the distinction
between odd and even numbers, which is common enough
to provide the basis of a game called "odd-and-even"
(ἀρτιάζειν, cf. Plato, *Lysis* 206 E). Thus this distinction —
besides that of male and female, and right and left —
provides Plato with the model for every "natural" *diairesis*
(cf. *Statesman* 262 A–E, *Phaedrus* 265 E).[65] But while this
distinction concerns the determinateness of numbers as
limited, i.e., as determinate in their quantity, *it is completely
independent of whatever "material" may happen to underlie any
particular case of counting*.

Thus the absence of any mention of either *arithmos* or
arithmoi in the definitions of arithmetic and logistic in the
Gorgias and in the *Charmides* not only expresses the fact that
the multitude of arbitrarily chosen assemblages of monads is
accessible to *episteme* only through the determinate *eide*
which can always be found for these assemblages, but it also
indicates that the characteristics of all possible kinds of
numbers, beginning with the odd and the even, are to be
found *indifferently in all countable things, be they objects of sense
or "pure" units*. These definitions of arithmetic and logistic
are therefore indeed independent of any distinction between
the theoretical or the practical use of either discipline,
although their formulation presupposes a *theoretical* point of
view. This point of view is indeed already implied in the fact
that strict definitions are at issue. But the rigor of these
definitions consists precisely in the fact that they articulate
only one of the two characteristics of the *arithmos*, namely
that all-pervasive limitedness which is rooted in an *eidos*, so
that they can cover the *whole* realm of that which is countable — *and yet avoid the indefiniteness which attends the term*
"arithmos" insofar as by itself it gives no indication of the
sort of collection concerned, i.e., of that which the number
is meant to be a number of. On the other hand, the rigour of
both definitions does not exclude them from the "common
understanding," since (1) *the understanding that the* arithmoi

with which episteme *deals are numbers of "pure" units is in no way presupposed* and (2) the two *eide* of the even and the odd appeal to a thoroughly familiar distinction. We shall see that beyond this these formulations are also suggested by certain ontological considerations.

We are now finally able to appreciate the meaning of the commentaries of Olympiodorus and of the *Gorgias* scholiast (Pp. 13 ff.). Arithmetic deals with numbers insofar as these present assemblages whose unity is rooted in the unity of a certain *eidos*, although this fact usually remains hidden from the person immersed in the practical activity of counting. As a theoretical discipline, at least, arithmetic studies each quantity and each multitude of monads which falls under a particular *eidos* only indirectly. "Logistic," on the other hand, be it practical or theoretical, aims of necessity — insofar as it is concerned with the mutual relations of numbers — directly at the "quantity," at the multitude, of those things which are in each case related to one another or computed, i.e., at the "material" which underlies each relation or calculation. But while the definition of logistic in Plato leaves completely undetermined whether this material belongs to the realm of the sensible or the purely noetic (an indeterminacy which insures the universality of the definition), Olympiodorus and the *Gorgias* scholiast are forced from the very beginning to regard the "hylic" monads, i.e., the monads which form the *hyle* of the numbers (cf. Domninus, Pp. 32 ff.) as sensible "units," since only these are amenable to that partitioning which exactitude of calculation requires (cf. Pp. 34 and 43). Thus both of the Neoplatonic commentators unwittingly enter an ontological realm which is no longer consonant with the Platonic definition of the monad. The next two sections, which deal with the ontological problems raised by *arithmoi*, will help to make this clearer.

7

The ontological conception of the arithmoi *in Plato*

W E HAVE so far avoided coming to terms with the onto-logical point of view which from the very first determined the form taken by the Greek doctrine of the *arithmoi*. Yet any attempt to understand Greek mathematics as a self-sufficient science must fail. It is impossible to disregard the ontological difficulties which fundamentally determine its problems, its presentation, and its development, especially in its beginnings. It is just as impossible, on the other hand, to understand Greek ontology without reference to its specifically "mathematical" orientation. For the character-ization of "mathematical" truths as *mathemata*, i.e., as things "to be learned," serves as model for all teachable and learnable knowledge — and it is knowledge so understood which determines the sphere of Greek ontological inquiry.

Since the framework of this study does not permit an exhaustive consideration of these problems in all their ramifications, only those factors will be emphasized which bear directly on our theme.

Since the appearance of the first fundamental studies by Julius Stenzel[66] a whole series of attempts has been made to clarify the connection between Greek mathematics and Greek philosophy which is so essential for the formation of mathematical concepts.[67] Especially Oskar Becker ("Die diairetische Erzeugung der platonischen Idealzahlen," *Quellen und Studien*, vol. I, pp. 464 ff.) has further developed Stenzel's lines of thought and has pointed out the central significance of the "monads" for an understanding of the Platonic doctrine of the so-called "ideal numbers." The article by J. Cook-Wilson, "On the platonist doctrine of the ἀσύμβλητοι ἀριθμοί," *The Classical Review*, XVIII (1904), pp. 247 ff., which has, it seems, not had enough attention, points in the same direction. In the present study, in which the same question is approached from a different point of view, and whose results, as far as the Platonic theory of number is concerned, concur with those of Becker, an attempt is made to discover the basic presuppositions of Greek arithmetic and the ontological arguments connected with it within the structure of the *arithmos* concept itself. Since the opinions of other authors on this subject[68] can be but occasionally touched on in the text, the following brief remarks may not be out of place:

With respect to the interpretation of the *arithmos* concept and its role in Greek science a certain discrepancy can be seen both in Stenzel and in Becker. Thus Stenzel stresses the intuitive character or perceptual immediacy ("die Anschaulichkeit") and the figure-like nature ("das Gestalthafte") of the Greek "number" concept, and fixes on the interplay of thinking, counting, and intuition ("das Wiederspiel ... von Denken und Zählen und Anschauung") as its guiding principle (*Zahl und Gestalt*[1], pp. vi and 43 f.). But he begins on a highly sophisticated level of thought, so that the fundamental phenomena cannot really emerge in their simple pregnancy. Following him, Becker speaks of a strange, figure-like, archaic significance ("uns fremden, gestalthaften, 'archaischen' Bedeutungssinn") of the *arithmos* which, he says, still appears in Aristotle (p. 491, cf. *Zahl und Gestalt*, pp. 5 and 29), and of the Greek *number-formations* ("Zahlgebilden"), as having a "certain intuitively apprehensible 'dimension'" ("gewissen anschaulichen 'Umfang'"); but in general he is, especially in the interpretation of the *arithmoi eidetikoi*, guided after all by *our* number concept, which has a totally different structure. In this whole context the relation of "number" to geometry and to geometric intuition plays a decisive role. Now it is without doubt true that there is a very close connection between the *fully developed* Greek arithmetic and geometry, as is attested by the theory of "figurate" numbers, by the structure of the theory of proportions and generally, by the illustrations used in arithmetic (or logistic). From this connection Zeuthen deduces the concept of "geometric algebra" which determines his whole view of Greek mathematics.[69] We shall have to show later how little this concept does justice to the Greek procedure.

However, one fundamental objection is to be raised immediately against stressing the "intuitive" character of the *arithmos* concept, namely that it arises from a point of view whose criteria are taken not from Greek, but from modern, symbolic, mathematics. But nothing hinders us from doing justice to the originality of ancient science by allowing ourselves to be guided only by those phenomena to which the Greek texts themselves point and which we are able to exhibit directly, our different orientation notwithstanding. Greek scientific arithmetic and logistic are founded on a "natural" attitude to everything countable as we meet it in daily life. This closeness to its "natural" basis is *never* betrayed in ancient science. It follows that, strictly speaking, it is not possible to call *arithmoi* "numbers." The peculiarity of the Greek concept of "number" lies therefore less in an "archaic" or "intuitive" character (which is not at all its primary property) than in the *kind of relation* it has to the "thing" it intends. In the last section we shall have to show how this relation of concept and intended content is subjected to a fundamental modification under the aegis of a new modern intentionality, and how thereby the modern concept of "*number*" first becomes possible. What we are interested in is not only the exemplification of some *general* subject matter, but also an attempt to determine more exactly the conceptual dimension which is characteristic of modern consciousness by examining one of its crucial aspects.

A
The science of the Pythagoreans

The development of the theory of *arithmoi* is undoubtedly in large part due to the men traditionally called "*Pythagoreans*." Their chief object was to understand the "order within the heavens" (Alex., *in metaph.*, Hayduck, 75, 15; Diels, I³, 347, 1 f. and 355, 29, based on the Aristotelian work on the Pythagoreans; cf. Eudemus in Simplicius' commentary on *De Caelo*, Heiberg, 471, 5 f.; Diels, I³, 19, 12 f.), i.e., to understand the *visible order of the visible whole*, as is clearly enough attested by Aristotle (*Metaphysics* A 8, 989 b 33 f.: "their discourse and concern is all about nature" [διαλέγονται . . . καὶ πραγματεύονται περὶ φύσεως πάντα]; see also the passage which follows; cf. *Metaphysics* A 5, 986 a 2 ff., *Physics* Γ 4, 203 a 7 f.: Δ 6, 213 b 22 ff.; *On the Heavens* Γ 1, 300 a 14 ff.); most modern scholars do, in fact, recognize

this.[70] The general point of view governing the efforts of the
Pythagoreans might be sketched out as follows: They saw
the true grounds of the things in this world in their *countableness*, inasmuch as the condition of being a "world" is
primarily determined by the presence of an "*ordered
arrangement*" (τάξις) — and this means a *well*-ordered
arrangement — while any order, in turn, rests on the fact
that the things ordered are delimited with respect to one
another and so become countable. Aristotle, who accuses
their "definitions" of superficiality, states the fundamental
principle of their procedure with complete clarity: "That to
which [in the order of things] the term in question *primarily*
belongs, this they consider to be the *being* of the thing."
(ᾧ πρώτῳ ὑπάρξειεν ὁ λεχθεὶς ὅρος, τοῦτ' εἶναι τὴν οὐσίαν τοῦ
πράγματος ἐνόμιζεν — *Metaphysics* A 5, 987 a 22 ff.[71]) But, in
accordance with Aristotle's statement, which is valid for *all*
of Greek cosmology, "the order proper to the objects of
sense [i.e., the order of the visible world] *is* nature" (ἡ γὰρ
τάξις ἡ οἰκεία τῶν αἰσθητῶν φύσις ἐστίν — *On the Heavens* Γ
2, 301 a 5 f.; see also *Metaphysics* Λ 10, 1075 a 11–23); in
other words, this order determines the *very being* of things,
and, furthermore, this order rests in the final analysis on the
possibility of *distinguishing* things, i.e., of *counting* them (cf.
Philolaos, fragment 4, Diels, I[3], 310, 8-10).We may, therefore, in accordance with the fundamental Pythagorean
principle mentioned earlier, conclude that the arithmetical
properties of things concern their *being itself* and that, in
truth, "the being of all things is number" (ἀριθμὸν εἶναι τὴν
οὐσίαν ἁπάντων — *Metaphysics* A 5, 987 a 19). All properties,
conditions, and modes of behavior would therefore be
reducible to those properties which things have in virtue of
their capability of being counted, to those affections (πάθη)
which are to be found in everything counted *as such;* the
ultimate "elements" and "sources" of all that is would be
identical with the elements and sources of their countableness
(*Metaphysics* A 5, 985 b 23–26, 29; 986 a 6, 15–17; A 8, 990 a

2 ff., 19). "And whatever they got hold of in numbers and conjunctions [of numbers][72] which agreed with the affections and parts of the heavens and the whole world order, this they collected and [fittingly] conjoined." (καὶ ὅσα εἶχον ὁμολο- γούμενα ἔν τε τοῖς ἀριθμοῖς καὶ ταῖς ἁρμονίαις πρὸς τὰ τοῦ οὐρανοῦ πάθη καὶ μέρη καὶ πρὸς τὴν ὅλην διακόσμησιν, ταῦτα συνάγοντες ἐφήρμοττον — 986 a 3–6.) They did this according to their fundamental principle, that is, by using the *pathe* and *logoi* (i.e., ratios) of *numbers* as the basis of their comparisons (*Metaphysics* N 3, 1090 a 20–25). Here lay that "super-ficiality" of their definitions which Aristotle had censured: They evidently considered a vague "structural" similarity a sufficient basis for speaking of an "imitation" of numbers by things (*Metaphysics* A 6, 987 b 11 f.; Theophrastus, *Meta-physics* 11 a 27 f.); so, for instance, they recognized retribution in "the just" and in retribution the "reciprocal ratio," whence "the just" came to be defined as "reciprocal ratio" (*Nicomachean Ethics* E 8, 1132 b 21 ff.; see also Euclid VI, Def. 2 and Nicomachus 13, 18 f.; cf. *Magna Moralia* A 1, 1182 a 14: justice — δικαιοσύνη — as "quadratic" number). The Pythagorean mode of definition is, then, characterized by the attempt to define the *being* of things by reducing and *assimilating* them to conditions "primarily" exhibited in the realm of counted collections as such;[73] the being of a particular thing would then be defined in terms of a certain *kind* of number or ratio. But this procedure can be driven even further: There is always a *first* and "smallest" number (or ratio) of a particular kind such that a particular property belongs to it *primarily*; this "first" number or ratio therefore represents the "root" (πυθμήν) of its kind. Thence the number *ten* acquires a special significance: together with *one*, it must (according to the Greeks, among all peoples) be regarded as a *fundamental element* of all counting, especially since the numbers comprised in the decad, together with their mutual relations, are themselves the most important "roots."[74] (Cf. *Theol. arithm.*, De Falco, 82 ff.=Diels, I³,

303, the fragment by Speusippus concerning the decad; also Aëtius I, 3, 8, Diels, I³, 349, 26 ff. = *Doxographi Graeci*, 281; Philolaos, fragm. 11, Diels, I³, 313, 5–9; furthermore [Arist.], *Problems* IE 3, 910 b 23 ff., where a Thracian tribe is mentioned as the only exception in regard to this mode of counting.) This Pythagorean method of identifying the being of things which have analogous characteristics thus extends to the very elements of the decad and is elaborated in the later Neopythagorean and Neoplatonic tradition so as to become the most important tenet of "arithmology" (to use a modern term coined for this purpose[75]). The cosmological approach remains, however, the controlling point of view throughout (cf. especially Aristotle, *Metaphysics* A 8, 990 a 18 ff.); the guiding thought of this tradition is always that of the perfect arrangement *(διακόσμησις)* or order *(τάξις)* of the Whole. Within the horizon of this tradition of inquiry the very countableness of things, which makes this *taxis* possible, is understood as a definite "being-in-order" of things, and thus the sequence of numbers is understood as the *original order of the being* of these things. Here the sequence of numbers represents not a linear chain whose links are all "of the same kind" but an "ordering" in the sense that each number precedes or follows in the order *of its being*, i.e., is related as prior and posterior *(πρότερον* and *ὕστερον* — cf. Aristotle, *Metaphysics* Δ 11, 1018 b 26–29).

Thus the science of the Pythagoreans is an *ontology of the cosmos*, a doctrine concerning the mode of being of the world and of the things comprised in it. This holds particularly for their "arithmetical" and "logistical" science, whose true object is the being of the very constituents of the world (cf. Archytus, fragment 1 and 4, Diels, I³, 330 f. and 337; Philolaos, fragment 6 and 11, Diels, I³, 310 f. and 313 f.). Whatever special motives may have led to such a conception — among others, it must have been an insight into the dependence of musical consonances on "logistic"[76] — the basis of its possibility is the "natural" conception of the *arithmos* as characterized earlier. Only when "number" means a number *of things*, when the *counted things themselves* are intended whenever a number is determined, is it possible to understand their being (precisely insofar as they are things) as "number." Aristotle stresses again and again that it is characteristic of the Pythagorean view that "they do *not* make number *separable* [from the things]" (οὐ χωριστὸν ποιοῦσι τὸν ἀριθμόν); this means that they do not go so far as to suppose the existence of "pure" numbers of "pure" units, although *they were the very men* who concerned themselves with numbers not for a practical but for a theoretical purpose, who conceived of the *arithmos* as *arithmos mathematikos*, as scientific number (*Metaphysics* M 6, 1080 b 16 ff.). For they, together with the other "physiologists," as Aristotle (*Metaphysics* A 8, 990 a 3 f.) says, were of the opinion that being extends just as far as sense perception, and they labored to prove that the mode of being of precisely such things as are perceived by sense is determined by "number," which meant, to be sure, that they were on the way toward discovering a mode of being of a higher order. We may conjecture that they saw the genesis of the world as a progressive *partitioning* of the first "whole" *one*, about whose origin they themselves, it seems, were not able to say anything conclusive (cf. *Metaphysics* M 6, 1080 b 20 f.; N 3, 1091 a 15 ff.), but which, at any rate, already contained as fundamental constituents, as "elements," the two fundamental kinds of the realm of number, the "odd and the even," namely "limit" and the "unlimited" (cf. *Metaphysics* A 5, 986 a 17 ff. and Theon 22, 5 ff.). This first "one," as well as all the subsequent "ones" which were the result of partition, i.e., the "numbers" themselves, they therefore regarded as having *bodily extension*: "They understood the monads to have magnitude" (τὰς μονάδας ὑπολαμβάνουσιν ἔχειν μέγεθος — *Metaphysics* M 6, 1080 b 19 f.). This allowed them to reduce the *measurableness* of things to their *countableness* —

until the discovery of "incommensurable" magnitudes proved the impossibility of this reduction.

Now in whatever way they imagined that the world and its bodies had been built "out of numbers" (ἐξ ἀριθμῶν), the structure of the *arithmos* concept enabled them to understand not only the *bodily monads* as the "material" of the being of things as things (which relates them to the atomists), but going further, to see in the *properties of the kinds* of numbers the models which things "imitate." Thus, by following the principle that the being of things with analogous characteristics is identical, they were able to define their being as *being number*. Cf. *Metaphysics* A 5, 986 a 16 f.: "considering numbers to be the source of the things that are, both *as their material and as their characteristics, as well as their states*" — τὸν ἀριθμὸν νομίζοντες ἀρχὴν εἶναι καὶ ὡς ὕλην τοῖς οὖσι καὶ ὡς πάθη τε καὶ ἕξεις; A 8, 990 a 18 f.: "How are we to take it that *the characteristics of number, and number itself,* are responsible [for the things which are and have come to be in the heavens]?" — πῶς δεῖ λαβεῖν αἴτια μὲν εἶναι τὰ τοῦ ἀριθμοῦ πάθη καὶ τὸν ἀριθμὸν ... ; N 6, 1093 a 11 f.: "Those [different things] would be the same with one another since *the same kind of number belongs to them.*" — ταὐτὰ ἂν ἦν ἀλλήλοις ἐκεῖνα τὸ αὐτὸ εἶδος ἀριθμοῦ ἔχοντα. Cf. also Philolaos, fragment 5, Diels, I³, 310, 13 f., referring to the *eide* of the odd, the even and even-times-odd: "There are many shapes belonging to each of the two forms" — ἑκατέρω δὲ τῶ εἴδεος πολλαὶ μορφαί, [ἃς ἕκαστον αὐταυτὸ σημαίνει]. From the Philolaos fragment, omitting the last clause, which is probably corrupt, it is, at any rate, clear that by the *morphai* of the *eide* are to be understood their subspecies, i.e., certain characteristics or properties of the kinds, about which Boeckh remarks[77] "that they were developed by the ancient arithmetician with a special industry which appears to us as petty" (. . . die alten Arithmetiker [hätten sie] mit besonderem, uns freilich kleinlich vorkommenden Fleiss entwickelt). The significance of these *pathe* or *morphai* or *eide* of numbers has never, since Boeckh, been properly acknowledged. Yet the *eide* of numbers and of their relations, together with their correlative roots (πυθμένες), not only form, as we have seen, the true object of Greek arithmetic but also the foundation of all cosmological speculation as originated by the Pythagoreans and developed through the centuries up to Kepler. In particular we shall have to examine the role given to this *eidos* concept in the mathematics of Vieta.[78]

B
Mathematics in Plato — *logistike* and *dianoia*

There can be no doubt that Plato's philosophy was decisively influenced by Pythagorean science, whatever the exact connection between Plato and the "Pythagoreans" may have been. So too those definitions of arithmetic and logistic which were the basis of the preceding reflections seem to point to a Pythagorean origin. We saw (P. 59) that these definitions in no way presupposed the existence of "pure" numbers; rather they referred to *everything* countable as such, and thus, above all, to the objects of sense in this world. The definitions seem to have preserved the original *cosmological* significance of these two sciences, especially since we must keep in mind that the "very first kinds" (πρώτιστα εἴδη) of number, the "odd" and the "even," represented for the Pythagoreans the "limit" and the "unlimited," i.e., the *archai* of *all* being, whose union, whose "mixture," first brings the world into being. This view, which is part of the cosmology of the *Philebus* and the *Timaeus*, is directly connected with the general theory of opposites of the Pythagoreans (cf. especially the table of opposites in Aristotle, *Metaphysics* A 5, 986 a 22 ff.). Nor should we overlook the symmetry of the pairs of opposites in the *Charmides*: even-odd (ἄρτιον — περιττόν, 166 A), heavy-light (βαρύ — κοῦφον, 166 B), and ignorance-knowledge (ἀνεπιστημοσύνη — ἐπιστήμη, 166 E). So also in the *Gorgias* the opposition of the *peritton* and the *artion* appears alongside that of just and unjust things (δίκαια — ἄδικα, 454 B, cf. also 460 E).[79] Here then, in the *Charmides* and in the *Gorgias*, the object of arithmetic and logistic is, it seems, conceived from an ontological point of view identical with that of the Pythagoreans (cf. Pp. 66 f.). Yet there exists an indissoluble tension between Pythagorean science and Platonic philosophy, which manifests itself right within the Platonic *opus* in the opposition between the Socratic dialogue and the "likely tale" (εἰκὼς μῦθος) representing the cosmos. To be sure, Plato knows a "bond" (δεσμός) which ties dialectic and cosmology together and affects *both* decisively. This bond is *mathematics*. Its fundamental signifi-cance for Plato lies precisely in its "middle position," in its character as a *bond*. But precisely by giving this "mediating" role to mathematics Plato assigns to mathematical objects a totally different place and a totally different mode of being than they can possibly have in Pythagorean science. Espe-cially in discussing numbers, Aristotle never tires of stressing that Plato, in opposition to the Pythagoreans, made them "separable" from objects of sense, so that they appear "*alongside* perceptible things" (παρὰ τὰ αἰσθητά) as a separate realm of being. This realm is represented, as we have seen, by the field of "pure" units, which are indivisible, of the same kind, and accessible only to thought. The new point of view Plato brings to mathematical investigations, his general interest in the superiority of the purely noetic over everything "somatic," sufficiently explains the stress which he lays on the existence of "pure" numbers, such as will be inaccessible to the senses. Yet it is not unimportant to note that the emphasis with which the thesis of "pure" monads is propounded is indicative of the fact that *arithmoi* were ordinarily, and as a matter of course, understood only as definite numbers of sensible objects. The thought of "pure" numbers separated from all body is originally so remote that it becomes the philosopher's task precisely to point out emphatically the fact that they are *independent* and *detached*, and to secure this fact against all doubt. The way this task is to be fulfilled is prescribed by the course which leads from our actual counting to the conception of "pure" monads (cf. Pp. 49 f.). In giving an account of this course, we grasp the fact that this is also the way to account for our ability to count and calculate in daily practice: the fact that we are able to count off a definite multitude of objects of sense is grounded in the existence of "nonsensible" monads which can be joined together to form the number in question and which our thinking, our *dianoia*, *really* intends when it counts or calculates such things. But whatever furnishes the *foundation* for something else (in our case, for the ordinary numbers), whatever makes the being of another possible, is also more meaningful, more powerful in its being, than this other. The being of that which is the foundation takes precedence over that which is so founded, i.e., it is prior (*πρότερον*), because the second cannot be without the first, but the first can be without the second. Thus the special mode of being which belongs to numbers of "pure" monads begins to be comprehensible in the light of their foundational role; it is characterized as that primordial independence and "detachment" which is rooted in the founding function — an approach typical of Plato (and of all later Platonism). Aristotle bears witness to it when in the explication of the various meanings of prior and posterior (*πρότερον — ὕστερον*, *Metaphysics* Δ 11) he remarks expressly, in commenting on that meaning of these terms which has just been explicated, that "Plato made use of this distinction" (*ᾗ διαιρέσει ἐχρήσατο Πλάτων* — 1019 a 2–4), as if to say that it is characteristic of Plato, that it plays an essential role for him. Indeed, it is connected directly with the much discussed "hypothesis" (literally: sup-position) theory of Plato, which the subject of this study now compels us to take up.

The total realm of that which is accessible to thinking, which "is and can be thought" (*νοητόν*), is divided by

Plato in the *Republic* (510 B ff.) into two domains, one of which is thus characterized: There are objects of sense which we understand as an *image of* (Abbild) something "other," since the examining soul is *compelled* to *sup-pose*, i.e., to make to underlie, this "other" which is precisely the *noeton* in question: "Here the soul must of necessity begin its search [for objects of thought] from *suppositions*." (τὸ μὲν αὐτοῦ [sc. τοῦ νοητοῦ] ... ψυχὴ ζητεῖν ἀναγκάζεται ἐξ ὑποθέσεων — 510 B.) In such cases we do examine certain things by means of the senses; yet we do not *intend* these things but rather that on which they are "founded" and which they *image*. This object, which underlies them as a foundation and images itself in them is, *as such*, the object of the *dianoia* (510 D–E). And the result of each act of thinking (διανοεῖσθαι) or reckoning (λογισμός) is that certain properties of things perceived by sense are understood — understood *from their foundations*.

This is, above all, the procedure of geometers and *logisticians* (510 C). The former draw certain figures and exhibit their properties; yet they do not intend the drawn figure itself but that which is imaged in this figure, e.g., the rectangle which is, in its purity, accessible only to thinking. Still, in order to grasp the particular content involved, constant reference to the drawn rectangle is quite necessary. Similarly, logisticians have before their eyes the "odd" and the "even" in the shape of certain countable objects which they reflect on, but these reflections, being pursued in thought, are aimed not at these particular objects but at the "pure" numbers or their *eide*, which are "supposed" in thinking and imaged in the objects. Here, too, reference to those multitudes which are accessible to the senses is necessary, though mere "signs" or points marked in the sand are sufficient.

In the conduct of the geometers and logisticians the efficacy of the *dianoia* is directly displayed. But what does this special intimacy with which the *dianoia* is related to the provinces of geometry and logistic mean? After all, the *dianoia* is by no means confined to these alone[80] but obviously has an essential, perhaps *the* essential, part in all human activity and self-orientation. We must not overlook the fact that the procedure by "hypothesis" stressed by Plato is *not* a specifically "scientific" method but is that original attitude of human reflection prior to all science which is revealed directly in speech as it exhibits and judges things.[81] Thus, compared to the study of nature embarked upon by the physiologists, that "second-best sailing" (δεύτερος πλοῦς) of Socrates, which consists of "taking refuge in reasonable speech" (εἰς τοὺς λόγους καταφυγεῖν — *Phaedo* 99 E), is indeed nothing else than a return to the *ordinary* attitude of the *dianoia*; it is for this reason that Plato can characterize the method of "hypothesis" as "simple and artless and perhaps naive" (ἁπλῶς καὶ ἀτέχνως καὶ ἴσως εὐήθως — *Phaedo* 100 D).[82] When engaged in reasonable speech under the guidance of the *dianoia*, we always suppose something "other" to underlie the objects we perceive, namely *noeta*; these, albeit appearing in the mirror of our senses, are the true objects of our study, though we may not even be aware of making such "suppositions." There is, however, a higher kind of reflection in which this "supposing" *is* raised to the rank of a conscious procedure; this is the origin of every science and every skill (cf. *Philebus* 16 C). For all science and all skill grows out of the natural activity of reflection when it attains the character of a fully developed "art" (τέχνη) which obeys definite rules. The "devices" of the *dianoia* that now become transparent and thereby learnable make completely explicit what the *dianoia* has in effect been accomplishing *prior to* any science. Conversely, the nature of this ordinary accomplishment of the *dianoia* can be grasped only through such a reflective understanding. And precisely those *technai* which are most highly developed, the *science of measurement* and above all, the *science of counting and calculation* (cf. *Euthyphro* 7 B–C), that "common thing of which all arts as well as all thinking processes and all sciences make use" (κοινόν, ᾧ πᾶσαι προσχρῶνται τέχναι τε καὶ διάνοιαι καὶ ἐπιστῆμαι — Republic 522 C) and without which any techne would lose its character as techne (Philebus 55 E; cf. Republic 602 D), permit us to grasp the true sense of the dianoia. This is why the dianoia is assigned to the realms of geometry and logistic: these are the realms within which its activity is exemplary. And that the ordinary activity of the dianoia does indeed refer back to a knowledge precisely of logistical matters is shown by those arguments in the Republic which establish the leading role played in education by mathematical subjects, especially the science of counting and calculation.

According to this argument there are objects of perception which directly satisfy the perceiving soul because they are in themselves sufficiently clear, so that there is no necessity to appeal to any authority beyond perception — there is no necessity to reflect in any way on such perceptions: "Some objects of sense do not invite thought to undertake an inquiry since they are sufficiently distinguished by sense." (τὰ μὲν ἐν ταῖς αἰσθήσεσιν οὐ παρακαλοῦντα τὴν νόησιν εἰς ἐπίσκεψιν, ὡς ἱκανῶς ὑπὸ τῆς αἰσθήσεως κρινόμενα — 523 A–B.) But there are also objects of perception which of necessity leave the perceiving soul unclear about their nature and which thus of necessity call up reflection to aid the soul in discovering it. The lack of clarity of such objects, it should be noted, is not such as to disappear with a better or sharper perception, for instance, when the object of perception is brought closer to the senses (523 B). Rather it has its basis in the very structure of the perceived object, insofar as the same object of perception may, in the same perception, also appear as its opposite, "since sense does not make clear one thing rather than its opposite" (ἐπειδὰν ἡ αἴσθησις μηδὲν μᾶλλον τοῦτο ἢ τὸ ἐναντίον δηλοῖ — 523 C). If, for instance, I look at the fingers on my hand, I perceive each of them directly as a finger. Plain perception is here the end of the matter: Within this perception there is nothing "problematical" — it is

complete in itself, it does not leave behind a feeling of a lack
of clarity such as might move me to reflect and to investigate
what there really is to the finger, *what* a finger might *be*
(523 D). The situation changes completely when I perceive
the characteristics of each of my fingers in turn, such as its
largeness or thickness or softness. Now I perceive in any one
finger *at one and the same time also* an opposite: its smallness,
its thinness, and its hardness. I therefore perceive opposites in
the same perception, and in the same object of perception,
e.g., softness and hardness or largeness and smallness. Here
something which clearly does not belong together is in an
obscure way "mixed," *as reflection immediately informs me*. In
order to resolve this difficulty, in order to *see* clearly, I must
of necessity go beyond "bare" perception and "call up"
and "awaken"[83] "reckoning and thought" (λογισμόν τε
καὶ νόησιν — 524 B). Thought here becomes active sponta-
neously, as it were, in direct succession to perception. It tells
me that I am dealing with something *twofold*, namely with
softness on the one hand and with hardness on the other, and
similarly with largeness on the one hand and with smallness
on the other (cf. *Theaetetus* 186 B). It makes me recognize
both together as "*two*," that is, as *this one* thing, namely softness (or largeness) and *that other one* thing, namely hardness
(or smallness) — not as "mixed" with one another but as
distinct: "*Each is one, but both are two*, and the two are
separated by thought . . . not fused together but distinct."
(ἓν ἑκάτερον, ἀμφότερα δὲ δύο, τά γε δύο κεχωρισμένα νοήσει . . .
οὐ συγκεχυμένα ἀλλὰ διωρισμένα — 524 B–C.)

But in which way does thought succeed in separating the
"mixture"? Clearly by way of *comparison* (cf. *Theaetetus*
186 B). Reflection informs me that this finger is small in
comparison with its left neighbor, and that this *same* finger is
large in comparison with its right neighbor. Or, as in the
Phaedo (102 B–C), Simmias is large, not insofar as he is
Simmias, but insofar as he can be compared with Socrates.
The same Simmias is, on the other hand, small — again, not

insofar as he is Simmias, but insofar as he can be compared with Phaedo; and so in reverse: Socrates is small "because he has smallness in relation to the largeness of the former," namely Simmias (ὅτι σμικρότητα ἔχει ὁ Σωκράτης πρὸς τὸ ἐκείνου μέγεθος), and Phaedo is large "because Phaedo has largeness in relation to the smallness of Simmias" (ὅτι μέγεθος ἔχει ὁ Φαίδων πρὸς τὴν Σιμμίου σμικρότητα). The largeness and the smallness of an object can, then, be recognized by thought as something twofold only when the object is recognized at one time as "being-larger-than..." and at another as "being-smaller-than...." Only by *relating* the one condition to the other is the *dianoia* enabled to suppose two distinct structures to underlie the "mixture" which assaults the senses.

The objects of the *dianoia*, the first kind of *noeta*, are therefore attained as a result of the fact that that which is accessible to the senses is, by reason of its "relational character" ("Bezüglichkeit"—Natorp), recognized as a manifold, a multiplicity. However, an awareness of the many *as many*, and thus also of the other *as other*, means precisely that the single constituents of the many are distinguished from one another and are at the same time, in all their distinctness, related *to one another*. The *dianoia* continually surveys and compares the many aspects which perception (αἴσθησις) offers it. Thus the *noeta* which it attains and which it causes to underlie each obscure *aistheton* also have relations to one another. The *dianoia* is never directed toward a single being as such; rather, its view always so encompasses a series of beings that the members of this series are carefully distinguished from, and thus simultaneously related to, each other.

Thus we see that the *dianoia* is based essentially on "account-giving and counting"[84] (λογίζεσθαι τε καὶ ἀριθμεῖν), namely on the ability to recognize many as *so* many, to see many *as many*, i.e., to distinguish the constituents of the many and at the same time to relate them to one another (see P. 20). It is for this reason that in the *Republic* the explication of the example of the finger leads directly to the question concerning "number" and the "one": "Well then, to which of these [i.e., the things inviting or not inviting reflection] do number and the one belong?" *(Τί οὖν; ἀριθμός τε καὶ τὸ ἓν ποτέρων δοκεῖ εἶναι; —* 524 D.) Glaucon is called on to "give an account analogous to what has been said previously" *(ἐκ τῶν προειρημένων ἀναλογίζου.)* Thus, if the perception of an object as *one* should at the same time include the perception of an "incompatibility" *(ἐναντίωμα),* in the sense that the *same* object appears as *one* and also as the *opposite of one,* "so that nothing appears any more as one than as the opposite" *(ὥστε μηδὲν μᾶλλον ἓν ἢ καὶ τοὐναντίον φαίνεσθαι —* 524 E; cf. *Theaetetus* 186 B), then here too the *dianoia* would have to intervene to remove the "obstacle" and to allow the examining soul to reach "clarity" *(σαφήνεια)* concerning the "one." Now this is, as Glaucon immediately notes, indeed the case: "For we see the same things at the same time as *one* and as *unlimited* in multitude." *(ἅμα γὰρ ταὐτὸν ὡς ἕν τε ὁρῶμεν καὶ ὡς ἄπειρα τὸ πλῆθος —* 525 A.)[85] But if this holds for one object, it holds also generally for "every number" of objects *(ξύμπας ἀριθμός),* because any "number" represents precisely a limited number of unit objects.[86] Thus in being compelled, in the course of distinguishing the various respects which permit the one object to be called many, to separate the unit as such, "the one itself" *(αὐτὸ τὸ ἕν),* from its opposite, namely from the unlimited many, the *dianoia* discovers the "one" *not by itself alone but among many ones,* relates the *one* one to the *other* ones, an activity which is, in effect, nothing but — counting. This is a proceeding of a perfectly ordinary kind. In all the daily routines of life we are dependent on just such interventions of the *dianoia.* And so the way of "learning about the one" *(ἡ περὶ τὸ ἓν μάθησις)* has already been entered upon. Led by the *dianoia,* the soul turns its attention to this, its very own way, "arousing thinking within herself" *(κινοῦσα ἐν ἑαυτῇ* τὴν ἔννοιαν) and discovers, within a field of infinitely many and homogeneous "pure" units, the "pure" numbers of these units; it raises its own relating activity to full explicitness by examining the relations of these numbers to one another, thus incidentally laying a foundation for the possibility of making calculations; it causes the *eide* of these relations of numbers, as well as of the numbers themselves, to "underlie" the objects of sense as "suppositions," "positing the 'odd' and the 'even'" *(ὑποθέμενοι τό τε περιττὸν καὶ τὸ ἄρτιον . . . — 510 C)* as the first of these — yet in spite of all this, it is unable to come face to face with the one *as it is in itself.* For the *dianoia* always deals with a *multitude* of ones; it cannot grasp the one except through an aggregate of ones, just as it cannot recognize one element of language, i.e., a single sound rendered by a letter, without the remaining sounds (*Philebus* 18 C), nor a single tone without the other tones. The *dianoia* cannot do this because, although it is directed toward *noeta*, it nevertheless always remains related to that *aisthesis* which first "called upon it" to clarify an obscure state of affairs. The *dianoia* effects its clarification by recognizing the opposition which underlies the obscurity of the *aistheton as* an opposition. For the recognition of an opposition as opposition is the proper function of the *dianoia* (cf. *Theaetetus* 186 B). Thus it at first always attains to noetic structures of oppositional character: "You are speaking of 'being' and 'non-being,' of 'similarity' and 'dissimilarity,' of the 'same' and the 'other,' and furthermore of the 'one' and the rest of 'number' which deals with these [i.e., with objects of sense]."[87] *(Οὐσίαν λέγεις καὶ τὸ μὴ εἶναι, καὶ ὁμοιότητα καὶ ἀνομοιότητα, καὶ τὸ ταὐτόν τε καὶ τὸ ἕτερον, ἔτι δὲ ἕν τε καὶ τὸν ἄλλον ἀριθμὸν περὶ αὐτῶν — Theaetetus* 185 C–D.) Thus it attains to the "beautiful" and the "ugly," to the "good" and the "foul" (186 A). The *dianoia* moves, from the very beginning, in the realm of opposition, and in conducting its "comparisons" *(ἀναλογίσματα — Theaetetus* 186 C) it discovers as the true "foundation" of this domain the realm of the "pure" relations of numbers, i.e., the ratios (λόγοι) and proportions (ἀναλογίαι) of the "pure" numbers, because every possible comparison is ultimately founded on these (cf. *Timaeus* 36 E–37 A). But even here the *dianoia* still remains dependent on *aisthesis* (cf. P. 72), and this applies, we must add, even to the *general* theory of proportions, because the work of the *dianoia* is everywhere "hypothetical" in character. It is true that it can "replace" the "more and less" (τὸ μᾶλλον καὶ τὸ ἧττον)[88] which always attaches to the realm of the *aistheta* by the *exact* relations of numbers, thereby accomplishing the most important step toward gaining that true *episteme* which no longer has any use for *aisthesis* and whose object is the realm of those other *noeta* which ascend to something "unsupposed" (ἀνυπόθετον — *Republic* 510 B, 511 B). But the *dianoia* itself is not able to appreciate the full range of significance of its accomplishment, because its own *noeta*, which it "supposes" to underlie the *aistheta*, appear altogether lucid and without further need of foundation. And yet this dianoetic quarry, as it is brought in especially by mathematicians, must first be handed over to the *dialecticians* for proper use (*Euthydemus* 290 C; *Republic* 531 C–534 E). Only dialectic can open up the realm of true being, can give the ground for the powers of the *dianoia* and can reveal Being and the One and the Good as they are — beyond all time and all opposition[89] — in themselves and in truth.

C
The *arithmos eidetikos*

In the *Hippias Major* the following strange fact is discussed at length (300 A–302 B): While in general a property which belongs to several things in common must be attributed also to each single one of them, so that a "common thing" (κοινόν) is something "which belongs to *both* [in this case to hearing — ἀκοή and sight — ὄψις] in common, as well as to *either* in its own right" (ὃ καὶ ἀμφοτέραις αὐταῖς ἔπεστι κοινῇ καὶ ἑκατέρᾳ ἰδίᾳ — cf. *Theaetetus* 185 A), there is also a *koinon* of such a kind that it does indeed belong to several things but *not* to each of these by itself. Hippias at first thinks that this state of affairs is impossible. He refers to the fact (300 E–301 A) that when something is said of him, Hippias, and of Socrates as holding for *both*, for instance, that "we are both" (ἀμφότεροί ἐσμεν) just, healthy, wounded, golden, silver, etc., then it is "entirely necessary" (μεγάλη ἀνάγκη) that each of these properties should also belong to *each one*. What is more, he now raises (301 B–C) the weighty objection against Socrates and "those with whom you usually talk" (ἐκεῖνοι, οἷς σὺ εἴωθας διαλέγεσθαι) that they do not look at the "wholes of things" (τὰ ὅλα τῶν πραγμάτων), but "you take apart the beautiful and each of the things there are and you pick at it in your talk until it is cut down to size" (κρούετε δὲ ἀπολαμβάνοντες τὸ καλὸν καὶ ἕκαστον τῶν ὄντων ἐν τοῖς λόγοις κατατέμνοντες). Hippias himself is clearly far from understanding the significance of this objection: It is nothing less than the *aporia*, the quandary, of the *Parmenides* (130 E–131 E) and of the *Philebus* (15 B 4–8) — namely how it is possible that one idea in its unity and wholeness is "distributed over" the many things which "partake" in it. The *aporia* thus formulated presents, *although merely on the level of the* dianoia, the problem of "participation" (μέθεξις), which reaches its full sharpness and force only when the relation of an idea of a higher order to the ideas under it, of a "genus" to its "species," is considered — that is, only within the realm of the ideas themselves. For as the elementary form of the problem of the "one and many" (see Note 85) is to the dianoetic *methexis* problem, so is the latter to the ontological problem of the "community" of ideas.[90] But Hippias is not able to grasp the meaning even of the merely dianoetic *methexis* problem since his objection applies much more to himself than to Socrates. Later he has to admit shamefacedly that there is indeed a *koinon* of the kind in question: Socrates and Hippias are *both* together *two*, yet *each* of them is not two, but only *one*; and conversely: what *each* of them is, namely *one*, that both together are *not*: "Each of us is one, but that very thing which *each* of us is, *both* of us are *not*; for we are not one but *two*." (... ἑκάτερος ἡμῶν εἶς ἐστί, τοῦτο δέ, ὃ ἑκάτερος ἡμῶν εἴη, οὐκ ἄρα εἴημεν ἀμφότεροι· οὐ γὰρ εἶς ἐσμέν, ἀλλὰ δύο — 301 D.) It follows directly from this that *both* together make an "even" number, while *each* of them taken separately is "odd" (cf. *Phaedo* 103 ff.). Socrates concludes this discussion with the words: "It is not then 'entirely necessary,' as you just now said, that whatever *both* are, that *each* is also, and what each is, that *both* are." (οὐκ ἄρα πᾶσα ἀνάγκη, ὡς νῦν δὴ ἔλεγες, ἃ ἂν ἀμφότεροι, καὶ ἑκάτερον, καὶ ἃ ἂν ἑκάτερος, καὶ ἀμφοτέρους εἶναι.) And to Hippias' last obstinate attempt to win the argument by saying that this might perhaps not be true for "such things" (τὰ τοιαῦτα), but that it still remains true for everything he had mentioned before, Socrates answers that this is sufficient for him—he is satisfied with that much. A little later (303 A ff.) the result of the argument turns out to be that besides "each" (ἑκάτερον) and "both" (ἀμφότερον), certain "irrational" magnitudes which when added make one "rational" magnitude[91] "and numerous other such things" (καὶ ἄλλα μυρία τοιαῦτα) fall under the domain of *koina* which Socrates has in mind. We easily see that this domain can be defined only within the mathematical realm and that numbers, above all, have this curious *koinon* character: *every number of things belongs to these things only in respect to their community, while each single thing taken by itself is one.*

From the foregoing we can draw a threefold conclusion: (1) There are two different kinds of *koinon*, of which one is represented by the "beautiful" (καλόν), the "just" (δίκαιον), etc., while the other can be shown to exist within the realm of quantity; (2) this second kind of *koinon* is most appropriately expressed in the fact that things are always a "*number of*" things and is signalized by Plato with the formulaic phrase: "each [is] one but both [are] two." (ἓν ἑκάτερον, ἀμφότερα δὲ δύο — *Hippias Major* 300 ff.; *Republic* 524 B; *Theaetetus* 146 E, 185 A, B, 203 D ff.; see P. 75 and Note 87; cf. also *Republic* 476 A, 479 C, 583 E and *Phaedo* 96 E–97 A; 101 B, C; furthermore *Parmenides* 143 C, D and *Sophist* 243 D, E; 250 A–D; (3) Plato himself relates these facts to the problem of *methexis*.[92] Now whatever the problems concerning the *koinon* of the first kind might be, it is, at least, clear that the understanding of the special *koinon* character of number is of crucial importance for the solution of the fundamental Platonic problem of the "community of the kinds" (κοινωνία τῶν εἰδῶν), i.e., of the ontological *methexis* problem. For this reason the dialogue called the *Sophist*, which brings this problem to the fore, is throughout engaged, although in a veiled way, with the curious kind of *koinonia* which shows itself in numbers.

The main task of the *Sophist* is that of exhibiting the foundations of the "possibility of being" of a sophist as identical with the ultimate foundation of every possible articulation of being itself. The fact that everything which is can be "duplicated" by an image (cf. *Timaeus* 52 C), an image which is, in some enigmatic way, precisely *not* that which it re-presents, so that it is at once *this* being and "*another*," is ultimately founded in the "mirror-like nature" of being itself: being itself has within itself the possibility of acting as the source of repetition, the ability to *counter-feit* itself, to *con-front* itself. If "imitation," "mirroring," "shadow," "similarity," in short, "copying," are to occur at all, being itself must have a primal character of "*image-ability*," a character which makes possible all "difference," all "inequality," and all "oppositionality," but also all "re-cognition." This primal character of being is, to anticipate ourselves, the effect of the "twofold in general," the "indeterminate dyad" (ἀόριστος δυάς). Through the dyad being is originally "alienated" from itself, is not only "itself" but also "another" than "itself." Within the human realm the *aoristos dyas* has its perfect "imaginal" embodiment in the *sophist*. Whether or not the sophist's being is a possible one must then, in the final analysis, be considered to depend on showing that the *aoristos dyas* is the *arche* of all duality and thus of all multiplicity. The sophist must therefore be understood through the primal phenomenon of "imageability," hard though it be to grasp; the mode of being of the "image" must occupy the center of inquiry.

Accordingly, the definition proper of the sophist begins with his — necessary — claim to omniscience, with the claim "to know everything" (πάντα ἐπίστασθαι — 233 A–C).[93] In order that the meaning of this senseless claim might be better understood, the "Stranger" gives a "clearer" example" (παράδειγμα σαφέστερον — 233 D), namely the claim to be able "to produce and do simply everything by one art" (ποιεῖν καὶ δρᾶν μιᾷ τέχνῃ συνάπαντα πράγματα). But such a claim can clearly be made only playfully — as Theaetetus says: "You are jesting somehow." (Παιδιὰν λέγεις τινά — 234 A.) So also should we judge the sophist's claim (A 7–10), especially if we do not at once deny it as fraudulent but take it seriously, i.e., if we mean to respond to it at all (cf. *Euthydemus* 277 D f.). More than this, if we want to do complete justice to this claim, we must understand it as the highest form of "play," namely as "*imitative play*," (234 B; cf. 231 A), for the whole activity of the sophist is to be understood as an "imitation of reasonable speech," (μίμησις . . . περὶ τοὺς λόγους), made "through the ears" (διὰ τῶν ὤτων — 234 C), analogous to the activity of the painter (cf. *Protagoras* 312 C) who can, with the aid of color, *produce* "images and namesakes of beings" (μιμήματα καὶ ὁμώνυμα τῶν ὄντων) for the eyes, and who can thus simulate a whole world (234 B; cf. *Statesman* 277 C and also 288 C). We must therefore begin by seeing the sophist as a "juggler and imitator" (γόης καὶ μιμητής — 235 A). But here, in the realm of *mimesis* itself, arises the question whether the sophist's imitation is a "true" or only an "apparent" one, i.e., whether it corresponds more to the relations of similarity in geometry where the "proportions" of the originals are always preserved, or to those of "the theory of perspective" (ὀπτική) or "scene painting" (σκηνογραφία) where they are distorted according to the laws of perspective. The question is whether the sophist has to do with a "likeness" (εἰκών) or with a mere "appearance" (φάντασμα — 236 A–C), an opposition which mirrors within the realm of *mimesis*, of "image-making" (εἰδωλοποιική), the opposition of being (εἶναι) and seeming (δοκεῖν — 236 E). This question is infinitely hard to decide because it begins by taking two things for granted: a "yes" and a "no," a "being" (ὄν) and a "non-being" (μὴ ὄν). How is this juxtaposition of the "*on*" and the "*me on*" possible? This is the ultimate question to which that concerning the possibility of the sophist's being leads; to pose this question means to ask about the "image-ability" of being in general. And the whole obscurity of what is sought becomes evident especially at the point where the "many-headed" sophist (cf. *Euthydemus* 297 C), by asking about the enigmatic nature of the being of "image" (εἴδωλον) which spans (239 D–240 A) "likeness" (εἰκών) and "mere appearance" (φάντασμα), mirrors himself within himself, as it were, in an infinite image. That he demands an answer such as comes "from reasonable speech" (ἐκ τῶν λόγων) marks him, to crown it all, as the highest "imitator" — for is not just this the demand of the "philosopher," of Socrates, who once entered on the "second-best sailing" (δεύτερος πλοῦς) by taking refuge in *logoi* (*Phaedo* 99 D–E; cf. also *Statesman* 285 E–286 A) and who now plays the silent listener? But the ability to determine the nature of the being of an "image" depends on the solution of the problem concerning the "non-," for in the image as such "being" and "*non*-being" are inextricably intertwined (240 A–C). Now characteristically the presentation of the "first and greatest quandary" (μεγίστη καὶ πρώτη ἀπορία — 238 A 2; D 1), which concerns non-being, already indicates the close connection between "speaking" (λέγειν) or "thinking" (διανοεῖσθαι) the me on on the one hand and the possibility of counting, i.e., the existence of arithmoi (238 A–239 B) on the other. "Then we posit all number whatsoever as belonging to being?" ('Ἀριθμὸν δὴ τὸν σύμπαντα τῶν ὄντων τίθεμεν;) asks the "Stranger." And Theaetetus answers: "At least if there is anything to be posited as being at all." (Εἴπερ γε καὶ ἄλλο τι θετέον ὡς ὄν.) This means that what is "countable" is always understood as "being," and "being" is always understood as "countable." We always speak of what is either in the "singular" or in the "plural" (leaving apart the "dual"): even one thing is only "one" among many things (cf. P. 77; also Parmenides 144 A). But the direct connection (cf. Pp. 76 f.) between "thinking" (διανοεῖσθαι) and "accounting for and counting" (λογίζεσθαι καὶ ἀριθμεῖν) becomes especially visible when we turn toward non-being or, going further, make "non-being" itself the object of study, for even then we speak of the me on and me onta, of non-being and non-beings; we articulate even that which defies all articulation, namely — nothing! It is most significant that this is precisely the reason why the Eleatic Stranger — "as ever, so also now" (καὶ γὰρ πάλαι καὶ τὰ νῦν) — declares himself no match for "non-being"; when he pretends to expect of his interlocutor, the young Theaetetus, "straight speaking about non-being" (ὀρθολογία περὶ τὸ μὴ ὄν), the deep and secret sense of this playful turn lies exactly in this — that the call "to make some correct pronouncement about this" (κατὰ τὸ ὀρθὸν φθέγξασθαί τι περὶ αὐτοῦ) is addressed to a mathematician, who, although incompetent not only in his youth but for all time to deal with the problem of the me on within his own realm, yet holds in his hand, as it were, the key to its solution.[94] In truth, neither of the two by themselves ("forget about you and me" — σὲ μὲν καὶ ἐμὲ χαίρειν ἐῶμεν — 239 C) but only both together ("at the same time from the mathematical side and from universal reasonings" — ἅμα ἐκ τῶν μαθημάτων . . . καὶ ἐκ τῶν λόγων τῶν καθόλου — Aristotle,

Metaphysics M 8, 1084 b 24 f.) can approach a solution. The subsequent parts of the dialogue indicate the way. "The present manner of inquiry" *(ὁ τρόπος τῆς νῦν σκέψεως* — 254 C) makes no attempt to give this solution "with complete clarity" *(πάσῃ σαφηνείᾳ,* cf. also *Republic* 435 D). What *is* done is to direct attention to varying aspects of that ever-same *aporia* whose very formulation already indicates the solution in Plato's mind.

First, we must keep in mind that "image" can only "be" if "non-being" and "being" can "*mix*" with one another — which holds just as much for the being of "seeming," of mere "appearance," of "lie," of the "false," and of "error." The problem of "non-being" simply cannot be detached from the problem of "being." More exactly: in asking about "non-being" at all, *we are already directed by the question "about being"* *(περὶ οὐσίας* — 251 C-D; "about that first and greatest founder" — *περὶ τοῦ μεγίστου τε καὶ ἀρχηγοῦ πρώτου* — 243 D; cf. also the traditional subtitle of the dialogue: "About Being" — *Περὶ τοῦ ὄντος),* just as we must of necessity come upon the "philosopher" in our search after the "sophist" (cf. 231 A-B; 253 C). But the converse also holds: the difficulty of the problem of "being" has an internal connection with the *aporia* of "non-being." At bottom we are dealing only with *one* difficulty: " 'being' and 'non-being' have equal parts in this quandary" *(ἐξ ἴσου τό τε ὂν καὶ τὸ μὴ ὂν ἀπορίας μετειλήφατον* — 250 E). Thus we are at the outset concerned with one question, which is in itself twofold. This is what the "ancients" as well as the "moderns" have failed to recognize; for this reason they are unable even so much as to see a difficulty which arises in all their solutions of the problem of being. We must ask particularly those who allow the "whole" to be more than only one, that is, those who reduce everything to *two* basic constituents, as for instance the "warm" and the "cold": "But what then are you addressing in both *when you say that both and also each is?*" *(τί ποτε ἄρα τοῦτ' ἐπ' ἀμφοῖν φθέγγεσθε,* λέγοντες ἄμφω καὶ ἑκάτερον εἶναι — 243 D–E.) Is this "being" "a third thing besides those two" (τρίτον παρὰ τὰ δύο ἐκεῖνα)? Then there would be—in contradiction to the thesis—*three* basic constituents. But neither can "being" coincide with *one of both*, for then only *this one* could be said to be, and consequently there would "be" only this *one* ."But do you then want to call *both together* ' *being* '?" ('Αλλ' ἆρά γε τὰ ἄμφω βούλεσθε καλεῖν ὄν;) But if they *are* only "together" then they "are" precisely only *together*; in the present case this would mean: the "warm" and the "cold" would no longer *be* by themselves separately, but there would be clearly only *one*, something "tepid," or, more generally, a "middle thing" (cf. P. 58). What is here under examination is then nothing but the special constitution of the *koinon* of which the *Hippias Major* speaks. That "both together" (ἄμφω) are indeed "one" and yet remain "two" (cf. *Parmenides* 143 C–D) cannot be demonstrated on this lowest level of ontological reflection because the two "substrates," the "warm" and the "cold" themselves, can be mixed. The subsequent discussion plays, in ever new variations, with the problem of the *koinon* which arises wherever there are "two" aspects (cf. especially 247 D), and finally comes to a head in the treatment of the relation of "rest" (στάσις) and "change" (κίνησις). It turns out (249 D) that "change" *as well as* "rest" must "both together" (συναμφότερα) be assigned to "being." *Kinesis* and *stasis* are "most opposite to one another" (ἐναντιώτατα ἀλλήλοις — 250 A 8 f.) and therefore completely uncombinable; yet *both* and *each* of them "*is*": "Stranger: And do you say that *both* and *each of these* alike *are*? Theaetetus: Yes, indeed I do." (ΞΕ. Καὶ μὴν εἶναί γε ὁμοίως φῇς ἀμφότερα αὐτὰ καὶ ἑκάτερον; ΘΕΑΙ. Φημὶ γὰρ οὖν.) As in the case of the "hot" (θερμόν) and the "cold" (ψυχρόν), only this much can be ascertained (250 B 2–D 4): to say that "change" and "rest" "both and each" (ἀμφότερα καὶ ἑκάτερον) "are" cannot mean that "being" coincides with one of them. But it is just as impossible, if they are "both [together said] to be" *(εἶναι ἀμφότερα)*, to posit their "being" as "a 'third' thing beside them" *(τρίτον τι παρὰ ταῦτα* — 250 B 7) or "outside both of them" *(ἐκτὸς τούτων ἀμφοτέρων* — 250 D 2), "taking both together and then disregarding them to look at the community of their being" *(συλλαβὼν καὶ ἀπιδὼν αὐτῶν πρὸς τὴν τῆς οὐσίας κοινωνίαν)*. For then being would be precisely not change and rest "together" *(συναμφότερον)*, but it would also be "according to its own nature" *(κατὰ τὴν αὐτοῦ φύσιν)* and therefore *neither* rest *nor* change, which appears to be the "most impossible of all things" *(πάντων ἀδυνατώτατον)* — for what is not at rest is surely changing, and what is not changing rests! The *aporia* of "being" is here left unresolved: "So then let us rest the matter here in all its difficulty." *(Τοῦτο μὲν τοίνυν ἐνταῦθα κείσθω διηπορημένον* — 250 E 5.) And yet it is just here that Plato stresses the internal connection of the *aporia* of "being" with that of "non-being" (250 D–E). The formulations are again chosen to allow the problem concerning "both together — each of both — neither of the two" to come to the fore: The strange *koinonia* among *on, kinesis,* and *stasis* is none other than that between "being" and "non-being."

Thus the relation of *stasis* to *kinesis* forms the nucleus of all the subsequent discussion. But from now on the conversation grows broader (251 A ff.); the point of departure is that "gift of the gods to human beings" (*Philebus* 16 C), namely the "astounding" (*ibid.*, 14 C) assertion that each thing is "'one' and 'many'" *(ἓν καὶ πολλά)* "at once" *(ἅμα)*. Next it is asked how the "many" are conjoined to form the "unity" of any being. This question is nothing but the generalization of the original problem of the "two" aspects, i.e., of the meaning of "at one and the same time" *(ἅμα)* or of "both" *(ἄμφω,* cf. *Phaedo* 96 E–97A). It is raised at first in the most general terms, namely in reference to any being, and then transformed into the narrower *ontological methexis* problem, into the question of the *koinonia ton eidon* (or, of *gene*). Thereupon three possibilities, and no more, arise (251 D; 252 E): (1) There is no *koinonia* at all. (2) All the *eide* are mutually related. (3) There is partial *koinonia*, in the sense that some *eide* can "mix" with each other but others not.[95] Since the two first possibilities are not in fact realizable, the third alone, of necessity, remains (most explicitly: 252 E; cf. also 256 C). *But the very formulation of this possibility indicates the* arithmos *structure of the* gene; for what is it but the division of the whole realm of *eide* into single groups or assemblages such that each *eidos*, which represents a unique eidetic "unit" (*ἑνάς*), i.e., a "*monas*" (*Philebus* 15 A–B), can be "thrown together" with the other ideas of the same assemblage, but not with the ideas of other assemblages? The *eide*, then, form assemblages of monads, i.e., *arithmoi* of a peculiar kind. The units of which the assemblages consist are not mathematical monads, for these are, as we have seen, completely similar and can therefore all be "thrown together" (Aristotle, *Metaphysics* M 7, 1081 a 5 f.: "capable of being thrown together and indifferent" — σνμβληταὶ καὶ ἀδιάφοροι, cf. Pp. 22 f., 46 and 50). While the numbers with which the arithmetician deals, the *arithmoi mathematikoi* or *monadikoi*, are capable of being counted up, i.e., added, so that, for instance, eight monads and ten monads make precisely eighteen monads *together*, the assemblages of *eide*, the "*arithmoi eidetikoi*," cannot enter into any "community" with one another. Their "monads" are *all* of different kind and can be brought "together" only "partially," namely only insofar as they happen to belong to one and the same assemblage, whereas insofar as they are "entirely bounded off" from one another (πάντῃ διωρισμέναι — *Sophist* 253 D 9) they are "incapable of being thrown together, in-comparable" (ἀσύμβλητοι).[96]

The notion of an "arithmetic" structure of the realm of ideas now permits a solution of the ontological *methexis* problem (cf. *Parmenides* 133 A). The monads which consti-tute an "eidetic number," i.e., an assemblage of ideas,[97] are nothing but a conjunction of *eide* which belong together. They belong together because they belong to one and the same *eidos* of a higher order, namely to a "class," a *genos*.[98] But all will together be able to "partake" in this *genos* (as, for instance, "human being," "horse," dog," etc., partake in "animal") *without "partitioning" it among the* (finitely) *many* eide *and without losing their indivisible unity only if the* genos *itself exhibits the mode of being of an* arithmos. Only the *arithmos* structure with its special *koinon* character is able to guarantee the essential traits of the community of *eide* demanded by dialectic; the indivisibility of the single "monads" which form the *arithmos* assemblage, the limitedness of this assemblage of monads as expressed in the joining of many monads into one assemblage, i.e., into one idea, and *the untouchable integrity of this higher idea as well.* What the single *eide* have "in common" is theirs only *in their community* and is not something which is to be found "*beside*" and "*outside*" (παρά and ἐκτός) them (cf. also *Philebus* 18 C–D). The unity and determinacy of the *arithmos* assemblage is here rooted in the content of the idea (ἰδέα), that content which the *logos* reaches in its characteristic activity of uncovering foundations "analytically." A special kind of number of a particular nature is not needed in this realm, as it was among the dianoetic numbers (cf. P. 56), to provide a foundation for this unity. In fact, it is impossible that any kinds of number corresponding to those of the dianoetic realm should exist here, since each *eidetic number is,* by virtue of its eidetic character, *unique in kind,* just as each of its "monads" has not only unity, but also uniqueness. For each idea is characterized by being always the same and simply singular, in contrast to the unlimitedly many homogeneous monads of the realm of mathematical number, which can be rearranged as often as desired into definite numbers (see P. 54). The "pure" mathematical monads are, to be sure, differentiated from the single objects of sense by being outside of change and time, but they are not different in this sense — that they occur in *multitudes* and are of the *same kind* (Aristotle, *Metaphysics* B 6, 1002 b 15 f.: [Mathematical objects] differ not at all in being many and of the same kind — τῷ δὲ πόλλ' ἄττα ὁμοειδῆ εἶναι οὐθὲν διαφέρει), whereas each *eidos* is, by contrast, unreproducible and truly *one* (*Metaphysics* A 6, 987 b 15 ff.: "Mathematical objects differ from objects of sense in being everlasting and unchanged, from the *eide*, on the other hand, in being many and alike, while an *eidos* is each by itself one only" — τὰ μαθηματικὰ . . . διαφέροντα τῶν μὲν αἰσθητῶν τῷ ἀΐδια καὶ ἀκίνητα εἶναι, τῶν δ'εἰδῶν τῷ τὰ μὲν πόλλ' ἄττα ὅμοια εἶναι τὸ δὲ εἶδος αὐτὸ ἓν ἕκαστον μόνον). In consequence, as Aristotle reports (e.g., *Metaphysics* A 6, 987 b 14 ff. and N 3, 1090 b 35 f.), there are three kinds of *arithmoi*: (1) the *arithmos eidetikos* — idea-number, (2) the *arithmos aisthetos* — sensible number, (3) and "between" (μεταξύ) these, the *arithmos mathematikos* or *monadikos* — mathematical or monadic number, which shares with the first its "purity" and "changelessness" and with the second its manyness and reproducibility. Here the "aisthetic" number represents nothing but the *things themselves* which happen to be present for *aisthesis* in this number. The mathematical numbers form an independent domain of objects of study which the *dianoia* reaches by noting that its own activity finds its exemplary fulfillment in "reckoning [i.e., account-giving] and counting" (λογίζεσθαι καὶ ἀριθμεῖν). The eidetic number, finally, indicates the *mode of being of the* noeton *as such — it defines the* eidos *ontologically as a being which has multiple relations to other* eide *in accordance with their particular nature and which is nevertheless in itself altogether indivisible.*

The Platonic theory of the *arithmoi eidetikoi* is known to us in these terms only from the Aristotelian polemic against it (cf., above all, *Metaphysics* M 6–8). It is questionable whether Plato sketched out more than the general framework of the theory. In his lecture *On the Good* ("Περὶ τ'ἀγαθοῦ")[99] he seems to have limited the realm of eidetic numbers to *ten* (cf. *Metaphysics* Λ 8, 1073 a 20; M 8, 1084 a 12 ff., 25 ff., and *Physics* Γ 6, 206 b 32 f., and elsewhere). In this he, as well as his successor Speusippus, remained true to the Pythagorean tradition; indeed, the eidetic numbers might, in their foundational function, be most easily compared to the Pythagorean "roots" *(πυθμένες)* of the realm of mathematical number. Now in understanding the *arithmoi* in the only way in which they can be understood in their very own, mathematical, domain, Aristotle exhibits the many contradictions which must arise from the transfer of the *universal* character of the countable as such to the *eide*, each of which has a special nature. *For Plato, however, it is precisely this unmathematical use of the* arithmos *structure which is essential.* For the *arithmoi eidetikoi* are intended to make intelligible not only the inner articulation of the realm of ideas but every possible articulation, every possible division and conjunction — in short, all *counting*. While the arithmetician and the logistician "suppose" certain *eide* to "underlie" the unlimitedly many monads of his domain in order to have "hypothetical" grounds on which they may be comprehended into single monadic assemblages (cf. Pp. 72 and 78), only the dialectician is able to give the true grounds for the existence of such *eide* of numbers and of each single number of pure units. Only because there are *eide* which belong together, whose community in each case forms a "kinship" which must, due to the "arithmetical" tie among its "members," be designated as *the* six or *the* ten, can there be arbitrarily many numbers, such as hexads or decads, in the realm of "pure" units as well as in the realm of sensibles; furthermore, only because of this can numbers exhibit such definite, unifying kinds (cf. Pp. 55 f.) as the "even-times-even" or the "triangular." Only the *arithmoi eidetikoi* make something of the nature of number possible in this our world. *They* provide the foundation for all counting and reckoning, first in virtue of their *particular* nature which is responsible for *the differences of genus and species* in things so that they may be comprehended under *a definite number*, and, beyond this, by being responsible for the *infinite variety* of things, which comes about through a "distorted" imitation of ontological *methexis* (cf. Pp. 82, 80 and 98–99; cf. *Philebus* 16 C–E; *Timaeus* 43). This foundational function guarantees their separate, independent and "absolute" being in relation not only to the *aistheta* but also to the "pure" mathematical numbers (see P. 71). What the Pythagoreans undertook with respect to the world of sense in which they believed that all beings were comprised (*Metaphysics* A 8, 990 a 3–5), Plato now undertakes to do with respect to the world opened up by the *logos*, the world of *noeta* which has true being. For him, as for them, the "numerical" being of the *noeta* means their ordered being, their *taxis*. The mathematical monads by themselves form a homogeneous field, but the "sequence" of monadic numbers (cf. P. 52) is grounded in the original *order* of the *eidetic* numbers. Every eidetic number is either "superior" or "inferior" in this order with respect to its "neighbor," so that a subsumption of all these numbers under one idea common to all, namely "number in general," is quite impossible.[100] This *taxis* of eidetic numbers is "logically" expressed in the relation of "being superior" or "inferior" in the order of *eide;* the relation of family "descent" between the higher and the lower ideas corresponds to the "*genetic*" *order of the eidetic numbers*. The "higher" the *genos*, i.e., the less articulated the eidetic number, the more original and "*comprehensive*" it is. In this order the "first" eidetic number is the eidetic "two"; it represents the *genos* of "*being*" *as such*, which comprehends the two *eide* "rest" and "change." That this last tenet, at least, is genuinely Platonic can be seen, if not "with complete clarity" yet clearly enough, from the *Sophist*.[101]

The third possibility, that of a "partial" *koinonia ton eidon* (cf. P. 89) is here investigated only for the "greatest *genera*" (μέγιστα γένη — 245 C), namely the following "five": being, rest, change, the same, the other (ὄν, στάσις, κίνησις,

ταὐτόν, θάτερον, cf. 254 E 4; 255 C 8; 255 E 8; 256 C–D).
This "count," the only one possible within the dianoetic-
dialogic method, does as little justice to the true ontological
state of affairs as the assertion that the sequence "sophist,"
"statesman," "philosopher" are three *gene* of equal importance to which, therefore, three dialogues should correspond
(217 A–B; cf. 254 B and *Stateman* 257 A–B). The purpose of
the inquiry concerning the *koinonia* among the "five"
greatest *gene* is, as is expressly stated, to grasp "being" as
well as "non-being" in a manner suited to the present mode
of examination, though for this very reason not completely
adequate: "So that if we are not able to grasp being and non-
being *with complete clarity*, we may yet at least *miss as little of
the account* as the present manner of inquiry allows." (ἵνα τό
τε ὂν καὶ μὴ ὂν εἰ μὴ πάσῃ σαφηνείᾳ δυνάμεθα λαβεῖν, ἀλλ' οὖν
λόγου γε ἐνδεεῖς μηδὲν γιγνώμεθα περὶ αὐτῶν, καθ' ὅσον ὁ τρόπος
ἐνδέχεται τῆς νῦν σκέψεως — 254 C; cf. P. 86) It is shown
(254 D–257 A) that there is indeed a "community" among
all these *gene*, although the kind of "community" is by no
means the same in all cases. The immediate basis of the
discussion is the *incompatibility* of *stasis* and *kinesis*, and this
fact must not be overlooked. Their "mixing" had already
been mentioned (cf. P. 87) in the strongest terms as
"impossible by the greatest necessity" (ταῖς μεγίσταις
ἀνάγκαις ἀδύνατον — 252 D). This is again stated in 254 D
7–9: "Stranger: And we do say that the two of them are not
to be mixed with one another. Theaetetus: We stress it."
(ΞΕ. Καὶ μὴν τώ γε δύο φαμὲν αὐτοῖν ἀμείκτω πρὸς ἀλλήλω. ΘΕΑΙ.
Σφόδρα γε.) But both "are", and from this follows the
"triad" of *stasis*, *kinesis*, and *on* (D 12), although it has already
been shown that the *on* is *not* to be understood as a "third
thing beside" or "outside these" (τρίτον παρὰ ταῦτα or
ἐκτὸς τούτων), since this would lead to the "most impossible
thing of all" (250 B–D, see P. 88). In respect to *on*, *kinesis*
and *stasis*, the *logos* fails! It fails because it must count
"three" when *in truth* there are only "two," namely *stasis*

and *kinesis*, which are "each one" and "both two"! (*ἑκάτερον ἕν* and *ἀμφότερα δύο*[102]—cf. also *Theaetetus* 185 B, *Timaeus* 38 B.) The *logos* cannot conclude the count with "two" because it says that *stasis* and *kinesis* "are" not only "together" but also "singly," while in the case of "two mathematical monads" it understands each of these by itself as *only one* and precisely *not as "two"* (cf. P. 81). *On, kinesis, and stasis, in spite of their "arithmetical" koinonia, cannot be "counted" at all* — this defines the "failure" of the *logos*.[103] The dianoetic understanding is clear only about this much, that each of the "three" things presupposed by it, insofar as each is "itself" exactly that which it is, and is grasped in its "self-sameness," is an "other" than the "two" others: "Then each of them is other than the two, but the same with itself." (*Οὐκοῦν αὐτῶν ἕκαστον τοῖν μὲν δυοῖν ἕτερόν ἐστιν, αὐτὸ δ'ἑαυτῷ ταὐτόν* — 254 D 14 f.) This crucial quandary of the *dianoia* is compounded by the introduction of yet a further "pair": "self-sameness" and "otherness" (*ταὐτόν — θάτερον*), but with these also emerges the solution which must suffice within the dianoetic realm (cf. 257 A 9–11): It consists in this, that "the other," *thateron*, analogously to the vowels among the letters (253 A 4–6), ranges "through all" (*διὰ πάντων* — 253 C 1) *gene* whatsoever (255 E 3 f.), and is, accordingly, both a connecting "bond" (*δεσμός*) and the agent "responsible for their division" (*αἴτιον τῆς διαιρέσεως* — 253 C 3), that is, "the other" is the "ultimate source" of all articulation whatsoever. *Thateron* is in itself twofold, for "the 'other' is always in relation to an 'other'" (*τὸ δέ γ'ἕτερον ἀεὶ πρὸς ἕτερον* — 255 D 1; cf. D 6 f.). This means that the possibility of "otherness" is dependent on the "self-sameness" of the participants in the relation of "otherness"; therefore *tauton*, too, pervades *all* the *gene*. This *koinonia* between *tauton* and *thateron*, which is nothing but another expression for the internal "twofoldness" of *thateron* itself,[104] permits the *dianoia* to understand the "duplicity" of "being," namely, that it means not only ever self-identical "rest," but conjointly also "change," and that this alone makes possible the "imaging" of being in "re-cognition," that is, of "knowing" and "being known" (γιγνώσκειν and γιγνώσκεσθαι —248 B ff.), and, beyond this, *all* image making (cf. *Cratylus* 439 E–340 A). "Conjointly" does not mean here, as for instance in the dialectic of Nicholas of Cusa or of Hegel, a "coincidence of opposites." Just as the *dianoia* finds in the realm of the "more and less" (μᾶλλον καὶ ἧττον) the "opposition," the "obstacle" (ἐναντίωμα) which first "awakens" it (see Pp. 75 ff.), so it must finally, at the end of its "dialectic" activity, come to see that the "conjunction" of *opposites* is in truth the "co-existence" of *elements other in kind.* "Otherness" makes possible an "arithmetic community" among eidetic monads which are not capable of "being mixed" although they "belong together"; the paradigm for such a *koinonia*, which is no longer accessible to the *logos* or, therefore, to counting, is the eidetic "two" which *consists of* stasis *and of* kinesis; *kinesis* is that which, in confrontation with *stasis*, is the "other" *without which even* stasis *itself cannot* "be," since precisely only "both together" amount to "being." This means that "being" itself is accompanied of necessity by a "not"; just as *stasis* is *not kinesis*, so *kinesis* is *not stasis.* "Otherness" turns out to be the ontological aspect of "non-being," which can never be separated from "being": "Of necessity, then, non-being is [immediately] involved in change and occurs throughout all the *gene* [of being]; for throughout these the nature of 'the other' works on each being to make it other and [thus] a non-being [namely not thus and such a being but another], and accordingly we rightly speak of all things whatsoever as 'non-beings,' and conversely, because things partake of being, as 'being' and 'beings.'" (Ἔστιν ἄρα ἐξ ἀνάγκης τὸ μὴ ὂν ἐπί τε κινήσεως εἶναι καὶ κατὰ πάντα τὰ γένη· κατὰ πάντα γὰρ ἡ θατέρου φύσις ἕτερον ἀπεργαζομένη τοῦ ὄντος ἕκαστον οὐκ ὂν ποιεῖ, καὶ σύμπαντα δὴ κατὰ ταὐτὰ οὕτως οὐκ ὄντα ὀρθῶς ἐροῦμεν, καὶ πάλιν ὅτι μετέχει τοῦ ὄντος, εἶναί τε καὶ ὄντα — 256 C–D.)[105]

The shadow of "non-being" necessarily attends all the "being" of that which is — just as the sophistic refutation, the *elenchos*, belongs to the essential business of the philosopher (cf. 230 A–231 B), just as the mutability of the cosmos perpetually returning into itself is the most immediate image of the true cosmos in its immutability. Everywhere "not being" is only "being *other*," "not something the *contrary of*, but *only other than*, being" (οὐκ ἐναντίον τι . . . τοῦ ὄντος ἀλλ' ἕτερον μόνον — 257 B; cf. 258 B), for the contrary of being would amount to the unthinkable as well as unspeakable "nothing"; what we everywhere have is an "opposition" of one being to an other being before us, "a confrontation of being with being" (ὄντος πρὸς ὂν ἀντίθεσις — 257 E 6). This is the reason for the possibility of a "mistake" or "interchange" of the "one" and the "other" or of "being" and "non-being," a possibility on which all "contradiction" (cf. 232 B), all "illusion," all "error," and every "lie" depend (260 B–264 B, also 266 D–E).

This "duplicity" of being is the ontological foundation and justification of the method of "division" (διαίρεσις). The "divisions" at the beginning and at the end of the *Sophist* (as also in the *Statesman*) are intended to hint repeatedly at this "duplicity" (cf. also *Statesman* 287 C; 306 C ff.; also *Charmides* 159 B ff.). They are intended at the same time to train the dialectician in "dividing according to genera" (κατὰ γένη διαιρεῖσθαι — cf. *Sophist* 253 D; *Statesman* 285 C–D), an activity in which he must be expert if he is to reach the primal "genetic" order of eidetic numbers.[106] In particular, the *Sophist* is intended to show that all articulated arrangements of *gene*, in other words, the "arithmetic" community of the ideas, can be understood only by means of "the other." Its very "nature" (φύσις) consists in "being broken up into parts," which makes it — certainly not accidentally — akin to the discerning *episteme*: "The nature of the other seems to me to be all broken up just like knowledge." ('Η θατέρου μοι φύσις φαίνεται κατακεκερματίσθαι καθάπερ ἐπιστήμη — 257 C 7 f.; 258 D–E; cf. *Parmenides* 142 E, 144 B and E; *Theaetetus* 146 C ff., *Meno* 79 A, C.) It is always, as it were, only a "part" *(μόριον)* of itself, namely an "*other of another*" *(ἕτερον ἑτέρου)*[107] — a "counter-part". In the context of the Platonic search for foundations (cf. P. 71) this *arche* of all doubleness must be recognized as the "twofold in general," the *aoristos dyas*.[108] This *dyas* is *aoristos* because it does not itself represent "two" beings of some particular kind such as are mutually delimited and univocally determined.[109] Rather, in endowing the being of each thing with "imageability," it "doubles" every thing, and so first allows it to come into "being" at all — it is "two-making" *(δυοποιός* — Aristotle, *Metaphysics* M 8, 1083 b 35 f.; M 7, 1082 a 15). Thus by a continual "duplication" of the *eide* — which the *logos* grasps in the "division" of the *gene* — it makes the "genetic" order of the eidetic numbers possible.[110] But it can do all this only because at the "head" of the *taxis* of eidetic numbers, at once concluding and introducing them, is the *One Itself* in its "absolute" priority (cf. P. 77). Since it is beyond all articulation, beyond the "two," and thus "beyond being itself" *(ἐπέκεινα τῆς οὐσίας* — *Republic* 509 B), it is not, like the mathematical unit, one one among many,[111] but rather the original, perfect, *all-comprehensive Whole* (cf. *Sophist* 244 D–245 D; also *Parmenides* 137 C, 142 D). As the "Whole" it is that which needs no "other" at all, that which is altogether "finished." In this sense it is "the perfect itself," namely, the model of every possible "relative" wholeness which is "delimited" in respect to an "other": it is the "Idea of the Good" *(ἰδέα τοῦ ἀγαθοῦ)*.

The doctrine of the *gene* as eidetic numbers must, finally, also furnish the foundation of an *eidetic logistic*. For instance, sound-mindedness *(σωφροσύνη)* and justice *(δικαιοσύνη)* in Book IV of the *Republic* (cf. I 337 A–C!; Aristotle, *Nicomachean Ethics*, E 6, 7), or the *taxis* of elemental materials in the *Timaeus*, can be understood only by means of *analogia*, proportion (cf. *Theaetetus* 186 A–B). So also the relation of the ontological to the dianoetic *methexis* problem (cf. P. 80), as well as the relation of original to copy in general, becomes comprehensible only in "logistic" terms. What is usually overlooked in discussions of the *methexis* question is the secondary, *the imaging, character of the whole* methexis *relation*, insofar as it concerns the dianoetic realm, i.e., the relation of one *eidos* to a series of *aistheta*. Only when these relations are reduced to relations of "community" within the realm of *eide*, can we see the *methexis* problem in its *original* form. But one of the possible solutions to this higher problem is precisely the conception of the *arithmos eidetikos*.

This solution at once gives the final answer to the problem of the "one and many": The *arithmos eidetikos* exhibits in itself the possibility of an immediate unification of the many. But this solution is bought, as we have seen, at the price of the transgression of the limits which are set for the *logos*, for from this point of view the ordinary mode of predication, such as: "the horse *is* an animal," "the dog *is* an animal," etc., is no longer understandable. Above all, the "natural" meaning intended when a multiplicity of things is called an "*arithmos*" is now lost. Here, therefore, are the points of departure for further development or correction of the Platonic doctrine by Speusippus and Xenocrates. But only Aristotle's critique exposes the root of this, as of all the other related difficulties, namely the postulate of the "separation" (χωρισμός) of all noetic formations, and in particular the *chorismos* of the *arithmoi monadikoi*, the numbers of "pure" monads.

8

The Aristotelian critique and the possibility of a theoretical logistic

THE PLATONIC *chorismos* thesis has its strongest support in mathematics. The exemplary *"mathema"* character of mathematical objects, their undeniably pure noetic nature, their "indifference" with regard to objects of sense — all this immediately indicates the possibility of the existence of noetic structures which are independent of and "detached," i.e. separated, from all that is somatic, just as the Platonic thesis affirms. Thus the obvious point of departure for the Aristotelian criticism of the Platonic school is the onto-logical standing attributed by them to the mathematical realm, in particular, to the numbers of pure units (cf. P. 71). It is not that Aristotle questions mathematical science itself — the Aristotelian theory of apodeictic science is too much indebted to procedures developed in mathematics for this — nor that he denies that mathematical inquiry has a special field of objects (as, for instance, arithmetic has the field of pure monads, cf. *Posterior Analytics* A 10, 76 b 4 f.); he is rather concerned with proving the Platonic conception of the *mode of being* of mathematical objects false, "so that our controversy will be not about their being but its mode" (ὥσθ᾽ ἡ ἀμφισβήτησις ἡμῖν ἔσται οὐ περὶ τοῦ εἶναι ἀλλὰ περὶ τοῦ τρόπου — *Metaphysics* M 1, 1076 a 36 f.).

Let us briefly recall the starting points of the Aristotelian criticism as they emerge in the course of an analysis of what is meant in ordinary speech. The fact that it becomes possible to call something a "whole" by indicating the "parts" of which the whole "consists," and that, therefore, "for declarative speech" *(τῷ λόγῳ)* the parts precede the whole does not mean that in respect to their "being" *(τῇ οὐσίᾳ)* these parts can lay claim to priority over the being of the whole (cf. *Metaphysics* M 2, 1077 b 1 ff. and Δ 11, 1018 b 34 ff.). To call an object which appears before us, e.g., a human being, a "white" human being presupposes the partial assertion "white," and yet in this particular assertion of "white" no other being is meant than just — this white human being (cf. on this *Metaphysics* Z 4, 1029 b 13 ff.). The "white" intended has no existence "outside" of this human being and "separate" from him. Its being is bound to the given whole, namely to "this white human being." It cannot exist if this human being of which it is predicated does not exist. Its being is therefore *dependent* on the being of the human being. *Exactly* in this same way does the assertion "three trees" presuppose the assertion "three," but what the assertion "three" intends has no existence "outside" of the trees of which there are said to be three. For the number of the trees, i.e., "three," has no proper, no independent, "nature" *(φύσις,* cf. *Metaphysics* M 6, 1080 a 15; 7, 1082 a 16; 8, 1083 b 22, and elsewhere). Their being "so many," just like their being, for instance, "green," is dependent on their being *trees*.

At the root of this Aristotelian conception lies the "natural" meaning of *arithmos*; the assertion that certain things are present "in a certain number" means only that *such* a thing is present in just this definite multitude: "To be present in number is to be some number of a [given] object." *(τὸ δ' εἶναι ἐν ἀριθμῷ ἐστὶ τὸ εἶναί τινα ἀριθμὸν τοῦ πράγματος* — *Physics* Δ 12, 221 b 14 f.; cf. Pp. 46 f.) But now, in determining the ontological standing of number, it becomes important not only to remember the "natural" significance of the *arithmos* but also to take into account its "dependence." Necessary though it be to presuppose numbers of "pure" units in order to understand the prior knowledge of numbers which is revealed in our daily calculating and counting, yet we must not, on that account, conceive of their being as independent, and separate or "absolute." The mode of being of the "pure" numbers can simply not be sufficiently determined from the point of view of the attempt to "account" for (speaking Platonically: of an "account of the responsible reason" for — λογισμὸς αἰτίας, cf. Meno 98 A) the possibility of counting and calculating.[112] Rather the dependence and bondage is indicative of the being of number. The whole difficulty is here precisely to bring this character of all possible numbers, and thus also of "pure" and "mathematical" number, into consonance with the purely noetic nature of the latter.

How will we be able, we must ask, to extract all the single parts *(μέρη)*, the single "constituents" of a thing which we get hold of in the *logos* one after another, e.g., "this" "round" "white" "column," from the concrete context into which they are fitted by reason of their being and to study each of them separately? Clearly only in this way: that in each case we *disregard* certain attributes of the thing in question, ignoring the nexus of being which links them all to one another. This "disregarding of . . ." is able to produce a new mode of seeing which permits something to come to light *in* the *aistheta* which, for all their variety and transitoriness, suffers no change but remains always in the same condition, thus fulfilling the demand that it can be an object of some science, of an *episteme*. In particular, the possibility of subjecting the numerical aspects (or the dimensions) of *aistheta* to an apodeictic discipline is based on this "disregarding of" every particular content. This, as Aristotle says, assures us "that there may be definitions and proofs [i.e., a science] of sensible magnitudes, though not insofar as these are objects of sense, but insofar as they are just such [namely so great or so many]" *(ὅτι ἐνδέχεται καὶ περὶ τῶν αἰσ-θητῶν μεγεθῶν εἶναι καὶ λόγους καὶ ἀποδείξεις, μὴ ᾗ δὲ αἰσθητὰ ἀλλ' ᾗ τοιαδί — Metaphysics M 3, 1077 b 20 ff.).* As objects of knowledge, these mathematical objects — the stereometric, planimetric, and arithmetic formations which are, as it were, read off the *aistheta* — are no longer subject to the senses, though they do not achieve any independent being, i.e., a being alongside that of the *aistheta*. Rather, their being remains dependent on the being of the objects of sense. However, insofar as the mathematical aspects become visible in their "purity," detached from all other content, they may be isolated within the "whole" and may be, without detriment to their dependence, "lifted off," as it were, from the "whole." For their more exact investigation this holds no danger. For: "if someone, positing things as separated from that which [otherwise] goes with them, examines them for that about them by which they are such [namely 'separable'], there will be no more falsification because of this than when someone draws something on the ground [for the purpose of demonstrating geometric theorems] and says that it has the length of a foot when it has not;[113] for the falsity is not in the premises [as such]." *(εἴ τις θέμενος κεχωρισμένα τῶν συμβεβηκότων σκοπεῖ τι περὶ τούτων ᾗ τοιαῦτα, οὐθὲν διὰ τοῦτο ψεῦδος ψεύσεται ὥσπερ οὐδ' ὅταν ἐν τῇ γῇ γράφῃ καὶ ποδιαίαν φῇ τὴν μὴ ποδιαίαν· οὐ γὰρ ἐν ταῖς προτάσεσι τὸ ψεῦδος — 1078 a 17 ff.)* Such a mode of study not only does not falsify, but it even allows the object of study to appear especially clearly: "Each thing may be viewed best in this way — if one posits that which is *not* separated [i.e., which has no separate existence] *as separate*, just as the arithmetician and the geometer do." *("Ἄριστα δ' ἂν οὕτω θεωρηθείη ἕκαστον, εἴ τις τὸ μὴ κεχωρισμένον θείη χωρίσας, ὅπερ ὁ ἀριθμητικὸς ποιεῖ καὶ ὁ γεωμέτρης — 1078 a 21 ff.)* This is the way in which these mathematical formations first become objects of *science*, which then has to accept their content, their "what," as given. Moreover, science simply has to "accept" (λαμβάνειν) the "being" of the various original formations, namely of the "one," the "line," the "plane," etc., and to "derive" from it the "being" of the rest, i.e., to display the noncontradictory connection of *all* the arithmetical and geometrical content which was given in advance (cf. *Posterior Analytics* A 10, 76 a 31–36). But to determine how this "being" itself is to be understood — this is no longer the task of mathematics but of "first philosophy" (πρώτη φιλοσοφία) alone (cf. *Metaphysics* K 4, 1061 b 25–27). It is her task to trace the mode of availability of the mathematical formations back to a *separation effected by reflective thought*. Aristotle's so-called "theory of abstraction" is, after all, not so much a "psychological" explication of certain cognitive processes as an attempt — fraught with heavy consequences for all later science — to give an adequate ontological description of noetic objects like the *mathematika*.

Science studies those objects, which in respect to their being are not "detached," *as if* they were detached or separated from objects of sense: "It thinks the mathematical objects which are not separate as separate when it thinks them." (τὰ μαθηματικὰ οὐ κεχωρισμένα ὡς κεχωρισμένα νοεῖ, ὅταν νοῇ ἐκεῖνα — *On the Soul* Γ 7, 431 b 15 f.) Accordingly the *mathematika* have their being "by abstraction" (ἐξ ἀφαιρέσεως), that is, their separate mode of being arises from their being "lifted off," "drawn off," "abstracted." This is why the "dependence" of mathematical formations works no detriment to their noetic character. For being "lifted off" is only a different expression for precisely that aforementioned "disregarding" of all the other content of things. In this "disregarding of..." the objects of sense wither away, as it were, and become "items" or "mere bodies." Thus these things are deprived of their *aisthetic* character and, to a great extent, robbed of their differences: "The mathematician makes those things which arise from abstraction his study, for he views them after having *drawn off* all that is sensible . . . , and he *leaves* only the [object of the question] 'how many?' and continuous magnitude." (. . . ὁ μαθηματικὸς περὶ τὰ ἐξ ἀφαιρέσεως τὴν θεωρίαν ποιεῖται, περιελὼν γὰρ πάντα τὰ αἰσθητὰ θεωρεῖ . . ., μόνον δὲ καταλείπει τὸ ποσὸν καὶ συνεχές . . . — *Metaphysics* K 3, 1061 a 28 ff.) If the reduction goes so far that things are no longer regarded even as "mere bodies" but only as "items," these things have been transformed into "neutral" monads. Just this "neutrality" of things which have withered away into mere countable "items" constitutes the "purity" of the "arithmetic" monads and turns them into the noetic material which underlies scientific study. Not *original* "detachment" but *subsequent* "indifference" characterizes the mode of being of pure numbers (*Metaphysics* M 2, 1077 a 15–18).

For Pythagorean science, as for Platonic philosophy, the basic problem of the theory of numbers was the problem of the "unity" of an *arithmos* assemblage — how the "many" can be understood as "one" at all. We saw that the whole structure of Greek theoretical arithmetic is conditioned by this approach. We saw, furthermore, that the Platonic doctrine of the *arithmoi eidetikoi* implies that the unity of numbers ultimately has its roots in the indivisible qualitative wholeness of a *genos* (cf. Pp. 90, 92 f., also Pp. 55 f.). Speaking generally, for Plato generic identity is the ultimate foundation of all possible unity. This is the very view which, above all, Aristotle attempts to refute in his attack on the *chorismos* thesis: Not only is the problem of the "unity" of the numbers or of the fact that soul and body are indeed one, or generally the problem of the relation of *eidos* to sensible object, *not* solved in this way: ". . . by reason *of what* the numbers are one, or the soul and the body, or generally, the *eidos* and the thing, no one tells us at all . . ." (. . . τίνι οἱ ἀριθμοὶ ἕν ἢ ἡ ψυχὴ καὶ τὸ σῶμα καὶ ὅλως τὸ εἶδος καὶ τὸ πρᾶγμα, οὐδὲν λέγει οὐδείς . . . — cf. *Metaphysics* Λ 10, 1075 b 34 ff.;

cf. also K 2, 1060 b 10–12); but also if generic identity entails unity, then "unity" is attributed to formations which, *strictly speaking, cannot be "one"* at all. Indeed, in speaking of "number," for instance, we mean precisely *more* than "one thing": "Some things are one by [immediate mutual] contact [of the parts], others by mingling, yet others by the disposition [of the parts]; none of this can possibly occur in the monads of which the dyad and the triad [consist], but just as two men are *not* one thing over and above *both* of them [being each by himself one], so is it necessarily also with [pure] monads."

(. . . τὰ μὲν ἀφῇ ἐστὶν ἓν τὰ δὲ μίξει τὰ δὲ θέσει· ὧν οὐδὲν ἐνδέχεται ὑπάρχειν ταῖς μονάσιν ἐξ ὧν ἡ δυὰς καὶ ἡ τριάς· ἀλλ᾽ ὥσπερ οἱ δύο ἄνθρωποι οὐχ ἕν τι παρ᾽ ἀμφοτέρους, οὕτως ἀνάγκη καὶ τὰς μονάδας

— M 7, 1082 a 20 ff.) What misleads us is precisely the supposed "detachment" of the pure monads, since under this supposition the possibility of collecting *two* monads in *one arithmos*-assemblage cannot but be understood as the effect of an original and therefore *independent eidos*, be it the "odd" *(ἄρτιον)*, be it the eidetic "two" (cf. P. 57 ff. and 93). For *these* monads, whose noetic character manifests itself in their "absolute" *indivisibility* (see Pp. 40 f.), in their unlimited multiplicity (see Pp. 51 and 54), and in their complete similarity (see Pp. 22 f.),[114] just do not offer any "natural" articulation (such as is found in the ever different and ever divisible objects of sense) which might serve as the *original source* of delimitation and unification productive of particular assemblages, i.e., of definite numbers. And since the pure "monads" are, in truth, only objects of sense reduced to mere countable "items," "they do not differ [from *aistheta*] because they are indivisible [while *aistheta* are divisible], for points [which, although they are also purely noetic formations, can nevertheless be immediately 'represented' by sensible signs] are also indivisible, and yet there is *nothing other aside from their 'being two'* [i.e., no 'one' thing different from themselves] which might be termed their 'twoness.'" (καὶ οὐχ ὅτι ἀδιαίρετοι, διοίσουσι διὰ τοῦτο· καὶ γὰρ αἱ στιγμαὶ ἀδιαίρετοι, ἀλλ' ὅμως παρὰ τὰς δύο οὐθὲν ἕτερον ἢ δυὰς αὐτῶν — Metaphysics 1082 a 24–26; cf. also M 9, 1085 a 23–31). A number is simply not *one* thing but a "heap" (σωρός) of things or monads (cf. Metaphysics H 3, 1044 a 4; M 8, 1084 b 21 f.; also above, P. 51).[115] "Being a number" is not a *koinon*, to be taken as a "whole" *above* and *alongside*, as it were, the parts of the "heap" (cf. Metaphysics H 6, 1045 a 8–10). "For a number is [only] that which has been counted or can be counted." (ἀριθμὸς γὰρ ἢ τὸ ἠριθμημένον ἢ τὸ ἀριθμητόν — Physics Δ 14, 223 a 24 f.; cf. Γ 5, 204 b 8.) Our preknowledge of all possible numbers does, to be sure, make available indifferent formations, each of which represents "a number by which we count" (ἀριθμὸς ᾧ ἀριθμοῦμεν — Physics Δ 11, 219 b 5 ff.; 220 b 4 f.), but it is a number which does *not* coincide with what is counted, and must not be called *one*. But just as this "availability" (which should, following Plato, be called a "stored possession" [κτῆσις] and not a "possession in use" [ἕξις] — Theaetetus, 197 B ff.; cf. Aristotle, Posterior Analytics A 1) is *revealed* only as the counting proceeds, so also it is *derived* from the experience of counting multitudes and of culling from them those indifferent formations "by abstraction" (ἐξ ἀφαιρέσεως). If we center our attention on this preknowledge itself, so that it is raised to the rank of a science (cf. Pp. 19 f. and 49 f.), we see that here too we are dealing with "heaps," namely with heaps of "pure" monads, which may be understood as "separable" but not as originally "detached." In this inexplicit preknowledge, numbers "are one" only as much as is anything whatever which extends "over the whole" (καθόλου, cf. Posterior Analytics B 19), i.e., which is general. As objects of explicit mathematical knowledge, the numbers of "pure" monads are as little "one thing" as any number of sensible objects (cf. also Metaphysics M 4, 1079 a 34–36).[116]

There remains, it is true, the question *by what* those numbers which are "heaps" are demarcated from one another, how it is possible to call one number just that, namely *one*. But this question should be posed only concerning objects which are *actually counted*. When we consider that all counting presupposes the *homogeneity* of that which is counted insofar as it is counted (see Pp. 46 and 53), we see not only that every number consists of many units, i.e., *is many* "ones" (ἕνα, see P. 51), but also that it is *kept together by a common measure*, namely by a particular unit which has become the basis of the count; this common measure first enables "many ones" to become "many": "For each number is 'many' because each is [made up of] 'ones' and because each is measured by [its own] 'one'." (πολλὰ γὰρ ἕκαστος ὁ ἀριθμὸς ὅτι ἕνα καὶ ὅτι μετρητὸς ἑνὶ ἕκαστος — *Metaphysics* I 6, 1056 b 23 f.; cf. *Physics* Δ 12, 220 b 20–22.) In *this* sense the "one" (or the *one* thing subjected to counting) makes counting and thus "counting-number" possible: In *this* sense it takes precedence over number and may be called its *arche* (cf. Pp. 53 f. and *Metaphysics* Δ 6, 1016 b 17–20; 15, 1021 a 12 f.; N 1, 1088 a 6–8). The priority of the one over number does not follow from a relationship of superiority of genus over species, but rather from the character of the one as "measure." And likewise the "unity" of a number is not of a generic sort but derives from the unity which each object has in virtue of being a "measure" of the count. We comprehend a number as *one* because we do our counting over one and the same thing, because our eyes remain fixed on *one and the same thing*.[117] And so in reverse — that things "are one" marks them as countable; it follows from this and only from this that "to be one is to be indivisible." (τὸ ἑνὶ εἶναι τὸ ἀδιαιρέτῳ ἐστὶν εἶναι — *Metaphysics* I 1, 1052 b 16; cf. I 3, 1054 a 23; furthermore, *Physics* Γ 7, 207 b 6 f., and P. 53.) For indivisibility belongs to things only insofar as they supply the *measure of a possible count*. And only insofar as it is understood as indivisible is a thing called "one": "Whatever is not subject to division is called one in respect of that by reason of which it is not divisible." (. . . ὅσα μὴ ἔχει διαίρεσιν, ᾗ μὴ ἔχει, ταύτῃ ἓν λέγεται . . . — Δ 6, 1016 b 4–6.) Thus the "one" is not a common property (κοινόν — I 1, 1053 a 14) but a measure (μέτρον — 1053 b 4 f.; 1052 b 18 ff.; Λ 7, 1072 a 33; N 1, 1087 b 33 ff.). And thus also the unity of a number of things is only the unity of its "measure," namely of the very object which is subjected to counting and is in that capacity indivisible, whence number can be very generally defined: "For number is a multitude measured by a unit." (ἔστι γὰρ ἀριθμὸς πλῆθος ἑνὶ μετρητόν — I 6, 1057 a 3 f.)

Now just as in every measurement, be it of weight, of speed, of area, of distance, etc., we count off the respective units of measurement (Δ 6, 1016 b 21 ff.; I 1, 1052 b 19 ff.; N 1, 1087 b 34–1088 a 2) — which is why we may call a mere count a "measurement" and the counted thing a "measure" — so we habitually reduce every count to a neutral expression. We do not say: one apple, two apples, three apples, but rather; one, two, three . . . (cf. M 7, 1082 b 35). Here we are already seeing objects as "reduced" structures, i.e., as indifferent, simply countable material (see Pp. 104 f.). In raising this procedure to the rank of a science by confining counting and calculating to "pure" monads,[118] we turn the property of "being one" and of "being indivisible" as such into an object of study. For the mathematical monas is nothing but the property of being a measure as such, which has been "lifted off" the objects. This is why the arithmetician understands the monas as the measure which is "totally indivisible" (πάντῃ ἀδιαίρετον — I 1, 1053 a 1 f.; Δ 6 1016 b 25), and consequently also as "completely exact" (ἀκριβέστατον — 1053 a 1). This is the basis of the universal "applicability" of the "pure" numbers: "For a human being as a human being is a thing one and indivisible; but [the arithmetician] has already posited the indivisible one [namely the "detached" monas, see above, P. 103], and then sees later what might belong to a human being that would make him indivisible" [i.e., what makes him subject to being counted or reckoned with as a 'unit']." *(ἓν μὲν γὰρ καὶ ἀδιαίρετον ὁ ἄνθρωπος ᾗ ἄνθρωπος· ὁ δ'ἔθετο ἓν ἀδιαίρετον, εἶτ' ἐθεώρησεν εἴ τι τῷ ἀνθρώπῳ συμβέβηκεν ᾗ ἀδιαίρετος* — M 3, 1078 a 23–25.) And exactly the same may clearly be said of any countable being whatsoever.

Aristotle's ontological view of *mathematika*, especially of the "pure" numbers, soon comes to influence the development of mathematical science itself. In respect to *theoretical arithmetic*, it is clear that the *eidos* concept will now have a considerably smaller significance than it did within the Pythagorean-Platonic framework. The *eide* of numbers can no longer be understood as structures which give unity and unambiguous articulation to the realm of numbers (cf. P. 54). If they are to have a place at all, the individual numbers themselves must henceforth be considered as *eide*, that is, in the "derived" sense (cf. *Metaphysics* Z 4, 1030 a 18–27): "When someone treats of the whole he must divide the genus into the *ultimate elements* which are indivisible *in kind*, for instance, number into *triad* and *dyad*. . . ." *(χρὴ δέ, ὅταν ὅλον τι πραγματεύηταί τις, διελεῖν τὸ γένος εἰς τὰ ἄτομα τῷ εἴδει τὰ πρῶτα, οἷον ἀριθμὸν εἰς τριάδα καὶ δυάδα . . .* — *Posterior Analytics* B 13, 96 b 15 ff.) The "even," the "odd," the "even-times-odd," etc., on the other hand, are now no more than the "peculiar characteristics" *(ἴδια πάθη)* of numbers (cf. Note 37; *Physics* B 2, 194 a 3–5). They represent merely a *quality*[119] *(ποιότης)* of numbers, for instance, the character of being a "composite" number, that is, a "plane" or a "solid" number (in contrast to a "prime number" and a "linear" number), while their being *(οὐσία)* is the multitude of units as such, for instance "six" — where "being" is again to be understood in the "derived" sense. Indeed, the *"what"* of each number insofar as it is a number is precisely that quantity which it indicates; thus "six" units are not in themselves "two times three" units or "three times two" units, for this indicates only their "composite quality," but "once six": "For the being of each number is what it is once, for instance that of six is not what it is twice or thrice but what it is once; for six is once six." (οὐσία γὰρ ἑκάστου [sc. ἀριθμοῦ] ὃ[for τὸ — Bonitz, Ross] ἅπαξ, οἷον τῶν ἓξ οὐχ ὃ δὶς ἢ τρὶς εἰσίν, ἀλλ᾽ ὃ ἅπαξ· ἓξ γὰρ ἅπαξ ἕξ — *Metaphysics* Δ 14, 1020 b 7 f.; also b 3–7.[120]) The "arithmetical" books of Euclid (VII, VIII, IX) directly mirror this ontological transformation. The geometric form of their presentation is suggested not only by a consideration of the great problem of "incommensurability" which forces a thorough "geometrization" on Greek mathematics (cf. P. 43). The "pure" units of which the numbers to be studied are composed are here understood precisely only as "*units of measurement*" such as can be represented most simply by straight lines which are *directly measurable* (rather than by points; cf. especially *Metaphysics* M 8, 1084 b 25–27), quite independently of whether they form a "linear" (prime), "plane," or "solid" number.[121] The same approach is indicated by Definitions 8, 9, 11, 12, 14 of the seventh book (namely of even-times-even, even-times-odd, odd-times-odd, prime and composite number: ἀριθμὸς ἀρτιάκις ἄρτιος, ἀρτιάκις περισσός, περισσάκις περισσός, πρῶτος and σύνθετος), which define the nature of each number with respect to the *measuring character* of its factors (cf. also Defs. 3 and 5) — this is the very approach which Nicomachus and Domninus, at least, make every effort to avoid.[122] It must be stressed that the "even-times-even," "even-times-odd," and "odd-times-even" numbers are, according to the Euclidean definitions, not mutually exclusive (cf. IX, 32–34).[123] In the light of Iamblichus' criticism of these definitions (Pistelli, 20 ff.) this shows very clearly that the object is no longer to provide a (more or less unambiguous) classification of all numbers but only to determine certain of their properties. For the same reason Euclid can join Aristotle (*Topics* Θ 2, 153 a 39 f.) in counting "two" among the "prime numbers" (πρῶτοι ἀριθμοί), a classification which is impossible for the Neoplatonic arithmeticians (see especially Theon 24, 4–8).[124]

From Aristotle's ontological conception insofar as it affects the problem of *theoretical logistic*, we can, however, draw another consequence, far more central in our context. We saw (Pp. 39 ff.) that the crucial difficulty of theoretical logistic as the theory of those mutual relations of numbers that provide the basis of all calculation lay in the concept of the monad, insofar as it is understood as an independent and, as such, simply indivisible object. *Aristotle's criticism obviates this difficulty* by showing that this "indivisibility" does not accrue to the *monas* as a self-subsisting *hen*, but by virtue of the *measuring character* of any such unit, be it of an aisthetic or a noetic nature. Only when the Aristotelian critique has taken effect can a whole series of "applied" sciences, such as were cultivated in the Alexandrian school, be justified *as* "sciences" (cf. *Physics* B 2, 194 a 7 ff.). The *Metrics* of Heron of Alexandria, for instance, starts out from the following observation: "In order, then, not to have to name feet or ells or their parts in each measurement, we will exhibit our numerical results as [reduced to indifferent] *monads*, for it is open to anyone to substitute for them whatever measure he wishes."[125] (ἵνα οὖν μὴ καθ᾿ ἑκάστην μέτρησιν πόδας ἢ πήχεις ἢ τὰ τούτων μέρη ὀναμάζωμεν, ἐπὶ μονάδων τοὺς ἀριθμοὺς ἐκθησόμεθα· ἐξὸν γὰρ αὐτὰς πρὸς ὃ βούλεταί τις μέτρον ὑποτίθεσθαι — Schöne, *Opera*, III, A, *Prooem.* 6, 4 ff.) Nothing now stands in the way of changing the unit of measurement in the course of the calculation and of transforming all the fractional parts of the original unit into "whole" numbers consisting of the new units of measurement. Thus even fractions can now be treated "scientifically." If we disregard for a moment the fact that Plato's demand for a theoretical logistic is in fact realized within a different context, namely in the general theory of proportions (cf. P. 44), there can be no doubt that only Aristotle's conception of *mathematika* makes possible that "theoretical logistic" which suffered from the dilemma of being at the same time postulated by the *chorismos* thesis and yet precluded from realization by that very thesis. We even possess a significant document which is able to give us a concrete notion of the *type* of theoretical logistic which can be built on Peripatetic foundations — the "arithmetical" textbook of *Diophantus*.

PART II

9

On the difference between ancient and modern conceptualization

THE RESULTS of the preceding investigation enable us to interpret the work of Diophantus in the light of the most general conceptual presuppositions of Greek arithmetic and logistic. This approach cannot but bring us into conflict with the standard presentation of Diophantus' *Arithmetic* as a rudimentary stage of modern algebra and hence of our symbolic mathematics in general. The latter interpretation of Diophantus is, as we shall see, intimately connected with the self-conception of modern mathematics since Vieta, Stevin, and Descartes. This new self-definition of the mathematical enterprise rests essentially upon a certain re-interpretation of ancient mathematics (as of the whole of ancient science). To clarify this reinterpretation means nothing less than to gain renewed access to Greek numerical doctrine, which is precisely what we have attempted to do in Part I of this study. Yet a better understanding of the sections to follow requires some further reflections on fundamental matters.

The difficulties in the way of an adequate understanding of the Greek doctrine of number lie above all (as hardly needs to be pointed out) in our own manner of dealing with concepts — in the nature of *our own intentionality*. [By intentionality is meant the mode in which our thought, and also our words, signify or intend their objects;* cf. P. 207.] The necessity of abstaining as far as possible from the use of modern concepts in the interpretation of ancient texts is therefore generally accepted, and even stressed. It is clear, to be sure, that the feasibility of an interpretation not based on modern presuppositions must always be limited; even if we succeed in ridding ourselves completely of present-day scientific terminology, it remains immensely difficult to leave that medium of ordinary intentionality which corresponds to our mode of thinking, a mode essentially established in the last four centuries. On the other hand, the ancient mode of thinking and conceiving is, after all, not totally "strange" or closed to us. Rather, the relation of our concepts to those of the ancients is oddly "*ruptured*" — our approach to an understanding of the world is rooted in the achievements of Greek science, but it has *broken loose* from the presuppositions which determined the Greek development. If we are to clarify our own conceptual presuppositions we must always keep in mind the difference in the circumstances surrounding our own science and that of the Greeks.

In Greek *episteme* the life of "cognition" and "knowledge" was recognized for the first time as an ultimate human possibility, one which enables men to disregard all the ends they might otherwise pursue, to devote themselves to contemplation in complete freedom and leisure, and to find their happiness in this very activity. This possibility is contrasted with the bondage imposed by the affairs of the day. Here science stands in original and immediate opposition to a nonscientific attitude which yet is its soil and in which it recognizes its own roots. In attempting to raise

* See J. Klein, "Phenomenology and the History of Science," *Philosophical Essays in Memory of Ed. Husserl* (Cambridge, Mass., 1940), pp. 143–163 [Trans.].

itself above this nonscientific attitude, science preserves intact these given foundations. It is therefore both possible and necessary to learn to see Greek science from the point of view of this, its "natural" basis. In its sum-total Greek science represents the whole complex of those *"natural" cognitions* which are implied in a prescientific activity moving within the realm of opinion and supported by a preconceptual understanding of the world.

Our science, the "new" science whose foundations were laid in the sixteenth and seventeenth century, is very differently circumstanced. Here the "natural" foundations are replaced by a *science already in existence*, whose principles are denied, whose methods are rejected, whose "knowledge" is mocked — but whose place within human life as a whole is placed beyond all doubt. *Scientia* herself appears as an inalienable human good, which may indeed become debased and distorted, but whose worth is beyond question. On the basis of this science, whose fundamental claim to validity is recognized, the edifice of the "new" science is now erected, but erected *in deliberate opposition to the concepts and methods* of the former. It is perfectly true that, as is pointed out again and again, the founders of the "new" science — men like Galileo, Stevin, Kepler, and Descartes — are carried by an original impulse which is quite foreign to the learned science of the schools. The scientific interest of these men and their precursors is kindled mostly by problems of applied mechanics and appplied optics, by problems of architecture, of machine construction, of painting, and of the newly discovered instrumental optics.[126] But it is nevertheless true that the conceptual frame for their new insights is derived from the traditional concepts. The very claim of passing on *true* science, *true* knowledge, induces the necessity of continual reorientation by reference to the traditional, firmly consolidated edifice of science. The "new" science shares with scholastic science the most general presuppositions of that "scientific attitude" which was developed by

Greek science in opposition to a "natural" existence. More-over, it returns to the sources of Greek science, which had been neglected by scholastic science, although it interprets both the presuppositions and the sources from a basis which is utterly foreign to ancient science. And this reinterpretation of the ancient body of doctrine, which brings with it a characteristic transformation of all ancient concepts, lies at the foundations not only of all concept formation in our science, but also of our ordinary intentionality, which is derived from the former.

Now that which especially characterizes the "new" science and influences its development is *the conception which it has of its own activity*. It conceives of itself as again taking up and further developing Greek science, i.e., as a recovery and elaboration of "natural" cognition. It sees itself not only as the science *of nature*, but as "*natural*" science — in opposition to *school* science. Whereas the "naturalness" of Greek science is determined precisely by the fact that it arises out of "natural" foundations, so that it is defined at the same time in terms of its distinction from, and its origin in, those foundations, the "naturalness" of modern science is an expression of its *polemical attitude toward school science*. This special posture of the "new" science fundamentally defines its horizon, delimits its methods, its general structure, and, most important, determines the conceptual character of its concepts.

In Greek science, concepts are formed in continual depen-dence on "natural," prescientific experience, from which the scientific concept is "abstracted." The meaning of this "abstraction," through which the conceptual character of any concept is determined, is *the* pressing ontological prob-lem of antiquity; it becomes schematized in the medieval problem of universals, and, in time, fades away completely. The "new" science, on the other hand, generally obtains its concepts through a process of polemic against the traditional concepts of school science. Such concepts no longer have that natural range of meaning available in ordinary discourse, by an appeal to which a truer sense can always be distinguished from a series of less precise meanings. No longer is the thing intended by the concept an object of *immediate* insight. Nothing but the internal connection of all the concepts, their mutual relatedness, their subordination to the total edifice of science, determines for each of them a *univocal* sense and makes accessible to the understanding their only relevant, specifically scientific, content. In evolving its own concepts in the course of combating school science, the new science ceases to interpret the concepts of Greek *episteme* preserved in the scholastic tradition from the point of view of their "natural" foundations; rather, it interprets them with reference to the function which each of these concepts has within the whole of science. Thus every one of the newly obtained concepts is determined by *reflection on the total context of that concept*. Every concept of the "new" science belongs to a new conceptual dimension. The special intentionality of each such concept is no longer a problem: it is indifferently the same for all concepts; it is a medium beyond reflection, in which the development of the scientific world takes place. The more recent endeavors to provide the edifice of science with a firm logical foundation do not alter this aspect of the situation.

In the present study we are dealing only with the history of concept formation within a certain scientific field, namely the mathematical. What has been said holds also for mathematical concepts. For the development of mathematics cannot be isolated from the general history of the modes of understanding — however strenuous an effort may be made to shape mathematics into a self-contained, independent discipline. To go even further — the nature of the modification which the mathematical science of the sixteenth and seventeenth century brings about in the conceptions of ancient mathematics is *exemplary* for the total design of human knowledge in later times: What we are referring to is the intimate connection between the mode of "*generaliza-tion*" of the "new" science and its character as an "*art.*"

The most characteristic expression of the above connection is to be found in the symbolic formalism and calculational techniques of modern mathematics. We shall see it displayed in all that follows. In particular, this intimate relation between the mode of generalization employed by the new science and its "art" character has determined the modern interpretation of ancient mathematics. Zeuthen was not the first to understand the ancient mode of presenting mathematical facts as a "geometric algebra,"[127] although he was the first to use this concept consistently. To anticipate, this interpretation can arise only on the basis of an insufficient distinction between the *generality of the method* and the *generality of the object* of investigation. Thus Zeuthen himself immediately relates his concept of "geometric algebra" to that of "general magnitude"; "geometric algebra," being a general method, has for its proper object precisely this "general magnitude."[128] Thereby the whole complex of problems which presented itself to the ancients because their "scientific" interest was centered on questions concerning the mode of being of mathematical objects is obviated at one stroke; ancient mathematics is characterized precisely by a tension between method and object. The objects in question (geometric figures and curves, their relations, proportions of commensurable and incommensurable geometric magnitudes, numbers, ratios) give the inquiry its direction, for they are both its point of departure and its end. The way they determine the method of inquiry is shown especially in the case of "existence" proofs, i.e., demonstrations that the "being" of a certain object is possible because devoid of self-contradiction (cf. P. 104). The problem of the "general" applicability of a method is therefore for the ancients the problem of the "generality" (καθόλου) of the mathematical objects themselves, and this problem *they* can solve only on the basis of an ontology of mathematical objects. In contrast to this, modern mathematics, and thereby also the modern interpretation of ancient mathematics, turns its attention first and last to *method as such*. It determines its objects *by reflecting on the way in which these objects become accessible through a general method*. Thus, arguing from the "generality" of the linear presentation in Euclid's "arithmetical" books, namely from the fact that with their aid *all arithmoi*, in all their possible relations to one another, can be grasped, its proponents reach the conclusion that this "generality" of presentation is intent upon "general magnitude." Now what is characteristic of this "general magnitude" is its indeterminateness, of which, as such, a concept can be formed only within the realm of symbolic procedure. But the Euclidean presentation is *not* symbolic. It always intends *determinate* numbers of units of measurement, and it does this *without any detour through a "general notion" or a concept of a "general magnitude."* In *illustrating* each determinate number of units of measurement by measures of distance it does *not* do two things which constitute the heart of the symbolic procedure: It does *not* identify the object represented with the means of its representation, and it does *not* replace the real determinateness of an object with a *possibility* of making it determinate, such as would be expressed by a sign which, instead of *illustrating* a determinate object, would *signify* possible determinacy. Although in Euclid the linear presentation of the "arithmetical" books is tailored to the requirements of the theme of the tenth book, with which the earlier books have a "systematic" connection (cf. P. 43), yet the objects of these books are, in spite of the *sameness* of method, *different*: Books VII–IX deal with numbers which are so defined as to be always commensurable (cf. VII, Def. 2, 14–16), while Book X, on the other hand, deals with magnitudes whose ratios cannot be reduced to number ratios (cf. X, 5–8) and which are incommensurable precisely on this account. And again, when in the arithmetical books an arithmetical, or more exactly, a logistical proposition is demonstrated *generally* with the aid of lines, this does not in the least mean that there exists either a general number or the concept of a "general," i.e., indeterminate, number corresponding to this general proof (cf. on this above all Aristotle, *Posterior Analytics* A 24, 85 a 31–b 3). In the second book of Pappus (Hultsch, I, p. 2 ff.), where he presents and comments on Apollonius' system of counting and calculating (see Part I, Note 36), we may see directly the manner in which the *general* "linear approach" (τὸ γραμμικόν) intends only *determinate* numbers (cf. the comments of Hultsch, III, 1213–1261).[129] Since here, as in Euclid, the single illustrative lines are additionally identified by a letter, the possibility arises of representing the numbers intended by those letters. This does not, however, in the least amount to the introduction of symbolic designations. Letters for indicating magnitudes and numbers seem to have been used already by Archytas (cf. Tannery, *Mém scient.*, III, p. 249; also Note 41). On these letters Tannery comments (*ibid.*, p. 249; cf. also III, p. 158): "La lettre remplace bien un nombre quelquonque . . . , mais seulement là où ce nombre est supposé placé; elle n'en symbolise pas la valeur et ne se prête pas aux opérations," i.e., *the letter does not symbolize the value of a number, and does not lend itself to being operated on.*[130] Aristotle, too, made use of such mathematical letters, e.g., in the *Physics* and in *On the Heavens*; and he even introduced them into his "logical" and "ethical" investigations. But such a letter is never a "symbol" in the sense that that which is signified by the symbol is in itself a "general" object.[131]

In the sections to follow we shall have to trace the conceptual transformation which permits the ancient *arithmos* to appear as "number" ("Zahl") — as opposed to "numbered assemblage" ("Anzahl") — *and concomitantly, as "general magnitude."* This transformation is heralded at the end of the Middle Ages by an increasing interest in the practical, i.e., applied, mathematical disciplines. In contrast to "theoretical" arithmetic and geometry (arithmetica et geometria *specula-*

tiva) such as especially Boethius, using Neopythagorean or Neoplatonic sources, passes on to the Middle Ages, the corresponding "practical" disciplines of logistic and mensuration (arithmetica et geometria *practica*) are now preferred (cf. Part I, Section 2.) These disciplines are consistently understood as "*artes*"; to learn them means to master the corresponding "rules of the art." "*Artful procedure*," the "practica" or "praxis" of calculating and measuring, is the object of this doctrine. Now at that significant moment when these disciplines first succeed in gaining recognition as part of the "official" science, it is precisely their character as "arts" which is thought to lend them their true theoretical dignity.[132] Conjointly with this, the structure of the objects with which mathematics deals is transformed. A new kind of generalization, which may be termed "symbol-generating abstraction," leads directly to the establishment of a new universal discipline, namely "general analytic," which holds a central place in the architectonic of the "new" science. And this whole process, which it is our task to describe in detail, is mainly initiated by the reintroduction and assimilation of the *Arithmetic* of Diophantus.

The Arithmetic *of Diophantus as theoretical logistic. The concept of* eidos *in Diophantus*

THE SIX BOOKS by Diophantus[133] which have come down to us under the title *Arithmetica* (*'Αριθμητικά*) — Diophantus himself speaks of thirteen books in the preface (16, 7) — essentially teach the solution of those computational problems which are known today as determinate and indeterminate equations of the first and second degree. In the course of his presentation Diophantus uses, besides various other signs, a series of abbreviations for unknowns and their powers which enter into the calculation itself, thus lending it an "algebraic" character. This is why Diophantus could come to be called — always with certain reservations — the "inventor" or "father" of our present-day algebra.[134] Nesselmann,[135] and later Tannery[136] and Heath[137] already stressed the fact that we have every reason to doubt Diophantus' originality, although no similar work from antiquity is known to us. The *material* assimilated in Diophantine problems is to be found already in Thymaridas (cf. P. 36), in Plato (*Laws* 819 B, C), in the *Charmides scholium* (cf. P. 12), in the "arithmetical" epigrams of the *Palatine Anthology*, and in Heron of Alexandria.[138] And although the *form* in which this material is presented by Diophantus cannot be documented before him, there is reason for the claim that he must have been related to his predecessors somewhat as Euclid was to the authors of the earlier *Elements* (Theodias, Leon, and Hippocrates of Chios), whose works were completely eclipsed by the Euclidean compilation.[139]

In addition, attempts have long been made to connect the "Alexandrian" Diophantus directly with the Egyptian tradition and even to understand the whole Diophantine technique as a development of the Egyptian Hau ("heap") calculations.[140] In contrast to this, Neugebauer has recently fitted Diophantus' work into a far more extensive context.[141] He sees in him the last offshoot primarily of the Babylonian "algebraic" tradition, a tradition which the Arabs later continue directly on the basis of direct oriental sources and not only through Greek mediation.[142] That the science of Diophantus exhibits certain non-Greek traits can hardly be denied.[143] But whatever the case for the more remote prehistory may be — the intentionality of a work which developed, as did this one, on Hellenistic soil must first of all be understood on the basis of *Greek* presuppositions. With respect to the "latent algebraic component in classical Greek mathematics" and the incisive difference between external structure and internal motivation which is implied by this, Neugebauer himself emphasized the "necessity of formulating the question in terms which might almost be said to belong to stylistic history."[144] Such a formulation presupposes a certain independence of the material mathematical content from the "external" form, which amounts, in the final analysis, to taking it for granted that the content is to be understood "algebraically." But our task consists precisely in bringing the content of Greek mathematics to light not by externally transposing it into another mode of presentation but rather by comprehending it in the one way which seemed comprehensible to the Greeks. Only then can we determine what kind of conceptual means Greek, in distinction from modern, mathematics employs. And similarly, only by making explicit the particular character of Greek intentionality, whose peculiar transformation in the sixteenth and seventeenth century is equivalent to the "introduction of a completely new means of expression for mathematical thinking,"[145] namely, a formal "algebraic" symbolism, can the specific conceptual character of the latter be understood.

Tannery's evaluation of the significance of Diophantus in the history of Greek mathematics is now generally accepted, and we must therefore briefly review it: Tannery sees Diophantus' achievement as original, or at least as based on only a few predecessors — it consists in the new conception which he had of the object of *logistic*. Even the title of his chief work: "*Arithmetika*" bears witness to this new conception, for only an *episteme*, a true science, would be thus designated.[146] Diophantus is therefore said to have raised logistic to the rank of a real science. This is why he adds the "demonstration" (ἀπόδειξις), i.e., the actual solution of each problem; the earlier logisticians failed to do this precisely because logistic was then not a true, "apodeictic," science. Above all, the "numbers" with which he calculates are, with a single exception (the thirtieth problem of the fifth book), "abstract" and not "concrete" as the "numbers" of the logisticians before him had been, at least for the most part. Later writers then misunderstood his work and interpreted his new "abstract" logistic in the spirit of the old "concrete" study. Thus an anonymous commentary on Nicomachus (printed in the second volume of the Tannery edition of Diophantus, p. 73) says: "For Diophantus, in the thirteen books of his *Arithmetic*, treats of the measured number [i.e., the number counted off *on things*]" (τὸν γὰρ μετρούμενον ἀριθμὸν Διόφαντος ἐν τοῖς δέκα καὶ τρισὶν αὐτοῦ βιβλίοις τῆς ἀριθμετικῆς παραδίδωσιν), while Nicomachus speaks of the measuring number, the number itself.[147]

In a later publication (1896),[148] Tannery attempted to give an explication of the "abstract" character of Diophantine problems, which does not, however, quite fit the exposition just mentioned, and which rests on highly controversial presuppositions. Basing his argument on the relation of Diophantus to Anatolius, the bishop of Laodicea, a relation supposedly implied by the Psellus fragment discovered by Tannery (printed in the second volume of the Tannery edition of Diophantus, pp. 37 ff.), he thought that he could identify the Dionysius to whom the Diophantine *Arithmetic* is dedicated with St. Dionysius, the bishop of Alexandria, and he concluded from this that the Diophantine textbook was intended for use in the Alexandrian catechumenical school; it followed that Diophantus must have been a Christian.[149] This thesis seemed to him to find particular support in the special form of the Diophantine *Arithmetic*. For the usual verbal problems composed in terms of pagan mythology, as we know them from the *Palatine Anthology* and the Archimedean *Cattle Problem*, would obviously be unsuitable for instruction at a Christian school. But to relinquish such verbal formulations would immediately produce that thoroughgoing "abstractness" which the Diophantine problems in fact have. (Note, in addition, that the verbal formulation of the thirtieth problem of the fifth book is not of a mythological sort, apart from the fact that this problem at the end of the book may perhaps be a later addition.)

This exposition does, in agreement with the afore-mentioned anonymous commentary, ultimately present the Diophantine *Arithmetic* as a "logistic" in the Neoplatonic sense which has sloughed off its "concrete" aspects only in a very external way; yet from this point of view the "scientific," "theoretical" character of Diophantus' work would not really be quite comprehensible. Clearly the opposition of "abstract" and "concrete" is insufficient for characterizing a work of this kind. It is, to be sure, characteristic of Diophantus that he calculates with *arithmoi* which are nothing but numbers of pure monads. But such numbers cannot, without further ado, simply be called "abstract" (see P. 48). What is crucial is rather to know how the being of these monads is understood, whether as independent and therefore incapable of any division, or as the result of reduction to "neutral" items which merely indicate the character of the "measure" and thus easily admit further partition. Now which of these two possibilities is here realized cannot be in doubt.

First of all, it is immediately clear not only from "Definition I" (2, 14 f.), which is reminiscent of Euclid VII, Def. 2 (cf. P. 51), and according to which "all numbers consist of a certain multitude of monads" (... πάντας τοὺς ἀριθμοὺς συγκειμένους ἐκ μόναδων πλήθους τινός ...) but also from the way in which Diophantus designates the numbers which are presupposed as "known" (the determinates — οἱ ὡρισμένοι — 6, 6 f.) in the calculation, that we are indeed dealing with numbers of pure monads. The letter which gives the number of monads usually has, in accordance with "Definition II" (6, 6–8), the sign for *monas (M̊)* as a prefix; for instance: M̊δ (four units) M̊͞ϛυ (six thousand and four hundred units), etc. Nesselmann[150] and, following him, Tannery[151] and Heath[152] have attempted to explain this notation as follows: Diophantus simply juxtaposes those expressions in an equation which we now conjoin by a plus sign and does not separate them by a special connecting sign; now if the sign M̊ were left out, the coefficients of an unknown would run together with the expression for any "known number" which might follow, so that, for instance, ϛ͞κM̊ε̄ (twenty unknowns + five monads) would, without the M̊ sign, look like this: ϛ͞κε̄, which could then easily be confused with ϛ͞κε̄ (twenty-five unknowns). We may object that in most cases such an amalgamation is not possible, and that in the rest it could easily have been rendered impossible by means of a special word or word sign, e.g., "and." Indeed our own plus sign (+) seems to be nothing but a very much simplified ligature of that same word in Latin (et).[153] The introduction of such a connecting sign must, on the contrary, have seemed unnecessary precisely because the M̅ sign had already excluded any such amalgamation. But the use of the sign itself was the result *of the meaning of the word* arithmos *in common speech, namely*: *a determinate number of determinate things* — in the extreme case this might be a number of units of measurement, or of "pure" or "neutral" monads (cf. Pp. 46 ff.).[154] This linguistic usage also occurs throughout the *Metrics* of Heron (cf. P. 112), where it has not yet, to judge from the surviving manuscript (of the eleventh or twelfth century),[155] coagulated into a univocal, fixed sign language. The same holds for the second book of Pappus (of which we have fragments—Hultsch, I, p. 2–28), where the system of calculation and the nomenclature of Apollonius are described (cf. Note 36). The same linguistic usage is, incidentally, the basis both of Apollonius' system of nomenclature and of Archimedes' exposition in a lost work addressed to Zeuxippos[156] as well as in the *Psammites*; both depend on the introduction of *units* of a higher order which, when counted, yield *numbers* of a higher order; they are called by Archimedes "numbers [formed] analogously *to those based on monads*" (ἀριθμοὶ ἀπὸ μονάδος ἀνάλογον — II[1], 270, 2 f. and 21 f.; 272, 4= II[2], 240, 2 and 20 f. and 29, etc.), by Apollonius simply "analogous numbers" or "analogues" (ἀνάλογοι [ἀριθμοί] or τὰ ἀνάλογα — Pappus, Hultsch, I, pp. 20, 13, and 20; 28, 13, and 21; 26, 4; cf. Hultsch, II, p. 1213); these then permit unusually large numbers of simple monads to be pronounced, written down, or made objects of calculation without much difficulty.[157] Even within the ordinary system of counting, the *myriad*, the unit-ten-thousand (μυριάς), forms a new and higher unit, and so, actually, already does the *chiliad*, the unit-thousand (χιλιάς), if the numerical adjectives thousand, two thousand, three thousand (χίλιοι, δισχίλιοι, τρισχίλιοι) and their written form are taken into account.[158] The systems of Archimedes and, above all, of Apollonius represent nothing but the consistent development of the Greek mode of thought and speech.

Although in this respect the Diophantine *Arithmetic* is obviously in line with the ordinary prevailing Greek understanding and treatment of number, it is, on the other hand, clear that it is founded on an *ontological* conception of monads which is altogether irreconcilable with that of the Neoplatonists. In the Diophantine text calculation with fractional parts of the unit of calculation, i.e., the division of this unit, is simply admitted without comment (most clearly in the formulations of the thirty-first problem of the fourth book and in problems 9, 10, 11, 12 of the fifth book);[159] so also the unknown and its powers can, according to "Definition III" (6, 9–21), appear as "denominators" of such fractions, which then bear their name "by derivation" (παρομοίως)[160] from these same "unknown" magnitudes, as for instance, the "cubic fraction," i.e., the reciprocal of a cubed unknown (τὸ κυβοστόν, from κύβος, cube) and the "square-times-square fraction," i.e., the reciprocal of an unknown taken to the fourth power (τὸ δυναμοδυναμοστόν, from δυναμοδύναμις, fourth power), etc. What is more, this handling of fractional parts of the unit clearly does not diminish the "theoretical" character of the whole work, although this ought to be the case in the light of Neoplatonic conceptions (cf. Pp. 43 and 44 f.). This "theoretical," apodeictic, character shows itself unambiguously in the fact that every problem, with the exception of those of the sixth book and of the thirtieth problem of the fifth book, is both posed and solved in terms of "pure" monads. Furthermore, the course of the solution, the *apodeixis* (cf. 256, 12; 430, 17; 208, 13, etc.),[161] is in each case strictly "methodical," i.e., transferable to other cases of the same type, although, since the calculation involves determinate numbers, it cannot always be effected without difficulty. In certain cases ever new solutions can indeed be found "*infinitely many times*" (ἀπειραχῶς) on the basis of an *apodeixis completed once* (cf. especially 184, 4 f. and

200, 21). Furthermore, the rules for the treatment of a first-degree equation, as well as of the *pure* quadratic and *pure* cubic equation, etc. ("Def. XI," 14, 11–20), the "method of backward calculation" (the expression is Nesselmann's, *Algebra der Griechen*, p. 370), which is generally called the method of the "false supposition," continually used by Diophantus, and finally also the rule for the solution of the so-called "double equation" (cf. 96, 9 ff.[162]) have a *general character*. The *Arithmetic* of Diophantus is thus indeed a theoretical work (cf. the expression "arithmetical theory" — ἀριθμητικὴ θεωρία — 4, 14) which does not scruple to introduce fractional parts of the unit of calculation into the inquiry. *This is possible only if the ontological understanding of the monas is Peripatetic in character* (see Pp. 104 ff.).[163] This means that in Diophantus, as in Heron (cf. P. 112), we are dealing with "neutral" monads obtained "by abstraction" (ἐξ ἀφαιρέσεως — whereby, incidentally, the thirtieth problem of the fifth book ceases to be an exception). But when, on the other hand, we consider the essential content of the Diophantine work as it appears in the formulations of the problems themselves, the thought that this is a discipline to which the Platonic definition of logistic (see P. 17) is immediately applicable can hardly be avoided.

In judging this work, we should indeed not allow ourselves to be guided by modern algebraic concepts. With these in mind, it appears to be Diophantus' peculiarity — corresponding to the primitive stage of algebra which he represents — to pose problems belonging to indeterminate analysis (which, mostly through the agency of Fermat, gave rise to the present-day concept of the so-called "Diophantine equations"), but always to transform these problems into determinate equations by means of arbitrary numerical assumptions which permit a univocal (although by no means exclusively integral) solution. Beyond this, the crucial weakness in Diophantine technique must, in the light of modern techniques of solution, appear to be that it is not directed

toward obtaining general solutions. Now Diophantus
does know problems and solutions of a "general" kind
(in his terms: "in the indeterminate [solution]" — ἐν τῷ
ἀορίστῳ or ἐν τῇ ἀορίστῳ [sc. λύσει]), namely those which
leave their object "*indeterminate*": "For to seek [the solution
of a problem] indeterminately means that a hypothetical
expression [must be constructed] such that, of whatever size
someone wishes the unknown to be, when he [chooses the
unknown and] sets it in the hypothetical expression, the
conditions of the problem [as formulated in the expression]
are preserved." (τὸ γὰρ ἀορίστως ζητεῖν ἐστιν ἵνα ἡ ὑπόστασις
τοιαύτη ᾖ, ἵνα ὅσου τις θέλει τὸν ς εἶναι, ἐπὶ τὰς ὑποστάσεις ποιήσας·
σώσῃ τὸ ἐπίταγμα—232, 6–8.) And similarly:"The procedure
[of finding the resolution] 'in the indeterminate [solution]'
is such that, of however many monads someone might wish
the unknown to be, when he sets [the indeterminate number]
in the hypothetical expression, the problem is [at once]
completed." (τὸ δὴ ἐν τῇ ἀορίστῳ [sc. λύσει λύειν] τοιοῦτόν ἐστιν,
ἵνα τὸν ς, ὅσων ἄν τις θέλει Μ̄ εἶναι, ἐπὶ τὰς ὑποστάσεις ποιήσας,
περάνῃ τὸ πρόβλημα — 278, 10–12.) But it appears *that such
problems and solutions have a merely auxiliary character.* This
holds not only for problems explicitly designated as such
(namely the lemmata to IV, 34, 35, 36 and lemma 1 to V, 7),
where it is required to find certain"indeterminate numbers"
(ἀριθμοὶ ἀόριστοι) subject to certain requirements, but for
problem IV, 19 as well, which is preliminary to the problem
following; furthermore, it holds for all problems within
which the "indeterminate" solution forms only a stage
preliminary to the final one, as is the case especially in
problems IV, 16, 17, 21; V, 18; and in a series of problems
in the sixth book.[164] Seen under the aspect of the conceptual
presuppositions of Greek numerical doctrine, Diophantus is
throughout concerned only *with finding wholly determinate
numbers* (and under certain circumstances also wholly determinate fractional parts of the unit of calculation) *such as have
to each other a certain particular relation given by the problem.*

Thus the Diophantine work does indeed represent a *theoretical logistic*, namely one founded on a Peripatetic theory of number relations. *Seen in its own terms*, it is not organized around types of equations and methods of solution — which is what modern interpreters usually look for[165] — but according to the possible relations that numbers, particularly "quadratic" *(τετράγωνοι)* and "cubic" *(κύβοι)* numbers and their "roots" *(πλευραί* — sides; cf. 2, 17–22; cf. also 16, 2–4), can bear to one another.

From this point of view, the *Arithmetic* of Diophantus is not very far from the "arithmetical" books of Euclid (VII, VIII, IX). Not only is the Euclidean linear presentation used in problem V, 10 (and, incidentally, also in the work on polygonal numbers), not only are all the problems of the sixth book directed toward finding right-angled triangles with "rational" sides (or "rational" angle-bisectors, cf. VI, 16), so that here, as in Heron, the monads appear as units of calculation immediately "abstracted" from units of measurement,[166] but Diophantus' whole work is related to these Euclidean books as a highly developed, detailed science to its "elementary" foundations. Like these, it therefore appears as an *arithmetical* work (cf. P. 43). To go even further — Euclid in general provides the background of Diophantine science. Whether the *Porisms (Πορίσματα)* to which Diophantus refers (316, 6 ff.; 320, 5 ff. and 358, 4 ff.) formed a part of the *Arithmetic* itself or were an independent work,[167] it is in any case hardly possible to avoid being reminded of the Euclidean work by the same name,[168] just as the concepts of "enunciation" *(πρότασις)*; of "reduction" *(ἀπάγειν)*, i.e., of one problem to another, prior one; and of "lemma" *(λῆμμα)* in Diophantus correspond to Euclidean terminology. (Cf. also the "restrictive condition" — προσδιορισμός — 36, 6; 340, 9 f., called in Euclid, Archimedes, and Apollonius διορισμός; cf. Proclus, *in Euclid*. 66, 20 ff., where the concept is traced back to Leon, and 202, 3 f.; furthermore, Pappus, Hultsch, II, 636, 15 f.; also "side" — πλευρά, i.e., root; cf. Euclid VII, Defs. 17, 18.) So also the Diophantine concept of the *arithmoi aoristoi* becomes transparent only in connection with the corresponding geometric concept, which Diophantus uses in the sixth book, namely "the triangle given 'in figure'" (τρίγωνον δεδομένον τῷ εἴδει — 396, 11 f. etc.; see Euclid, *Data*, Def. 3), which is a right triangle given only by shape, and therefore only in terms of its angles and the proportions of its sides. (In these problems of the sixth book the sole point is to determine the proportions correctly, from which, taken together with the further conditions of the problem, a univocal solution follows immediately.) Although it is especially characteristic for Diophantus that he emancipates himself extensively from the geometric mode of presentation and understands even geometric concepts "arithmetically,"[169] we should not, on the other hand, overlook his connection with the strict mathematical tradition which goes under Euclid's name and which is, as we saw (Pp. 110 ff.), influenced in essential respects by the Peripatetic doctrine concerning *mathemata*. Here Suidas' notice is of interest, according to which Hypatia, daughter of Theon of Alexandria, the editor and commentator of Euclid, wrote *A Memorandum on Diophantus* (ὑπόμνημα εἰς Διόφαντον).[170] Note that Hypatia herself (died A.D. 415) was a member of the Alexandrian school which combined the cultivation of scholarship and of the various scientific and, especially, mathematical disciplines, with a definite leaning toward Aristotelianism, albeit permeated by Neoplatonic dogma.

Diophantus' use of "fractions" should by no means lead us to conclude that on the level of Diophantine logistic the concept of the *arithmos* comprehends the whole of what we call the "realm of rational number." In order to understand a fraction as a "nonintegral rational number," *arithmos*, i.e., the "numbered assemblage," must itself already have been understood as "number" in the modern sense. We shall have to investigate the presuppositions of such an understanding in the last two sections. In Diophantus, at any rate, *arithmos* means — as it does in the whole of Greek mathematics — nothing but "a number of. . . ." Correspondingly, by a fraction Diophantus means nothing but *a number of fractional parts.*[171] The "magnitude" of such a fractional part corresponds to "how many" (i.e., what multitude of) partitions the *monas* undergoes. The fractional part therefore takes its name from this multitude as "the fraction derivatively named" (τὸ ὁμώνυμον μόριον); compare Euclid VII, 37: "If a number be measured by a certain number, the number measured will have a part having the same name as the measuring number." *Ibid.*, 38: "If a number have any part whatever it will be measured by a number having the same name as the part." (᾽Εὰν ἀριθμὸς ὑπό τινος ἀριθμοῦ μετρῆται, ὁ μετρούμενος ὁμώνυμον μέρος ἕξει τῷ μετροῦντι. ᾽Εὰν ἀριθμὸς μέρος ἔχῃ ὁτιοῦν, ὑπὸ ὁμωνύμου ἀριθμοῦ μετρηθήσεται τῷ μέρει. Cf. also Diophantus, "Def. III," 6, 9 ff.; "Def. V," 8, 11 f.; "Def. VII," 8, 16 ff., cf. also P. 43.) The expressions "the reciprocal square," "the reciprocal cube" (τὸ δυναμοστόν, τὸ κυβοστόν) are formed analogously (see P. 132). Now the larger the *arithmos* which does the partitioning, the smaller the "fraction" (μόριον) which arises from the partition, and just as numbers increase infinitely in the course of counting as well as in partitioning (2, 16), so the fractional parts of the units diminish infinitely. Thus the anonymous *scholium* to Iamblichus, Pistelli, 127 (second volume of the Tannery edition of Diophantus, p. 72) says: "So Diophantus in his *Fractional Parts*; for fractions [involve] progress in diminution carried to infinity."[172] (οὕτως ὁ Διόφαντος ἐν τοῖς Μοριαστικοῖς· μόρια γὰρ τὴν εἰς ἔλαττον τῶν μονάδων πρόοδον εἰς τὸ ἄπειρον.) In this sense the *monas* does indeed form "the borderline between [the realm of] number and fractions" (ἀριθμοῦ καὶ μορίων μεθόριον), as the passage in Iamlichus (Pistelli, 11, 9–11), to which the *scholium* refers, has it.[173] *In its character as measure,* the *monas* itself is not affected by any partition.[174] What is divided is only a bodily, or perhaps a geometric, i.e., linear, measuring "item" which *is* in itself divisible (cf. Pp. 39 f.; 104; 109; 112). When a fractional part of a unit is multiplied as many times as the homonymous number indicates, the whole original unit is reconstituted ("Def. V," 8, 11 f.). On the other hand, the fractional parts of an original unit may be transformed into a new unit by *changing the measure* (cf. P. 112), which is precisely what is done by Diophantus in problems I, 23, 24, 25; III, 14; VI, 2 and 16.[175] Moreover, Diophantus makes an effort to state the solutions of those problems where the introduction of a new measure would contradict the meaning of the problem in terms of only *one* fractional part (ἓν μόριον) of the *monas* (as is seen especially clearly in problems IV, 34, 35; cf. also IV, 22, 37, etc.).

It is, in fact, quite impossible to arrive at the concept of "negative" and "irrational" number by way of Diophantus' concept of number. Therefore, in order to avoid "negative" and "irrational" solutions, a "restrictive condition" (προσδιορισμός) is introduced in each case. An equation which leads to a "negative" or an "irrational" solution is "impossible" (ἀδύνατος) or "not enunciable" (οὐ ῥητή — i.e., ir-rational), just as it is "absurd" (ἄτοπον) to state such solutions, particularly to state an "ir-rational number" (ἀριθμὸς οὐ ῥητός, cf. 312, 17–19; 250, 14 with 251, note; 424, 12 f.; 204, 19 f.; 208, 7; 210, 1 f.; 212, 6 f.; 264, 12 f.; 270, 4–6) — an irrational *arithmos* is precisely not an *arithmos* at all.[176] Nor can there be a realm of "negative" numbers. It would sooner be possible to comprehend the fractional parts of the unit (in accordance with the definition in Iamblichus making "one" a boundary — μεθόριον — between numbers and fractional parts) under one separate realm, which would represent, as it were. the downward continuation of the realm of numbers beyond one. It appears from a series of formulations where the fractional parts (μόρια) and, incidentally, also the one itself, are understood as a kind of *arithmoi* that this notion is indeed not foreign to

Diophantus (cf. especially problems IV, 12, 22, 23, 24, 26 and also Note 159). This, however, changes nothing in the fundamental understanding of "fractions" set forth above, namely that they are fractional parts of whatever the unit of calculation happens to be.

What has been said up to this point about Diophantus' concept of number was shown to follow directly from that aspect of the *arithmos* which implies that it is a "number *of something*" (ἀριθμός τινος, cf. Pp. 48, 50). To understand what follows, it is, however, no less essential to see that the other aspect of the *arithmos*, namely that the objects intended by it are precisely *so and so many*, is also preserved in Diophantus, as indeed in the whole of Greek mathematics. Here we are confronted with the important question whether Diophantine logistic may not contain within itself the possibility of a symbolic calculating technique. Since Vieta this question has been, on principle, answered positively, even, or rather especially, by those who see the Diophantine science merely as the primitive "preliminary stage" of modern algebra. From the point of view of modern algebra only a single additional step seems necessary to perfect Diophantine logistic: the thoroughgoing substitution of "general" numerical expressions for the "determinate numbers," of *symbolic* for *numerical* values — a step which was, subsequent to a great deal of progress in the treatment of equations in general, finally taken by Vieta. This development, which leads from Diophantus over the Arabs and the Italian algebraists of the sixteenth century to Vieta, is understood as being strictly in line with the demands, as it were, of the subject itself. This view finds its chief support in two circumstances: (1) Diophantus' calculation with the *sign* of the magnitude sought, where this sign is introduced into the very process of solution as it develops step by step (Hultsch, Pauly-Wissowa, see "Diophantus," pp. 1059 f.); (2) the existence of a *general treatment* of mathematical, and especially also of arithmetical, subject matter in Euclid, a treatment which is exemplary for Greek mathematics as a whole and which scholars are generally inclined to understand as "geometric algebra." The second point has already been discussed (Pp. 112 ff.). Here we need only comment on the first point, the use of the sign for the unknown in the Diophantine text itself. [We follow Heath's typography in using final sigma to represent this sign; see *Diophantus of Alexandria*,[2] p. 37.]

In its special Diophantine sense of an *unknown* number, the *arithmos* is defined as "having in itself an *indeterminate* multitude of monads" (ἔχων ἐν ἑαυτῷ πλῆθος μονάδων ἀόριστον) and is rendered by the sign ς (6, 4 f. — on the text cf. the second volume of the Tannery edition, pp. ix f.). The multitude of monads which the unknown number contains is, however, indeterminate only "for us" (πρὸς ἡμᾶς). The whole point of each problem lies precisely in this — that a *completely determinate number of monads completes the solution of each problem.* When the unknown or its sign is introduced into the process of solution, it is precisely *not* indeterminacy in the sense of "potential" determinacy which is intended. Just as the indeterminate solution is always only provisional (see P. 134) and as the concept of the *arithmoi aoristoi* becomes fully understandable only on the basis of *figures* "similar" to one another (i.e., given only in shape and not determinate in size), so also is the unknown to be understood as an "indeterminate multitude" (πλῆθος ἀόριστον) only from the point of view of the completed solution, namely as "provisionally indeterminate," and as a number *which is about to be exactly determined in its multitude*, like these other *arithmoi aoristoi*. Furthermore, it is not permissible simply to call the expressions *dynamis, kybos, dynamodynamis, dynamokybos, kybokybos* "powers of the unknown." Rather, we must attend to the way in which these terms are introduced by Diophantus himself. After mentioning the series of numbers going to infinity (2, 16) he lists as "*among these*" (ἐν τούτοις — 2, 17) the quadratic and cubic numbers and furthermore those numbers which result from "squaring" the quadratic, from multiplying a quadratic with a corresponding cubic (i.e., one which has the same root as the quadratic), and finally from the "quadrature" of a cubic number.[177] It is by intertwining these and their roots in various ways, Diophantus goes on to say, that the texture of most "arithmetical" problems is knit (4, 7–10). Now if, as is usual, the numbers are marked by brief names indicating their kind, then they may well be called the elementary constituents of "arithmetical" science (4, 12–14). These names are, he says, *dynamis, kybos, dynamodynamis, dynamokubos, kybokybos,* i.e., square, cube, square-times-square, square-times-cube, cube-times-cube, each of which is in addition rendered by a "sign" ($\sigma\eta\mu\epsilon\hat{\iota}o\nu$), namely $\Delta^\gamma, K^\gamma, \Delta^\gamma\Delta$, $\Delta K^\gamma, K^\gamma K$, in the solutions which follow (4, 14–6, 2). In this whole context no unknown or indeterminate number is so much as mentioned. *Only at this point* is the expression *arithmos* in the sense of an "indeterminate multitude of monads" ($\pi\lambda\hat{\eta}\theta\sigma\varsigma$ $\mu\sigma\nu\acute{\alpha}\delta\omega\nu$ $\acute{\alpha}\acute{o}\rho\iota\sigma\tau\sigma\nu$), and its sign, ς, as well as the sign for the immutable *monas*, $\overset{\text{м}}{M}$, introduced (6, 3–8). Thus, although within the course of each solution the numbers rendered by $\Delta^\gamma, K^\gamma, \Delta^\gamma\Delta, \Delta K^\gamma$, and $K^\gamma K$ have as their "root" (cf. 2, 19 f.; also Euclid VII, Defs. 17, 18) precisely the ς for each case, it is by no means their original character to be merely the powers of an unknown or "indeterminate" number. For even as powers of the ς they represent only transformations of a *determinate* number of monads (cf. Aristotle, *Posterior Analytics* XII, 78 a 18). Exactly the same thing holds true also for the fractional parts homonymous with these, namely the "fraction named after 'number,' 'square,' 'cube,' etc. . . . [i.e., the reciprocals of the terms named earlier]" ($\mu\acute{o}\rho\iota o\nu$ $\acute{\alpha}\rho\iota\theta\mu\sigma\sigma\tau\acute{o}\nu$, $\delta\upsilon\nu\alpha\mu\sigma\sigma\tau\acute{o}\nu$, $\kappa\upsilon\beta\sigma\sigma\tau\acute{o}\nu$, $\delta\upsilon\nu\alpha\mu\sigma\delta\upsilon\nu\alpha\mu\sigma\sigma\tau\acute{o}\nu$, $\delta\upsilon\nu\alpha\mu\sigma\kappa\upsilon\beta\sigma\sigma\tau\acute{o}\nu$, $\kappa\upsilon\beta\sigma\kappa\upsilon\beta\sigma\sigma\tau\acute{o}\nu$ — 6, 9–21; cf. P. 137). It is for this reason that the sign χ, which denotes reciprocals (6, 20 f.), is not confined to the fractional parts named but is applicable to all root fractions (cf. Note 171, also second volume of the Tannery edition, pp. xlii f.). There is, then, no sharp boundary between "the determinate" (οἱ ὡρισμένοι — 6, 6 f.) and the "indeterminate" (ἀόριστοι) numbers.[178] Maximus Planudes (end of thirteenth century) is therefore perfectly justified in elucidating the facts implied in the rules for multiplication (IV–VIII, 8, 11–12, 18) by means of determinate numbers (in his *scholia* to the two first books of Diophantus, second volume of the Tannery edition, pp. 127 ff.), although these facts are especially emphasized by Diophantus only for dealing with the special numbers which are represented by the signs $Δ^γ$, $K^γ$, $Δ^{γι}$, $K^{γι}$, etc. The fact that the multiplication of *such* a number with the homonymous fractional part of the unit yields *one* ("Def. V") is exclusively based on the fact that this holds for the multiplication of *any* number with its homonymous fractional part (cf. P. 138). That the multiplication of several fractional parts of the unit again yields fractional parts which are homonymous with the product of the numbers homonymous with the original fractional parts ("Def. VII") likewise holds true of all numbers and *therefore* also of those special multitudes of monads which occur within the course of the solution and which are, at that point, still "indefinite." We shall presently have to discuss "Definition VI," which concerns multiplication with the unit itself, from another point of view. In any case, all these rules for multiplication are not "definitions"[179] in the Euclidean, and certainly not in the modern, sense; that is, rules by which certain relations among newly introduced magnitudes, *and thus the magnitudes themselves*, are originally determined; they are rather the rules of calculation already in use within the *ordinary* realm of numbers, which are here, in deference to the unfamiliar notation, merely explicitly confirmed. What is more, these rules can be inferred from the nomenclature itself: "They will be clear to you because they are pretty well expressed in the name itself" (ἔσονται δέ σοι καταφανεῖς διὰ τὸ προσδεδηλῶσθαι σχεδὸν διὰ τῆς ὀνομασίας — 6, 24 f.), for the relevant expressions show directly the multiplicative components of the numbers which are intended.

Now what, more exactly, does this nomenclature mean? It concerns, as we have said before, the "kind," the *eidos*, of numbers. In Diophantus we are, to be sure, no longer dealing with the original Pythagorean-Platonic *eidos* which is responsible for the characteristics of a "kind" because it gives unity to, and thereby originally makes possible, the being of numbers (cf. Pp. 54 ff.; 68 f.). And yet in Diophantus too the *eidos* of a number means — as in the Neoplatonic arithmeticians[180] — the characteristic of its kind which a number shares with other numbers, or by which it is, in turn, separated from them so that a classification of numbers can be obtained, except that this characterizing *eidos* is here understood rather in the Peripatetic-Euclidean manner, as a *mere property* of the various numbers (cf. Pp. 110 f.). All numbers which exhibit one and the same such *eidos* belong to one and the same "class," exactly as, for instance, all hundreds belong to one and the same *taxis* (cf. Pappus, Hultsch, I, pp. 20, 14 f.; Domninus, *Anecdota Graeca* pp. 415, 16 ff.), although it may well be that the same number is included in some other respect in another class. The classification which is based on so attenuated a concept of the *eidos*[181] has therefore no ontological but merely an instrumental significance. For instance, all numbers which can be rendered with the aid of the signs ς, $\varDelta^\gamma$, $K^\gamma \ldots$, ς^x, $\varDelta^{\gamma x} \ldots$ represent, in respect to their multiplicative composition, one *eidos* each, while all numbers which are written with the aid of the sign $\mathring{M}$ likewise form an *eidos* by themselves (cf. e.g., 114, 1 f.: "the rest of the classes . . . of the unknowns and units" — τὰ λοιπὰ . . . εἴδη τῶν ϛ καὶ τῶν $\mathring{M}$). Thus the sign x is said to be a sign "distinguishing the class" (διαστέλλουσα τὸ εἶδος — 6, 21) in that it allows the *eidos* in question (e.g., $K^{\gamma x}$) to be distinguished from the homonymous *eidos* (K^γ). Thus, with reference to the classes enumerated, the division of one number by another of another class is mentioned:

"the partitionings of the afore-mentioned classes" (οἱ μερισμοὶ τῶν προκειμένων εἰδῶν — 14, 2); and practice in the additive, subtractive, and multiplicative combination of numbers of the same or of different classes is recommended (14, 3–10). So also the general rule (14, 11–23) for the treatment of equations of the first order (and of pure quadratic and pure cubic equations, etc.) has precisely this object — to order and combine the numbers *according to their class membership* until finally both sides (μέρη, i.e., parts) of the equation are reduced to numbers of *one class*, that is, "until one class becomes equal to one class" (ἕως ἂν ἓν εἶδος ἑνὶ εἴδει ἴσον γένηται — 14, 14).[182] Now it is of the greatest significance that here *the* eide *themselves are always mentioned* and not the numbers, each of which belongs to a certain *eidos*. This is especially clear in the case of "Definition VI," according to which, when the *eidos* is multiplied by the unit, the "*eidos* so multiplied ... remains the same *eidos*" (τὸ πολυπλασιαζόμενον εἶδος ... αὐτὸ τὸ εἶδος ἔσται — 8, 14 f.). And the problems are spoken of explicitly as "having their material for the most part concentrated *in the classes as such*" (πλείστην ἔχοντες τὴν ἐπ᾽ αὐτοῖς τοῖς εἴδεσι συνηθροισμένην ὕλην — 14, 25–27), i.e., the presentation and solution of the problems takes place essentially *in terms of the* eide *themselves. But this does not mean that the* eide *as such are numbers.* Rather they represent, as we have shown earlier, a "characteristic of the kind" of each determinate number. The fact that multiplication by the unit does not change the *eidos* means only that each number (or fractional part of the unit) taken "once" remains untouched in its multitude *and consequently also in its* "*kind*," while, on the other hand, every multiplicative change of "kind," such as, for instance, "Definition VIII" (10, 1–12, 18) describes, is based on a change of the multiplicative composition of that particular multitude of monads. In other words, here too we must distinguish strictly between the *procedure* and the *object*; while the procedure is applied to the *eide* which are as such independent of each "multitude of monads" (πλῆθος μονάδων) and in this sense "general" (καθόλου), the object intended is in each case a determinate number of monads. Here, in the realm of mathematics, we thus find mirrored, albeit in a disjointed fashion which is no longer appropriate to the original phenomenon, the relation between *eidos* and single object — in the language of the schools, between "secondary" (δευτέρα) and "primary substance" (πρώτη οὐσία).[183] On its home ground, that is, within Greek intentionality, this relation is taken as a "matter of course"; in Aristotelian ontology it is forcefully brought to the fore: As the single object is *just that* which we name in its *eidos*, and is therefore accessible to "scientific" treatment through it alone, so the numbers which the Diophantine problems are supposed to yield are available to the "scientific" grasp only in their "eidetic" nature; and this is precisely because in each *eidos* a completely determinate number of monads is always *intended*. Different as the Diophantine *Arithmetic* may seem when contrasted with the Neoplatonic "arithmetical" science, on this point it is altogether comparable; it too deals with the multitude of monads essentially only in the medium of the *eide* (cf. Pp. 13 ff.; 32 ff.; 59 f.).

What distinguishes the Diophantine *eide* from those others is, aside from the fact of their being limited to a few "figurate" ones, such as arise by multiplication and correspond to the "figurate" numbers,[184] their purely *instrumental* significance. Within the calculating procedure itself the *eide* are used as special units of calculation exactly in the same sense in which units of a higher order occur in Archimedes and Apollonius (cf. P. 131) and in which the myriad is used in other places (cf. in Diophantus himself, 332, 8 f.: "1 second-degree myriad, and 8-thousand 7-hundred 4-ty and 7 first-degree myriads and 4-thousand 5-hundred and 60 monads"[185] — β′ $\overset{\prime}{M}$ᾱ καὶ α′ $\overline{,ηψμζ}$ καὶ $\dot{M},δφξ$ = δευτέρων μυριάδων ᾱ καὶ πρώτων $\overline{,ηψμζ}$ καὶ μονάδων $\overline{,δφξ}$). So also the sign for the ordinary unit, $\overset{\circ}{M}$, is introduced as a complete parallel for the sign ς: "and its [i.e., the number's] sign is ς. There is also another sign, the invariant sign of determinates, the monad ... $\overset{\circ}{M}$" (... ἔστιν αὐτοῦ [sc. ἀριθμοῦ] σημεῖον τὸ ς. ἔστι δὲ καὶ ἕτερον σημεῖον τὸ ἀμετάθετον τῶν ὡρισμένων ἡ μονὰς ... $\overset{\circ}{M}$ — 6, 5–8); and the "classes of unlike multitude" (εἴδη μὴ ὁμοπληθῆ), that is, the classes not "present" and "lacking" in equal number, e.g., $Δ^γ\bar{η}$ ⋏ $Δ^γ\bar{β}$ (eight "squares" less two "squares") are spoken of in the same sense as elsewhere the different numbers of ordinary monads (14, 5 f. and 12). This is possible precisely because every eidos means, in every single case, just like the myriad, a definite number of units.

But we must not forget that all the signs which Diophantus uses are merely word abbreviations. This is true, in particular, of the sign for "lacking," ⋏ (12, 21), and of the sign for the unknown number, ς, which (as Heath has convincingly shown) represents nothing but a ligature for αρ (= ἀριθμός; cf. Heath, Diophantus of Alexandria, pp. 32–37;[186] on ⋏, see ibid. pp. 42 f.). For this reason Nesselmann (Algebra der Griechen, p. 302) called the procedure practiced by Diophantus a "syncopated algebra" which, he said, forms the transition from the early "rhetorical" to the modern "symbolic" algebra (according to Nesselmann even Vieta's mode of calculation belongs to the stage of syncopated algebra). Léon Rodet (Sur les notations numériques et algébriques antérieurement au XVI siècle [1881], p. 69) has, with some justice, opposed this tripartite division with the thesis that only two types of algebra should be recognized, namely "l'algèbre des abbréviations et des données numériques" and "l'algèbre symbolique." It is to the first type, that of the algebra of abbreviations and numerically given constituents, that the Diophantine "Arithmetic" unquestionably belongs. Of the second type Rodet says that "it was not born before someone had had the idea of representing what is given in a problem in a general form by means of a symbol, and of similarly symbolizing each of the operations by a special sign." In this way, he says, *general* solutions may then be achieved, be they numbers, figures, or natural phenomena. But in a historical study this should itself be the true problem: How was it possible at all that in the face of a *conceptually self-sufficient and complete* calculating procedure, such as the Diophantine *Arithmetic* is *par excellence*, the "idea" of a symbolic algebra was conceived? Or, more exactly: *What transformation did a concept like that of* arithmos *have to undergo in order that a "symbolic" calculating technique might grow out of the Diophantine tradition?* The *Arithmetic* of Diophantus may, to be sure, itself refer back to a pre- and *non*-Greek, perhaps even a "symbolic," technique of counting (cf. P. 127). The ease with which Diophantus carries out the multiplication of expressions which are composed of numbers of different "kinds," the matter of-course fashion in which he handles such expressions in general, the way, furthermore, in which he teaches multiplication with "lacking" magnitudes, without, however, considering negative numbers possible as such,[187] and, finally, the purely instrumental use he makes of the *eidos* concept — all this does indeed show an inner tension between the "material" treated and the character of the concepts forced on it. But for the origin of modern algebra and its formal language it was precisely the *direct* assimilation of Diophantus' work and of its concepts which was crucial. The indirect route over the Arabs, by which the origin of modern algebra can also be traced, does, to be sure, likewise lead back to Greek sources, namely to Diophantus and Anatolius,[188] but the tradition advancing along this route includes an independent pre-Greek element. From Leonardo of Pisa (beginning of the thirteenth century), *via* the "cossic" school and up to Michael Stifel (1544), Cardano (1545), Tartaglia (1556–1560), and Petrus Nonius (Pedro Nuñez, 1567), "algebra," maintaining itself apart from the traditional disciplines of the schools, struggles for a place in the system of western science. While, however, the "algebra" which has Arabic sources is continually elaborated in respect to techniques of calculation, for instance by the introduction of "negative," "irrational," and even so-called "imaginary" magnitudes (numeri "absurdi" or "ficti," "irrationales" or "surdi," "impossibiles" or "sophistici"), by the solution of cubic equations, and in its whole mode of operating with numbers and number signs, its self-understanding fails to keep pace with these technical advances. This algebraic school becomes conscious of its own "scientific" character and of the novelty of its "number" concept only at the moment of direct contact with the corresponding Greek science, i.e., with the *Arithmetic* of Diophantus. Under the influence of the impression made by a reading of the Diophantine manuscript, Bombelli changes the "technical" character of his manuscript, which was probably already written by 1550.[189] The third book of his *Algebra*, which appeared in a new version in 1572, now contains also (in a modified, partly "generalized" form) the majority of the problems in the first five books of Diophantus. Here the original verbal "cloaking" of the problems, which had been the practice in previous "algebraic" works, is abandoned in favor of a "pure" form taken over from none other than Diophantus.[190] In place of the term "censo," which renders the Arabic word for "means" corresponding to the Greek *dynamis*, the expression "potenza"[191] is used; for the term "cosa" (Latin: res, i.e., "the [unknown] thing"), the expression "tanto" (corresponding to the Greek *arithmos*: "so and so many") is substituted.[192] In 1575 there appears the first Latin translation of Diophantus by Xylander, which contains also the *scholia* of Maximus Planudes; in 1577 the algebraic work of Gosselin, which includes a treatment of the equations of Diophantus. Finally Diophantus is put into modern "symbolic" form in 1585 by Stevin and in 1591 by Vieta. Although Stevin, who depends directly on Bombelli for his terminology and symbolism, already makes the new *"number" concept* which is the basis of the "symbolic" procedure completely explicit, it is Vieta who, by means of the introduction of a general mathematical symbolism, actually realizes the fundamental transformation of the conceptual foundations. We shall therefore now turn to the latter.

II

The formalism of Vieta and the transformation of the arithmos *concept*

A
The life of Vieta and the general characteristics of his work

THE MATHEMATICAL investigations of Vieta (1540–1603) are in every respect characteristic for the situation of the "new" science at the turn of the sixteenth century. This is especially true of the place which these investigations take within Vieta's own life. Born a Catholic, but personally unengaged in questions of faith, by profession a jurist, he entered the service of the Huguenot Antoinette d'Aubeterre (who held the seigniory of Soubise) at age twenty-four, and, as her legal adviser, became involved in the dispute between the Catholics and the Calvinists[193] which shook France to the foundations. On the side he supervised the education of Catherine of Parthenay, later of Rohan (the daughter of Antoinette d'Aubeterre), whose loyal friend and adviser he remained all his life. Of the lectures designed for her use (1564–1568), which comprehended all areas of knowledge, only one has come down to us. It appeared for the first time in 1637 in a French translation under the title *Principes de cosmographie*[194] and indicates Vieta's special interest in cosmological and astronomical questions. His chief work, *Harmonicum coeleste* (*Harmonic Construction of the Heavens*), which never appeared in print, was still available in manuscript in the nineteenth century (Libri is responsible for its later disappearance)[195] and seems to have been an outgrowth of these first didactic essays.[196,197] *All the mathematical investigations of Vieta are closely connected with his cosmological and astronomical work.* The *Canon mathematicus, seu ad triangula* (*Mathematical Canon, or About Triangles*, 1579), which is based on the works of Regiomontanus and Rheticus,[198] was intended to form the preparatory "trigonometric" part of the main work.[199] The second part of the *Canon*, the *Universalium Inspectionum ad Canonem mathematicum liber singularis* (*A Book of General Tables attached to the Mathematical Canon*), gives, among other things, the *computational methods* used in the construction of the canon (constructio Canonis), and teaches, in particular, the computation of plane and spherical triangles with the aid of the *general* trigonometric relations which exist between the different determining components of such triangles. These relations are collected by Vieta in the form of tables which allow one to read off directly the relevant proportion (analogia) obtaining between three "known" and one unknown component of the triangle.[200] This work is directly related to problems in the formulation of equations as treated in the contemporary "algebraic" works of Cardano, Tartaglia, Nonius, Bombelli, and Gosselin. References by Peletier and Petrus Ramus,[201] as well as Xylander's translation, must certainly have introduced Vieta to Diophantus' *Arithmetic*,[202] which he undoubtedly came to know also in the original.[203] And from this study of the Diophantine work eventually grew his symbolic algebra, whose fundamental characteristics he sketched out programmatically in his work *In artem analyticen Isagoge* (*Introduction to the Analytical Art*), which appeared in 1591. [For a translation of this work see the Appendix,

Pp. 315–353.] Of the writings announced there, only a part appeared during Vieta's lifetime. Called to high official positions, adviser of Henry III and Henry IV,[204] he hardly found time, aside from occasional writings,[205] to give his mathematical works the final form suitable for publication.[206]

It is important to be clear about the fact that modern mathematics is guided from the outset by cosmological-astronomical interests. This is true not only of Vieta, but of Kepler, Descartes, Barrow, Newton, etc. In this respect the "new" science repeats the course of ancient science (cf. Part I Section 7A). But the manner in which the founders of modern science set about attaining a mathematical comprehension of the world's structure betrays, from the outset, a different conception of the world, a different understanding of the world's being, than that which had belonged to the ancients. They were not, for the most part, themselves aware of their own conceptual presuppositions; thence came the tension within the science which they founded: Eventually, the ancient legacy came into conflict with the modern mode of cognition which rests upon a new ontological understanding. In the nineteenth century, this conflict then led to a new "formalized" foundation of infinitesimal analysis, and today its effects are seen in the struggle to fix the principles of mathematical physics. In germ it is already present in Vieta.

In contrast to many of his contemporaries, for instance, Petrus Ramus and Bacon, Vieta's comprehensive humanistic education does not lead him into open and explicit opposition to the traditional science. But he resembles them in turning, by preference, toward the neglected or unknown sources of ancient literary tradition. He wishes to be in every respect the loyal preserver, rediscoverer and continuator of our ancient teachers.[207] The *Harmonicum coeleste*, whose time of origin coincides with Kepler's *Mysterium cosmographicum* (1596), has the task of renewing *in Ptolemy's spirit*, his *Mathematical Composition* (Μαθηματικὴ σύνταξις), the *Alma-*

gest. Vieta seems to accept the Copernican thesis because it retains the methods of Ptolemy's science.[208, 209] Similarly, in his remaining mathematical writings he is always concerned not only with borrowing his terms directly from ancient terminology (or at least, when he must invent new terms, to match it as closely as possible) but also with interpreting all "innovations," once they have been introduced, as a mere development of the tradition. All "innovation" is for him, as for so many of his contemporaries, "*renovation*" (renovatio, restitutio; cf. L. Olschki, *Galilei und seine Zeit* [1927] P. 147; so also in the writings of Tartaglia, Commandinus, Galileo, Viviani). In the *Letter Prefatory* to the *Isagoge* (Appendix, P. 318), which is addressed to Catherine of Parthenay, he says characteristically: "Those things which are new are wont in the beginning to be set forth rudely and formlessly and must then be polished and perfected in succeeding centuries. Behold, the art which I present is *new*, but in truth *so old*, so spoiled and defiled by the barbarians, that I considered it necessary, in order to introduce an entirely new form into it, to think out and publish a new vocabulary, having gotten rid of all its pseudo-technical terms, lest it should retain its filth and continue to stink in the old way...." (Quae nova sunt solent a principio proponi rudia et informia, succedentibus deinde seculis expolienda et perficienda. Ecce ars quam profero nova est, aut demum ita vetusta et a barbaris defaedata et conspurcata, ut novam omnino formam ei inducere, et ablegatis omnibus suis pseudo-categorematis, ne quid suae spurcitiei retineret, et veternum redoleret, excogitare necesse habuerim et emittere nova vocabula. . . .) Thus in the edition of 1591 the *Isagoge* is described as a part of the "*Opus restitutae Mathematicae Analyseos*" (*Opus of the restored Mathematical Analysis*) or of the "*Algebra nova.*"[210] The very expressions "algebra" or "algebra et almucabala" clearly belong for Vieta to this barbarian "pseudo-terminology." Descartes, incidentally, later follows him in this, speaking in his *Regulae* of that "art which they by a barbaric name call algebra" (ars quam barbaro nomine Algebram vocant — Adam-Tannery, X, 377).[211] Passing over this "Ars magna,"[212] which is surrounded by mystery and recalls the dark art of the alchemists, Vieta claims to have been the first to have discovered the "previously buried genuine gold" (aurum fossile et probum) of the *ancient* mathematicians, which they had guarded jealously, and whose possession allows Vieta to solve not only, as people did before him, "this and that problem" singly, but precisely to manage problems of this kind in any desired amount — "by tenths and twenties" (decadas et eicadas). The severe and pedantically learned sobriety of his thinking[213] is complemented by a consciousness of the immense power which this discovery gives to mankind and which he, anticipating centuries to come, expresses in the proud parole at the end of his *Isagoge*: "*To leave no problem unsolved*" — "*Nullum non problema solvere.*"

B

Vieta's point of departure: the concept of synthetic *apodeixis* in Pappus and in Diophantus

In order better to understand Vieta's relationship to the ancients, a relationship typical for the sixteenth and seventeenth century, we turn to his presentation of it in the *Isogoge* itself. The two main Greek sources on which Vieta's work draws appear immediately in the opening chapter: (1) Pappus' seventh book, which was destined to continue to play so significant a role in the development of modern mathematics,[214] and (2) Diophantus' *Arithmetic*.[215] There is in mathematics, Vieta says, a special procedure for discovery, "a certain way of investigating the truth" (veritatis inquirendae via quaedam) which, so it is claimed, was first discovered by Plato.[216] Theon of Alexandria gave this procedure the name of "analysis" and defined it precisely, namely as a process beginning with "the assumption of what is sought as though it were granted, and by means of the consequences [proceeding to] a truth [which was in fact already] granted" (adsumptio quaesiti tanquam concessi per consequentia ad verum concessum), just "as in converse" (ut contra) he defined "synthesis" as a process beginning with "the assumption of what is granted and by means of the consequences [proceeding to] the conclusion and comprehension of what is sought" (adsumptio concessi per consequentia ad quaesiti finem et comprehensionem). These definitions,[217] which are here ascribed to Theon, also occur in Pappus in a modified and clarified form, namely at the beginning of his seventh book (Hultsch, II, P. 634, 11 ff.).[218] In a *scholium* to Euclid[217] it is shown with reference to the first five theorems of the thirteenth book how the "synthesis" results in each case from the preceding "analysis" *by means of conversion* (analysis and synthesis both proceeding "without drawing the figure" — ἄνευ καταγραφῆς — Heiberg-Menge, Pp. 366, 4; 368, 16). And Pappus, who mentions the aforesaid procedure with reference to the so-called *Treasury of Analysis* (ἀναλυόμενος τόπος), emphatically stresses the relationship of conversion. Now according to the application to be made of the "analysis" — whether it is to be applied to the discovery of the *proof of a "theorem"* or to the *solution* (i.e., construction) *of a "problem"* — Pappus distinguishes *two kinds, two gene, of analysis*: "the one is for searching for the truth [i.e., *zetetic*, from ζητέω, to search], which is called *theoretical*, and the other is for supplying what is required [i.e., *poristic* from πορίζω, 'to supply'], which is called *problematical* (τὸ μὲν ζητητικὸν τἀληθοῦς, ὃ καλεῖται θεωρητικόν, τὸ δὲ ποριστικὸν τοῦ προταθέντος, ὃ καλεῖται προβληματικόν — Hultsch, II, P. 634, 24–26). Thus in the first case the conversion of the analysis, the "synthesis," represents a direct *apodeixis*, while in the second case it consists first of a geometric construction (κατασκευή), or sometimes a porism (πορισμός, i.e., the production or finding of something already implicit in the figure; Hultsch, II, p. 650, 16 ff.; cf. Note 46), upon which the *apodeixis* then follows.[219] However, in elucidating the difference between "theoretical" and "problematical" analysis, Pappus at both times calls the synthesis simply an "*apodeixis*": "And in reverse, the proof is the converse of the analysis." (636, 5 f.: καὶ ἡ ἀπόδειξις ἀντίστροφος τῇ ἀναλύσει, also 636, 12 f.: καὶ πάλιν ἡ ἀπόδειξις ἀντίστροφος τῇ ἀναλύσει.) So also in Diophantus—and this was clearly crucial for Vieta's terminology — the conversion of each solution, namely the "test proof" which is intended to show that the numbers found do, indeed, fulfill the conditions, that they "do the problem" (ποιοῦσι τὸ πρόβλημα), is called *apodeixis*; the words "and the proof is clear" (καὶ ἡ ἀπόδειξις φανερά) form the conclusion of a whole series of his problems (cf. second volume of the Tannery edition, Index, see ἀπόδειξις).[220] On the other hand, we must recall that it is, after all, particularly characteristic of the Diophantine procedure to operate with the *quaesitum*, namely with the number sought in each case, *as with something already given* or "granted" (concessum). To construct an *equation* means nothing but to put the conditions of a problem into a form which enables us to ignore whether the magnitudes occurring in the problem are "known" or "unknown." The consequences (consequentia) to be drawn from such an equation, that is, its stepwise transformation into a canonical form[221] (its standard form, as we would say), finally lead, by means of computation, to the finding of the number sought, i.e., to the "true" number which is only then, at the end, "granted" (verum concessum). Should an "impossible" number (see P. 138) result from the final computation, the problem itself is taken to have been badly posed, that is, impossible; as Pappus says: "If we come on [a solution] agreed to be impossible, the problem will also be impossible." (ἐὰν δὲ ἀδυνάτῳ ὁμολογουμένῳ ἐντύχωμεν, ἀδύνατον ἔσται καὶ τὸ πρόβλημα — Hultsch, P. 636, 13 f.) In that case it is in need of a condition of possiblity (διορισμός), or in Pappus' words (636, 15 f.), an "additional specification for when, and how and in how many ways the problem will be possible" (προδιαστολὴ τοῦ πότε καὶ πῶς καὶ ποσαχῶς δυνατὸν ἔσται [καὶ] τὸ πρόβλημα); Diophantus calls this condition a *prosdiorismos* (cf. P. 135; P. 138).[222]

The point of departure for Vieta's "renovation" is then this: He conjoins these facts, which are presented by Pappus only in reference to *geometric* theorems and problems (such as are found in Euclid, Apollonius, and Aristaeus), with the procedure of the Diophantine *Arithmetic*. *On the basis of Pappus' exposition, Vieta calls this procedure "analysis," "ars analytice"* — "analytic art." He is completely aware of the purely geometric character of the analysis intended by Pappus. In the *Appendicula* I to *Apollonius Gallus* (ed. van Schooten, pp. 339 ff.), confronted with certain problems which Regiomontanus[223] could solve "algebraically, as he says" (algebrice, ut loquitur) but not geometrically, he provides their geometric construction and notes by way of introduction that these geometric constructions are *also* of importance: "But algebra, as Theon, Apollonius, Pappus and the other ancient analysts passed it down, is *generally geometrical* and always exhibits the magnitudes which are sought either *as an object* [i.e., in a visible construction] or *in number*, or else it will be an 'irrational or absurd problem' [which here simply means 'impossible' — ἀδύνατον[224]]." (At Algebra, quam tradidere Theon, Apollonius, Pappus et alii veteres Analystae, omnino Geometrica est, et magnitudines, de quibus quaeritur, sive re, sive numero statim exhibet, aut erit ἄρρητον ἢ ἄλογον πρόβλημα.) He is, then, of the opinion that the *geometric* failure of Regiomontanus is in this case due to the fact that he did not know the genuine "algebra" at all. At the end of his work "*Ad Problema, quod omnibus mathematicis totius orbis construendum proposuit Adrianus Romanus, Responsum* (*An Answer to the Problem in Construction which Adrianus Romanus Proposed to All the Mathematicians of the Whole World* — 1595), he sets himself a problem which he takes from the seventh book of Pappus, solves it under the name of "Apollonius Gallus" (cf. Note 214), and notes in conclusion: "But the problems which Regiomontanus solved algebraically he confesses *he* could sometimes not construct *geometrically. But was that not because algebra was up to that time practiced impurely?* Friends of learning, embrace a *new* algebra; farewell, and look to the just and the good." (Sed quae problemata Algebrice absolvit Regiomontanus, is se non posse aliquando[223] Geometrice construere fatetur. An non ideo quia Algebra fuit hactenus tractata impure? Novam amplectimini, φιλομαθεῖς, valete, et aequi bonique consulite — van Schooten, p. 324). This "new" and "purified" algebra, which is represented by the "ars analytice," is for Vieta quite as much "geometric" as "arithmetical." Thus he conjectures that a generalized procedure which is not confined to figures and numbers lies not only behind the geometric analysis of the ancients but also, as we shall see, behind the Diophantine *Arithmetic* — although for Vieta the ultimate aim of this procedure is indeed to find geometric constructions and numbers; in the latter case, this means finding "possible" numbers, that is, according to the passage from *Apollonius Gallus*, such numbers as have a direct geometric interpretation.

Vieta's conception of a "pure," "*general*" algebra which will be equally applicable to geometric magnitudes and numbers is met half-way by the *general theory of proportions* of Eudoxus as transmitted in the fifth book of Euclid. Aristotle had already used the Eudoxian theory of proportions, together with the "common notions" (κοιναὶ ἔννοιαι), as the classical example of a discipline which has a "*general object*" and is not bound to a specific realm of objects: "Among the mathematical [sciences] ... geometry and astronomy *deal with a nature of a certain kind, while the general science is common and about all things.*" (... ἐν ταῖς μαθηματικαῖς ... ἡ μὲν γεωμετρία καὶ ἀστρολογία περί τινα φύσιν εἰσίν, ἡ δὲ καθόλου πασῶν κοινή — *Metaphysics* E 1, 1026 a 25–27.) So also in the *Posterior Analytics*: "Just as it used to be proved separately that a proportion can be alternated [cf. Euclid V,

Def. 12 and Prop. 16] insofar as it is a proportion of numbers and lines and solids and times [the last with reference to astronomy, cf. the previous quotation and *Metaphysics* M 2, 1077 a 1–4; 12], *so it now can be shown for all in one demonstration*; but because all of these — numbers, lengths, time, solids — did not have any one name and differed from one another in species they were taken separately, but *now* [using the model of Eudoxus] *it is proved generally*; for it belongs to them [to be alternable] not insofar as they are lines or numbers, *but insofar as they are in general supposed to have this property.*" (... καὶ τὸ ἀνάλογον ὅτι ἐναλλάξ, ᾗ ἀριθμοὶ καὶ ᾗ γραμμαὶ καὶ ᾗ στερεὰ καὶ ᾗ χρόνοι, ὥσπερ ἐδείκνυτό ποτε χωρίς, ἐνδεχόμενόν γε κατὰ πάντων μιᾷ ἀποδείξει δειχθῆναι· ἀλλὰ διὰ τὸ μὴ εἶναι ὠνομασμένον τι πάντα ταῦτα ἕν, ἀριθμοὶ μήκη χρόνος στερεά, καὶ εἴδει διαφέρειν ἀλλήλων, χωρὶς ἐλαμβάνετο. νῦν δὲ καθόλου δείκνυται· οὐ γὰρ ᾗ γραμμαὶ ἢ ᾗ ἀριθμοὶ ὑπῆρχεν, ἀλλ' ᾗ τοδί, ὃ καθόλου ὑποτίθενται ὑπάρχειν — A 5, 74 a 17–25; cf. P. 162.) Proclus, likewise (*in Euclid.*, p. 7 ff.; 18 ff.), speaks of the "one science" (μία ἐπιστήμη) which gathers all mathematical knowledge into one, and of its "common theorems" (τὰ κοινὰ θεωρήματα) which can, it is true, be studied "in numbers and magnitudes and motions" (ἐν ἀριθμοῖς καὶ ἐν μεγέθεσι καὶ ἐν κινήσεσι), but which are identical neither with arithmetical nor with geometric nor with astronomical theorems. To this science belong, above all, the theorems of the theory of ratios and proportions (cf. Euclid V, Defs. 14, 15, 16, 12, Props. 17, 18, corollary of 19, 16; and Nicomachus I, 17–II, 4; II, 21–29; see also Pp. 26 f.): "the [theorems] *about proportions*, their compositions and divisions, their conversions and alternations, furthermore [theorems] about *all ratios*, such as about their multiples, their superparticulars, their superpartients and their opposites, *and the theorems established generally and in common for the equal and the unequal simply*" (τά τε τῶν ἀναλογιῶν καὶ τὰ τῶν συνθέσεων καὶ διαιρέσεων καὶ τῶν ἀναστροφῶν καὶ ἐναλλαγῶν, ἔτι δὲ τὰ τῶν λόγων πάντων οἷον πολλαπλασίων καὶ ἐπιμορίων καὶ ἐπιμερῶν καὶ τῶν τούτοις ἀντικειμένων καὶ ἁπλῶς τὸ περὶ τὸ ἴσον καὶ ἄνισον καθόλου θεωρούμενα καὶ κοινῶς — Proclus, in Euclid., 7, 22–27). Proclus means theorems which are not tied to "figures or numbers or motions [of heavenly bodies]" (ἐν σχήμασιν ἢ ἀριθμοῖς ἢ κινήσεσιν) but in which "a certain common nature" (φύσις τις κοινή) is grasped as such, "itself by itself" (αὐτὸ καθ᾽ αὑτό). In just this connection Proclus also mentions the procedure of "analysis" and "synthesis" *common to all mathematical disciplines*: "the road from things better known to things sought and the reversal [of the process, so as to go] from the latter to the former, which they call *analysis and synthesis*" (ἡ ἀπὸ τῶν γνωριμωτέρων ὁδὸς ἐπὶ τὰ ζητούμενα καὶ ἡ ἐκ τούτων ἐπ᾽ ἐκεῖνα μετάβασις, ἃς καλοῦσιν ἀναλύσεις καὶ συνθέσεις — 8, 5–8). With this tradition in mind Vieta maintains (in the conclusion of the second chapter of the *Isagoge*) that every "equation" (aequalitas) is a "solution of a *proportion*" (resolutio proportionis), and correspondingly, every proportion is the "construction of an equation" (constitutio aequalitatis). Therefore he always speaks of "equations" and "proportions" *together*, e.g., in Chapter II, end: "A proportion can be called the construction of an equation, and an equation the solution of a proportion." (Proportio potest dici constitutio aequalitatis: aequalitas, resolutio proportionis.) That is to say, "pure" algebra is for him not only a "general theory of equations," but at the same time a "general theory of proportions."[225] That is why he amalgamates some of the "common notions" enumerated in Euclid's *Elements* Book I with some definitions and theorems of the "generalized" Book V, of the geometric Books II and VI, and of the "arithmetical" Books VII and VIII to form his "stipulations for equations and proportions" (Symbola aequalitatum et proportionum — Chap. II),[226] which are to serve as the general and firm foundations (firmamenta) "by means of which the equations and proportions are obtained as conclusions" (quibus aequalitates et proportiones concluduntur — Chapter I).[227]

Here Vieta does indeed prove himself the faithful preserver and interpreter of the traditional doctrine. But the
crucial difference with respect to the ancient "general
treatment" (καθόλου πραγματεία) appears in the conception
which Vieta has of its *object*, although here too he refers
directly to the tradition. His point of view is (1) suggested by
the comparison which can be drawn between the role of
"analysis" in geometry on the one hand and in the Diophantine *Arithmetic* on the other, (2) most strongly supported
by the use of the *eidos* concept in Diophantus, (3) extensively
influenced by Proclus' position on the general theory of
proportion; yet it can be thoroughly understood only on the
basis of the fundamental reinterpretation which the ancient
mode of intentionality experiences in modern mathematics
and which reaches its most characteristic expression in the
transformed understanding of "*arithmos*."

C

The reinterpretation of the
Diophantine procedure by Vieta

1

The procedure for solutions "in the indeterminate form"
as an analogue to geometric analysis

For ancient science, the existence of a "general object" is
by no means a simple consequence of the existence of a
"general theory." Thus Aristotle says: "The general propositions in mathematics [namely the 'axioms,' i.e., the
'common notions,'228 but also all theorems of the Eudoxian
theory of proportions] *are not about separate things* which
exist *outside of and alongside* the [geometric] magnitudes and
numbers, but are just about these; not, however, insofar as
they are such as to have a magnitude or to be divisible [into
discrete units]." (... καὶ τὰ καθόλου ἐν τοῖς μαθήμασιν οὐ περὶ

κεχωρισμένων ἐστὶ παρὰ τὰ μεγέθη καὶ τοὺς ἀριθμούς, ἀλλὰ περὶ
τούτων μέν, οὐχ ᾗ δὲ τοιαῦτα οἷα ἔχειν μέγεθος ἢ εἶναι διαιρετά
. . . — *Metaphysics* M 3, 1077 b 17–20.) There cannot be a
specific mathematical object which is "neither a [determi-
nate] number [of monads] nor [indivisible, geometric]
points, nor an [arbitrarily divisible geometric] magnitude,
nor a [determinate period of] time" *(οὔτε ἀριθμός ἐστιν οὔτε*
στιγμαὶ οὔτε μέγεθος οὔτε χρόνος — *Metaphysics* M 2, 1077 a
12). This is, moreover, not some special Aristotelian dogma
but rather a *generally granted* premise of Aristotelian argumentation: "and if this is [granted to be] impossible . . ." *(εἰ δὲ*
τοῦτο ἀδύνατον . . . — 1077 a 12 f.; see also *Posterior*
Analytics A 24, 85 a 31–b 3). And the Platonic tradition likewise, as it makes itself heard one last time in Proclus, assigns
no special *mathematical* object to the "general study" *(καθόλου*
πραγματεία). This may, to be sure, be seen as an attempt
to understand the possibility of any "ratio," "proportion,"
and "harmony" on the basis of a "common property"
(κοινόν) of a *primordial kind* (cf. Pp. 98 f.), but as soon as
this attempt is realized the realm of the properly "mathematical" has been far superseded. This same fundamental
conception of the objects of mathematics certainly obtains in
Diophantus, namely insofar as in his problems and solutions
he admits, as we have seen (Pp. 134; 144 f.), only *determinate*
numbers of monads. But it is precisely on account of this
that the relationship between ("problematical") analysis and
synthesis which is traditional in geometry undergoes a
significant change in the *Arithmetic* of Diophantus. In the
solution of *geometric* problems the required *construction* forms
the *first part of the synthesis* (cf. P. 155); the *apodeixis* which
follows has to use the relations between the "given"
magnitudes, relations which are themselves "given" *from the*
beginning, together with those brought to light by the
construction, to prove that this construction satisfies the
conditions of the problem. In the "*arithmetical*" problems of
Diophantus, however, *the last step of the analysis*, namely the

final computation which produces the number sought, *is at the same time also the first step of the synthesis* — the final *computation* actually corresponds to the geometric "*construction*." The conversion of the process of solution in Diophantus thus corresponds only to the second part of the "problematical" synthesis in geometry (cf. Pp. 155 f.).[229] Now if the "analytical" manner of finding solutions in Diophantus and geometric ("problematical") analysis are understood as completely parallel procedures, as Vieta indeed understands them, then a sharper line must be drawn between the *transformations* of equations and the *computation* of the numbers sought than that which Diophantus generally does draw. In other words, the calculation ending "*in the indeterminate*" (the solution ἐν τῷ ἀορίστῳ, cf. Pp. 134 f.), *which Diophantus himself uses only as an auxiliary procedure, must be understood as the true analogue to geometric ("problematical") analysis*. Such an "indeterminate" solution then permits an arbitrary multitude of "determinate" solutions on the basis of numbers assumed at will. But for Diophantus there are only limited possibilities for the use of this procedure, because he still associates it with the assumption of determinate *numbers*. Here the comparison with *geometric* analysis causes Vieta to go beyond Diophantus.

2

The generalization of the *eidos* concept and its transformation into the "symbolic" concept of the *species*

When Hankel (*Zur Geschichte der Mathematik* . . . , p. 148) complains of the "Greek ethnic peculiarity" which loads down every analytical solution in geometry with the "useless ballast" of a "completely worked out synthesis which says everything over again in reverse order," he misses the essential difference between analysis and synthesis. Analysis merely shows the *possibility* of a proof or a construction. A theorem has been "proved" only when the facts in question have *actually* been "derived" from the "given" relations between the "given" magnitudes. The construction of a figure determined by certain definite conditions has taken place only when this figure has actually been drawn using the magnitudes "given" with *just these determinate dimensions*. The "giveness" of which analysis makes use should, by contrast, be understood only as a "*possible* givenness." Thus, while analysis is immediately concerned with the generality of the procedure, synthesis is, in accordance with the fundamental Greek conception of the objects of mathematics, obliged to "realize" this general procedure in an *unequivocally determinate object* (cf. Pp. 122 f.). This "possible givenness" appears in geometric analysis in the fact that the construction which is regarded as already effected (the "quaesitum tanquam concessum") does not need to use the "given" magnitudes as unequivocally determinate *but only as having the character of being "given"* (cf. Aristotle, *De Memoria* 450 a 1 ff.). How can this situation be transferred to "arithmetical" analysis? Clearly in this way — that the numbers "given" in a problem are also regarded only in their character of being given, and not as just these determinate numbers. In other words, to assimilate the arithmetical to the geometric analysis completely, the "given" numbers must likewise be allowed a certain indeterminateness which should, in fact, be limited only by the "condition of possibility" *(διορισμός)* of the problem. Thereupon a solution "in the indeterminate form" would be possible for every case, and three stages could always be distinguished in the process of solution: (1) the construction of the equation, (2) the transformations to which it is subjected until it has acquired a canonical form which immediately supplies the "indeterminate" solution, and (3) the numerical exploitation of the last, i.e., the computation of unequivocally determinate numbers which fulfill the conditions set for the problem. Here only the first two stages represent the properly "analytic" procedure. The last stage actually already belongs to the synthesis, since it corresponds to the geometric "construction." But how is it possible to allow "given" and therefore "determinate" numbers to play an "indeterminate" part? At this point the Diophantine model may be appealed to: Just as Diophantus represents the unknown number — although it is "in itself" likewise "determinate" (cf. Pp. 140 f.) — by its *eidos*, its "*species*," which leaves the question "how many?" provisionally indeterminate, so *every number can be expressed by its "species."* As soon as the *eide* of the unknown and its powers appear in Diophantus (cf. Pp. 143 ff.) as new units of calculation, while the real computation takes place in terms of determinate numbers, the "calculation," which is performed exclusively with number "species," ought to be entirely shifted into the domain of the "indeterminate." This crucial last step is also taken by Vieta. He takes it in the consciousness of merely confirming a practice long in use among the ancients (and found especially in Diophantus), although not sufficiently clarified by them. In the course of taking this last step he is forced to reinterpret the tradition at essential points.

Having established geometric ("problematical") analysis as a parallel to the Diophantine procedure in the manner shown, Vieta arrives at the conception of a mode of calculation which is carried out completely in terms of "species" of numbers and calls it "*logistice speciosa*" (in contrast to calculation with determinate numbers, which is "logistice numerosa"; cf. *Isagoge*, Chap. IV, beginning).[230] Consequently "logistice speciosa" is originally very closely connected with the Diophantine procedure, which, in turn, forms the "arithmetical" analogue to geometric analysis. But at the same time — and this is of symptomatic significance for the concept formation of modern mathematics — Vieta devotes the "logistice speciosa" to the service of "*pure*" algebra, understood as the most comprehensive possible "analytic" art, indifferently applicable to numbers and to geometric magnitudes (cf. Pp. 158–161). In the course of this, the *eidos* concept, the concept of the "species," *undergoes a universalizing extension while preserving its tie to the realm of numbers. In the light of this general procedure, the species,* or as Vieta also says, the "forms of things" (formae rerum: — Chapter IV, beginning),[231] *represent "general" magnitudes simply.*

This extension of the Diophantine *eidos* concept has an immediate effect on the new organization of the "analytic" art which Vieta undertakes (Chap. I).[232] To the two kinds of analysis mentioned by Pappus, the "theoretical" and the "problematical," Vieta adds still a third which he calls *rhetic* ($\dot{\rho}\eta\tau\iota\kappa\dot{\eta}$), i.e., "telling," or sometimes *exegetic* ($\dot{\epsilon}\xi\eta\gamma\eta\tau\iota\kappa\dot{\eta}$), i.e., "exhibiting," analysis. For the first two kinds he significantly borrows from Pappus' text not the expressions "theoretical" ($\theta\epsilon\omega\rho\eta\tau\iota\kappa\dot{o}\nu$) and "problematical" ($\pi\rho o\beta\lambda\eta\mu\alpha\tau\iota\kappa\dot{o}\nu$), but *zetic* ($\zeta\eta\tau\eta\tau\iota\kappa\dot{o}\nu$ [sc. $\tau\dot{\alpha}\lambda\eta\theta o\hat{v}s$]), i.e., "seeking [the truth]," and *poristic* ($\pi o\rho\iota\sigma\tau\iota\kappa\dot{o}\nu$[sc. $\tau o\hat{v}\pi\rho o\tau\alpha\theta\dot{\epsilon}\nu\tau os$]), i.e., "productive [of the proposed theorem]." While for Pappus the difference between "theoretical" and "problematic" analysis is a difference in the kind of object presented in a "theorem" and in a "problem" (cf. P. 155), Vieta is able to distinguish only very externally between "zetic" and "poristic" analysis, because in concentrating his reflections on procedure he no longer differentiates between "theorems" and "problems," or, more exactly, because *he sees all theorems as problems.* He is interested less in the "truths" themselves which are to be found than in the *finding of "correct finding."* This is also the source of his general definition of the "analytic art" as the "theory for finding [what is sought] in mathematics [in general]." "Zetetic" and "poristic" — Vieta explicitly notes that properly speaking these alone are intended in Theon's definition of analysis (cf. Pp. 154 f.) — both represent those two first, properly analytical, stages of the process of solution mentioned previously. "Zetetic" is defined by Vieta as the procedure "through which the equation or the proportion is found which is to be constructed by the aid of the given magnitudes with a view to the magnitude sought"; "poristic" is defined as the procedure "through which by means of the equation or proportion the truth of the theorem [!] set up [in them] is investigated."[233] In both cases "magnitude" (magnitudo) is spoken of in the most general sense. Now the new kind of analysis introduced by Vieta is defined as the procedure "through which the magnitude sought is *itself* produced out of the equation or proportion set up [in canonical form]." This required magnitude "itself" is either an unequivocally determinate number, and therefore communicable *in speech*, or a *visible* geometric magnitude, unequivocally measurable. Hence arises the double name of this third kind of analysis whose business it is to effect the *computation* of "arithmetical" as well as the *construction* of "geometric" magnitudes starting from canonically ordered equations; it is called *rhetic* ($\dot{\rho}\eta\tau\iota\kappa\dot{\eta}$) with respect to the numbers to which it leads and which can be expressed by the ordinary number names of our *language*; it is called *exegetic* ($\dot{\epsilon}\xi\eta\gamma\eta\tau\iota\kappa\dot{\eta}$) in respect to the geometric magnitudes which it makes directly available to *sight*.[234] "Rhetic" or "exegetic" thus represents the third (and final) stage in the solution of an equation, which, as we have seen (Pp. 164 f.), is actually already a part of the synthesis and which is nevertheless understood by Vieta as an analytical procedure.[235] In this last stage the "analyst" must become either a "geometer" or a "logistician," i.e., calculator, in the ordinary sense.[236] As *geometer* he does his work by finding the "solution" of "another, though similar" problem, insofar as he "repeats" in his construction in reverse order, "synthetically," that analysis (resolutio) which is performed on *other* "given" magnitudes (cf. P. 164), namely the "purely" algebraic analysis. In doing this he may "hide" the purely algebraic "analytic" work which preceded, and pretend to have solved the problem immediately by a "synthetic" process, attacking it analytically only later (to assist those interested in the calculations, as it were) by starting from the equation obtained in the synthetic construction.[237]

The "geometric solution" (effectio Geometrica) is thus
effected in such a way that "it does not derive and justify the
synthesis from the equation but the equation from the
synthetic construction — while the very synthesis speaks for
itself" (compositionem operis non ex aequalitate, sed
aequalitatem ex compositione arguit, et demonstrat: ipsa
vero compositio seipsam — Chapter VII).[238] As logistician
he finds numerical solutions — be it exactly, be it by approximation procedures — which may be either simple equations (resolutio potestatum purarum — analysis of *pure*
powers, i.e., simple equations whose unknown has the same
exponent everywhere), or composite equations, or any others
desired (resolutio potestatum adfectarum — solution of
conjoined powers, i.e., *impure* equations whose unknown has
different exponents for different occurrences).[239] In no case
must he forget to make clear, by means of a specimen, the
artful device employed.[240] "Rhetic" or "exegetic" thus
comprises a series of "rules" (praecepta) and is therefore to
be considered the most important part of "analytic" — for
these rules first confer on the "analytic art" its character as an
"art" — while "zetic" and "poristic" consist essentially of
"examples" (exempla): "Rhetic and exegetic . . . must be
considered to be most powerfully pertinent to the establishment of the art, since the two remaining [parts of analytic]
provide *examples* rather than *rules*, as we must by rights grant
the logicians." (ῥητικὴ ἢ ἐξηγητική . . . censenda est . . .
potissimum ad artis ordinationem pertinere, cum reliquae
duae [sc. partes Analytices] exemplorum sint potius quam
praeceptorum, ut logicis iure concedendum est.)[241]

This organization of "analytic" by Vieta shows that
"general analytic" is understood as nothing more than the
indispensable auxiliary *means* to the solution of geometric
and numerical problems, but not as a complete and self-
sufficient discipline. "General analytic" is first of all a
"technique," to use a modern expression — its aim is neither
to open up a domain of truths of fairly manageable extent

nor to solve a certain number of problems; rather it intends to be an *instrument* for the solution of *problems in general*. Its "material" is not a single problem or a series of single problems but, as Vieta himself says (cf. Pp. 154 and 185), the problem of the general ability to solve problems; it is the "problem of problems" (problema problematum), *the art of finding, or the finding of finding*. Vieta's "general analytic" is an "organon," an instrument in the realm of mathematical finding in the same sense as the Aristotelian "logical" works, above all, the (*Prior* and *Posterior*) "*Analytics*" are an *Organon* in the realm of all possible knowledge whatever. But when with Descartes the "general analytic" takes over the role of the ancient fundamental ontological discipline, an effort is made to let it supersede the traditional logic completely, as we shall see later. The battle between the proponents of Peripatetic syllogistic and mathematical analysis[242] about the primacy of their respective views concerning the framework of the world, which is thus initiated in the seventeenth century, is still being waged today, now under the guise of the conflict between "formal logic" and "mathematical logic" or "logistic," although its ontological presuppositions have been completely obscured. In Vieta himself, the "analytic art" as a *whole*, that is, all three "kinds," stands in the service of his cosmological and astronomical investigations, which are concerned mainly with the numerical exploitation of solutions and therefore with "rhetic" and "exegetic." Hence arises his effort to develop an approximation procedure, which may well claim a place within the framework of an art that is considered a "tool" for "finding."[243] This is the reason also for the priority of "rhetic" or "exegetic" over the two other "kinds" of analysis (cf. P. 168). If, for the reasons just given, we disregard "poristic," only "zetetic" employs the "logistice speciosa"; therefore, strictly speaking, it alone represents "general analytic," the "new algebra." But we have already seen (P. 165) that Vieta's "logistice speciosa" has a double function: It is understood on the one hand as the procedure of *general* "pure" algebra and on the other as the procedure analogous to geometric analysis and directly related to the Diophantine *Arithmetic*. The two interpretations are simultaneously possible only because the "general" algebra has the role of a merely auxiliary procedure. Connected with this fact is Vieta's assumption that Diophantus solved his "arithmetical" problems with the aid of "zetetic" as understood by Vieta himself, but that he considered it better to "hide" this. Thus at the end of the fifth chapter of the *Isagoge*, "On the Laws of Zetetic" (De legibus Zeteticis), Vieta says: "Zetetic was employed most subtly of all by Diophantus in those books which were written on the subject of arithmetic. *But he exhibited it as though it were founded only on numbers and not also on species — although he himself used them — so that his subtlety and skill might be more admired*, since things that appear very subtle and abstruse to one who calculates in numbers [i.e., practices *logistice numerosa*] appear very familiar and immediately obvious to one who calculates in species [i.e., practices *logistice speciosa*]."[244] The art of the "ancient analysts" appears (Chap. I) as "tedium" (oscitantia) because it pretends to work with numbers instead of proceeding "with the skill proper to it" (ex arte propria) and calculating throughout in species. When Vieta speaks of the "species calculation which is to be *newly* introduced" (logistice sub specie noviter inducenda — *ibid.*), he means only the re-activation of an art known and approved in antiquity.[245] He finds support for this not only in the instrumental use of the *eidos* concept in Diophantus but also in the "indeterminate" solutions which the latter used as an *auxiliary procedure* (cf. Pp. 134 and 163) and in which Vieta apparently sees the traces of the *true* Diophantine art, namely of a *"pure" analytic* conceived as a *"general" mathematical auxiliary technique*.[246]

With this interpretation of the Diophantine *Arithmetic*, which views it exclusively as an "artful" procedure, and has but a small interest in the determinate results of the solutions which (to speak with Descartes) are "sterile truths," Vieta prescribed to historical research the approach which governs it to this very day (cf. P. 135). This view comes to expression in the matter-of-course acceptance within modern consciousness of the revolution in the ancient mode of forming concepts and of interpreting the world, which first took tangible shape when Vieta founded his "general analytic."

Now what, precisely, are the conceptual presuppositions which make such an interpretation of Diophantine "logistic" possible? *What* does Vieta understand by the species which form the object of the "general analytic" and *in which way* does he understand them? Moritz Cantor (II², pp. 631 and 519) was of the opinion that the "species" are "signs of spatial structures which render them perceptible to the senses," i.e., "magnitudes, not numbers." He appealed for support to Vieta's linguistic usage, for in the fourth chapter of the *Isagoge* he designates the operations to be performed on the species, namely those of "multiplication" and "division" (mentioned in the same words in the second chapter) by "ducere" (sc. magnitudinem in magnitudinem — "to compose a magnitude with a magnitude") and "adplicare" (sc. magnitudinem magnitudini — "to apply a magnitude to a magnitude"). But these terms, Cantor observed, originate in *geometric* representations, and hence Tartaglia in his *General trattato di numeri et misure* (1556–1560) still assigned the terms "multiplicare" and "partire" to the numerical but "ducere" and "misurare" to the geometric realm (the latter being intended to apply also to "fractions," because continuous division is possible only in geometric objects). Now unquestionably the "species" have for Vieta their traditional geometric implications, since their nomenclature is (as it was indeed already in Diophantus, cf. Pp. 142; 145) reminiscent of the ancient doctrine of "*figurate*" numbers, a fact which was of course known to so great an expert on ancient mathematical literature as Vieta was from his reading of Nicomachus and Theon of Smyrna.[247] (Here too we should remember the relationship of the Diophantine concept to the corresponding *geometric* concept of the "triangle given in figure" — τρίγωνον δεδομένον τῷ εἴδει, cf. P. 136.) This may have been one reason the more for him to conjecture that behind geometric analysis lurked precisely the same "general" analytic art which, in his opinion, lay at the basis of the Diophantine *Arithmetic*. But far more essential for the structure of the species concept than these geometric reminiscences are: (1) the generalization which the *eidos* concept undergoes at Vieta's hands, a generalization through which the species become the objects of a "general" mathematical discipline which can be identified neither with geometry nor with arithmetic, and (2) the direct connection with the "logistice numerosa," i.e., with "calculation," which is nevertheless retained by Vieta; in its original sense the *logistice numerosa* presupposed a homogeneous field of monads (cf. Pp. 41; 46; 49; 53) and was consequently dependent on "*arithmoi*" and their relations. After Vieta has given the "most venerable and significant" ladder of "*genera*" of unknown magnitudes, "a venerable series or scale of magnitudes ascending or descending from genus to genus by their own power in [continuous] proportion"[248] (solemnis magnitudinem ex genere ad genus vi sua proportionaliter adscendentium vel descendentium series seu scala — Chap. I, end), and has complemented it, in the third chapter, with the corresponding series of known magnitudes to be related to the unknowns,[249] he goes on in the fourth chapter to lay down the fundamental rules, "*the canonical rules of species calculation*" (logistices speciosae canonica praecepta). These correspond to the rules for addition, subtraction, and multiplication used for instruction in ordinary calculation.[250] Every species is represented by a letter (the vowels being assigned to the unknown, the consonants to the known magnitudes; cf. Chap. V, 5),[251] to which are joined the designation of the degree or genus, beginning with the second degree. A calculation carried out in species in the "general analytic" of Vieta then looks, for example, like this:

$$\frac{A \text{ cubus} - B \text{ solido } 3,}{C \text{ in } E \text{ quadratum}}$$ read: *A* cubed minus 3 times *B* solid divided by *C* times *E* squared

$$\left(\text{in modern notation:} \frac{x^3 - 3b}{cy^2}\right).$$

The fundamental "*law of homogeneity*" (lex homogeneorum), says Vieta (Chap. III, beginning), according to which only magnitudes of "like genus" can be compared (i.e., can appear in the same equation) with one another, must be kept in mind throughout; we must be careful "to compare homogeneous with homogeneous [magnitudes]" (homogenea homogeneis comparari). This law represents the "first and eternal law of equations or proportions" (prima et perpetua lex aequalitatum seu proportionum) and says, in modern terminology, that all the numbers of an equation must have the same dimension. According to it, only such magnitudes can be related by way of addition or subtraction as belong to the same or the corresponding "rung," although this does not hold for multiplication and division.[252] Vieta's "law of homogeneity" has therefore no immediate connection with Adrastus'[253] explication of Euclid V, Def. 3, which Vieta quotes, for according to this definition a *ratio* can exist only between "homogeneous" magnitudes: "A ratio is a sort of relation between two magnitudes of like kind with respect to size." (Λόγος ἐστὶ δύο μεγεθῶν ὁμογενῶν ἡ κατὰ πηλικότητά ποιὰ σχέσις.) It is certainly not unusual for ancient mathematicians to "compare," for instance, a ratio of *lengths* with a ratio of *planes*, and to bring both into one "*proportion.*"[254] Vieta's law of homogeneity is concerned rather with the fundamental fact that every "calculation," since it does, after all, ultimately depend on "counting off" the basic units, presupposes a field of *homogeneous monads*. For Diophantine "logistic" this demand is fulfilled as a matter of course, because it already operates within such a field of "pure" monads — the known and unknown "magnitudes" which are there united in an equation represent, each and every one, an "*arithmos* of monads" (cf. Pp. 130 f. and 139 f.).[255] For the "logistice speciosa" this fundamental presupposition needs to be especially stressed; hence the emphasis with which Vieta, in contrast to the "ancient analysts" (veteres Analystae), expounds the "lex homogeneorum" as the foundation of the "analytical art" (cf. also Chap. I, end, and Chap. V, 4). Thus it appears that the concept of the species is for Vieta, its universality notwithstanding, irrevocably dependent on the concept of "*arithmos.*" *He preserves the character of the* "arithmos" *as a* "*number of...*" *in a peculiarly transformed manner.* While every *arithmos* intends *immediately the things or the units themselves* whose number it happens to be, his letter sign intends *directly the general character of being a number* which belongs to every possible number, that is to say, it intends "number in general" immediately, but the things or units which are at hand in each number only mediately. In the language of the schools: The letter sign designates the intentional object of a "second intention" (intentio secunda), namely of a concept which itself directly intends another *concept* and not a *being*. Furthermore — and this is the truly decisive turn — this general character of number or, what amounts to the same thing, this "general number" in all its indeterminateness, that is, in its merely possible determinateness, is accorded a certain independence which permits it to be the subject of "calculational" operations. This is achieved by adjoining the "rung" designations, whose interconnection according to precise rules indicates the particular homogeneous field underlying each equation which is constructed.[256] The "rung" designation, which taken independently corresponds to the Diophantine *eidos*, thus transforms the object of the intentio secunda, namely the "general number" intended by the letter sign, *into the object of an* intentio prima, of a "*first intention*," namely of a "being" which is directly apprehensible and whose counterpart in the realm of ordinary calculation is, for instance, "two monads," "three monads." But this means that the "being" of the species in Vieta, i.e., the "being" of the objects of "general analytic," is to be understood neither as independent in the Pythagorean and Platonic sense nor as attained "by abstraction" (ἐξ ἀφαιρέσεως), i.e., as "reduced" in the Aristotelian sense, but as *symbolic. The species are in themselves symbolic formations — namely formations whose merely potential objectivity is understood as an actual objectivity*. They are, therefore, comprehensible only within the language *of symbolic formalism*[257] which is fully enunciated first in Vieta as alone capable of representing the "*finding of finding*," namely "zetetic." Therewith the most important tool of mathematical natural science, the "*formula*," first becomes possible (cf. Note 239),[258] but, above all, a new way of "understanding," inaccessible to ancient *episteme* is thus opened up.

When we look back to the Pythagorean and Platonic concept of the *eidos* of an *arithmos* as that which first makes the unified being of each number possible (cf. Pp. 55 f.) and compare it with the concept of the species developed above, we may say that the *ontological* independence of the *eidos*, having taken a detour through the *instrumental* use made of it by Diophantus (cf. Pp. 143 f. and 170), here finally arrives at its *symbolic* realization. This heralds a general conceptual transformation which extends over the whole of modern science. It concerns first and foremost the concept of *arithmos* itself.

As soon as "general number" is conceived and represented in the medium of species as an "object" in itself, that is, symbolically, the modern concept of "number" is born. Usually its development is explained by a reference to its ever increasing "abstractness." But this facile and easily misunderstood manner of speaking leaves its true and complicated structure completely in the dark. The modern concept of "number," as it underlies symbolic calculi, is itself, as is that which it intends, *symbolic in nature — it is identical with Vieta's concept of species.* This appears most clearly in the species of the first degree, where the designation of the "rung" is not appended to the letter sign and may therefore be said to coincide with it. But it is obviously impossible to see "numbers" in the isolated letter signs "A" or "B," except through the syntactical rules which Vieta states in the fourth chapter of the *Isogoge,* contrasting them with the operational rules of the "logistice numerosa." These rules therefore represent the first modern *axiom system*; they create the systematic context which originally "defines" the object to which they apply. And yet — and here lies the germ of future difficulties — these rules are directly derived from "calculations" with numbers of monads. This means that a species can ultimately retain a numerical character even in its transformed mode and hence is able to become a "number," namely an object of "calculational" operations, only *because the ancient "numbered assemblages of monads" are themselves interpreted as "numbers,"* which means that they are conceived from the point of view of their symbolic representation. This reinterpretation has to this day remained the foundation of our understanding of ancient "arithmetic" and "logistic."[259]

That is why after Vieta and under his immediate influence (as indeed already before him) the "*numeri* algebraici" (or "algebrici", or "cossici") are posited along with the "numeri simplices" (or "vulgares").[260] We shall have to investigate this further, especially in Stevin. In this respect, the position of Bachet de Méziriac, the first editor of Diophantus, is characteristic: In his commentary on "Definitions" III–VIII of the proemium of the Diophantine *Arithmetic* (cf. Note 179) he attacks the opinion of Xylander and his model, the scholiast Maximus Planudes (cf. Pp. 142

and 148). According to Bachet (edition of 1621, p. 6), Diophantus is here speaking not of "absolute numbers and fractions" (de numeris et fractionibus absolutis), "which is completely beyond his design" (quod a scopo illius prorsus alienum est), but only of the species as such, or of "fractiones algebricae." "Definition VI," in particular, according to which an *eidos* remains the same *eidos* when multiplied by one, since multiplication by the unit *leaves every number and consequently also its* eidos *unchanged* (cf. Pp. 142; 144), is interpreted by Bachet to the effect that the species in each case does not change in the course of multiplication with *arbitrarily many units*, that is, with ordinary numbers. He adds an elucidation supported by a cautious appeal to Diophantus himself: Just as the multiplication of ordinary number by one produces that very same number, so the multiplication of a species with arbitrarily many units, that is, with an ordinary number, produces just the same species.[261] It is clear that Bachet is dealing with *two* realms of "numbers," for he assigns the role of the unit in the realm of ordinary "numbers" to the ordinary "numbers" in the algebraic realm. But this means that for Bachet the unit itself is not so much an elementary component of every number as a "numerical coefficient."[262] No less significant for this new conception of "number" and "unit" is the way in which Bachet declares the Diophantine sign $\dot{M}$ for known numbers (cf. P. 130) superfluous: "Who, in fact, when he hears the number 6 does not at once *think* of six units? *Why then is it necessary to say 'six units' when it suffices to say 'six'?*" (Ecquis enim cum audit numerum 6. non statim cogitat sex unitates? Quid ergo necesse est sex unitates dicere, cum sufficiat dicere, sex? — *ibid.*, p. 4.)[263] This distinction between the "dicere" (saying) and the "cogitare" (thinking), which antiquity knows well enough in practice but never as a principle,[264] is expressed in the *symbolic* notation, and holds both for the formal language of algebra and for the representation of the numbers themselves: The "numbered assemblages" are now conceived as "ordinary numbers," and for the everyday understanding, which in effect determines the formation of concepts, these coincide with the "number" *sign* as such, especially in the course of calculational operations.

The new "number" concept which already controlled, although not explicitly, the algebraic expositions and investigations of Stifel, Cardano, Tartaglia, etc.,[265] can now also be used to justify speaking of "fractions" as "nonintegral rational numbers," of "irrational" numbers, etc.[266] Finally this "number" concept is able to obliterate the distinction, so fundamental for antiquity, between "continuous" geometric magnitudes and numbers divisible into "discrete" units (cf. Pp. 10 and 53); this permits, for the first time, a "scientific" understanding of the approximation methods handled so masterfully by Vieta.[267] In the last section we shall have to look further into these consequences of the symbolic number concept.

3

The reinterpretation of the *katholou pragmateia* as *Mathesis Universalis* in the sense of *ars analytice*

The species conceived as algebraic "numbers" thus forms the object of the "logistice speciosa" which "zetetic" has to employ. A specimen of "logistice speciosa" is offered in the *Zeteticorum libri quinque* (*Five Books on Zetetics*, 1593); Vieta contrasts them directly with the Diophantine *Arithmetic* which, as we saw, in his opinion stays only too much within the limits of the "logistice numerosa." In order to stress the parallelism of the two works Vieta concludes the fifth book of his *Zetetics* with the same problem which forms the conclusion of the fifth book of the Diophantine *Arithmetic*: "Diophantus presented this in the last problem of Book V. Therefore let the fifth book of our *Zetetics* also have this end." (... Retulit Diophantus quaestione ultima libri V.

Quare et Zeteticorum quintus noster finem hic accipito.) In other parts of the book, too, he takes series of problems from the Diophantine work, and in the solutions of IV, 1–3 he appeals explicitly to the "analysis Diophantaea." But we must not forget that zetetic analysis includes the "general" analytic art which transcends the opposition of "geometry" and "arithmetic" and is therefore superior in rank to the Diophantine *Arithmetic*. We must not forget that the species serve as the objects of this "general" analytic, as well, namely *in their numerical character*. This is precisely what constitutes the crucial difference between Vieta's conception of a "general" mathematical discipline and the ancient idea of a *katholou pragmateia*, a "general treatment" (καθόλου πραγματεία). The "numerical" character of the species, while preserving the position of this *pragmateia* within the system of "science," lends it a completely new sense.

In the context of ancient science, the "general" propositions — the axioms on the one hand and the propositions of the general theory of proportions on the other — have the closest possible connection with the theme of the "highest" discipline, be this characterized with Plato as "dialectic" or with Aristotle as "first philosophy." For Aristotle they form not only the stock examples of the manner in which "first philosophy" (πρώτη φιλοσοφία) treats of the theme peculiar to it (cf. Pp. 158 f. and 161), but they form a part of that very theme. He poses, and answers affirmatively, the question whether the examination of "being" and the examination of the mathematical axioms are to be referred to one single science, namely to "first philosophy": "We must say whether concern with the so-called axioms of mathematics [cf. *Posterior Analytics* A 10, 76 b 14 f.] and with 'being' belongs to one or to different sciences. *It is clear that the examination of these things belongs to one [science], which is that of the philosopher too*," and also: "these too are the study of him who comes to know 'being' insofar as it is 'being.'" (Λεκτέον δὲ πότερον μιᾶς ἢ ἑτέρας ἐπιστήμης περί τε τῶν ἐν τοῖς

μαθήμασι καλουμένων ἀξιωμάτων καὶ περὶ τῆς οὐσίας. φανερὸν δὴ
ὅτι μιᾶς τε καὶ τῆς τοῦ φιλοσόφου καὶ ἡ περὶ τούτων ἐστὶ σκέψις —
Metaphysics Γ 3, 1005 a 19–22; τοῦ περὶ τὸ ὂν ᾗ ὂν
γνωρίζοντος καὶ περὶ τούτων ἐστὶν ἡ θεωρία — ibid. a 28 f.)
Whereas the mathematician makes use of "general"
propositions with reference to those objects which happen
to be before him, the "philosopher" deals with them as
primary, that is to say, not as they can be derived from lines or
angles or numbers but as they are *common to all* that is
countable or measurable *in general* and consequently as they
determine the generic character of countable, measurable,
and weighable things insofar as they are "quantitative"
beings: "Since even the mathematician uses the common
notions *in a special way* it should belong to first philosophy to
investigate their *origins*." (ἐπεὶ δὲ καὶ ὁ μαθηματικὸς χρῆται τοῖς
κοινοῖς ἰδίως, καὶ τὰς τούτων ἀρχὰς ἂν εἴη θεωρῆσαι τῆς πρώτης
φιλοσοφίας — Metaphysics K 4, 1061 b 17 ff.) This is here
illustrated by the axiom: "If equals are subtracted from
equals the remainders are equal" (ἀπὸ τῶν ἴσων ἴσων
ἀφαιρεθέντων ἴσα τὰ λειπόμενα), which represents this sort of
"common notion applicable to all that is quantitative"
(κοινὸν ἐπὶ πάντων τῶν ποσῶν, cf. Posterior Analytics A 10, 76 b
20 f.; A 11, 77 a 30 f.; Euclid 1, Common Notion 3).
Similarly Proclus, referring to this Aristotelian doctrine and
stressing the general theory of proportion, asks who would
be the man that would, in contrast to a "geometer" and an
"arithmetician," demonstrate, for instance, the proposition
"that alternation [of the middle members] again produces
proportion" (ὅτι ἀνάλογον καὶ τὸ ἐναλλάξ, cf. P. 159) using
either geometric magnitudes *or* numbers (in Euclid. 9, 2–8),
i.e., what man might study alternation and the like in
"general" and "in itself": "Who then comes to know
alternation, be it in magnitudes, be it in numbers, in his own
behalf . . . ?" (τίς οὖν ὁ καθ' ἑαυτὸν γνωρίζων τὸ ἐναλλὰξ εἴτε
ἐν μεγέθεσιν εἴτε ἐν ἀριθμοῖς . . .; — 9, 8 f.) His answer is that
this manner of inquiry belongs to an independent science

which is far *superior* in rank to geometry and arithmetic: "the knowledge of these things belongs to a science prior by far" (πολλῷ πρότερον ἡ ἐκείνων γνῶσις ἐστὶν ἐπιστήμη — 9, 14 ff.; cf. 8, 24–26); and the "ascent from more partial to more universal understandings" (ἄνοδος ἀπὸ τῶν μερικωτέρων [sc. γνώσεων] ἐπὶ τὰς ὁλικωτέρας) is that by which "we climb up to the very science of 'being' insofar as it is 'being'" (ἕως ἂν ἐπ' αὐτὴν ἀναδράμωμεν τὴν τοῦ ὄντος, ᾗ ὄν ἐστιν, ἐπιστήμην, cf. *Republic* 534 E). In the sixteenth and seventeenth century these words of Proclus acquired tremendous significance, since, as they came to be widely known, first through Grynaeus' edition of the Euclid commentary (1533), and then chiefly through Barocius' translation (1560), they were normally understood as a reference to the *Mathesis universalis*. For Proclus is referring precisely to this superior science when he mentions "common theorems which are both general and have arisen from one science *which comprehends all mathematical knowledge together in one.*" (τὰ κοινὰ ... θεωρήματα καὶ ἁπλᾶ καὶ τῆς μιᾶς ἐπιστήμης ἔγγονα τῆς πάσας ὁμοῦ τὰς μαθηματικὰς γνώσεις ἐν ἑνὶ ἐπεχούσης ... — 7, 18 f.; Barocius translates: "communia . . . Theoremata, et simplicia, et ab una scientia orta, quae cunctas simul Mathematicas cognitiones in unum continet . . ."; cf. P. 159) Barocius adds here, as in other places, the marginal note "*Divina Scientia,*" clearly basing himself on the subsequent Proclus passage: "It reveals in the reasonings proper to it *the truth about the gods* and the vision of things that are."[268] (. . . τὴν περὶ θεῶν ἀλήθειαν καὶ τὴν περὶ τῶν ὄντων θεωρίαν ἐν τοῖς οἰκείοις ἐκφαίνει λογισμοῖς — 20, 5 f.) Now it is significant that as early as 1577, Gosselin (*De Arte magna*, p. 3ʳ) had called *algebra* the "queen of sciences" (regina scientiarum) and a "*divina ars.*"[269] And even Descartes, pursuing those reflections which lead him to recognize in Pappus and Diophantus the traces of a "true science" (cf. Note 215), is still reminded of Proclus' thinking in the Euclid commentary. In the *Regulae* (IV, Ad.–Tann., X, 377 f.), he inquires "what precisely it might be that all understand by that name, [i.e., "Mathesis"] and why not only those just mentioned [arithmetic and geometry], but also astronomy, music, optics, mechanics, and many other sciences [cf. the enumeration in Proclus, 38, 11 f.] are said to be *parts* of mathematics" (quidnam praecise per illud nomen omnes intelligant, et quare non modo iam dictae [sc. Arithmetica et Geometria], sed Astronomia etiam, Musica, Optica, Mechanica, aliaequae complures, Mathematicae partes dicantur; cf. Proclus 35, 17–42, 8, where he speaks of the "parts of mathematics" — μέρη τῆς μαθηματικῆς, in particular 38, 13 ff.; 41, 4 and 24 f.; 42, 6–8). He continues, clearly attacking Proclus or Barocius: "Here it is certainly not sufficient to look at the *origin of the word*." (His enim vocis originem spectare non sufficit — cf. Barocius' heading of Proclus 44, 25 ff.: Chap. XV. "*Whence the name of mathematics arose*" — Mathematices nomen unde sit ortum.) "For since the name 'Mathesis' means no more than [scientific] *discipline*, these might indeed, with as good a right as geometry itself, be called mathematical." (Nam cum Matheseos nomen idem tantum sonet quod disciplina, non minori iure, quam Geometria ipsa, Mathematicae vocarentur; cf. Barocius: "When the Pythagoreans had seen that all that is called *mathesis, that is, discipline*, is recollection ..." — cum perspexissent [sc. Pythagorei] quidem, quod omnis quae Mathesis, hoc est disciplina appellatur, reminiscentia est ... = Proclus 45, 5–7: ... τῶν Πυθαγορείων κατειδότων μὲν ὅτι πᾶσα ἡ καλουμένη μάθησις ἀνάμνησίς ἐστιν ....) Yet anyone with even the slightest schooling can easily distinguish the mathematical aspect of any problem from that relating to the other sciences, and after attentive study it becomes clear that all those things and only those things in which "order and measure" (ordo et mensura) can be observed form the object of *mathesis*, but that it is not of importance whether one deals with numbers, figures, heavenly bodies, tones, etc. As Descartes says: "And thence it became clear that there ought to be *some general science* which would explain everything that could be investigated in respect to order and measure when these are not ascribed to any special material, and that this same science was named — *using a word not newly appropriated but old and of accepted usage* — *Universal Mathematics*, since in it was contained everything on account of which other sciences are called *parts* of mathematics." (Ac proinde generalem quamdam esse debere scientiam [innotuit], quae id omne explicet, quod circa ordinem et mensuram nulli speciali materiae addictam quaeri potest, eamdemque, non ascititio vocabulo, sed iam inveterato atque usu recepto, Mathesim universalem nominari, quoniam in hac continetur illud omne, propter quod aliae scientiae et Mathematicae partes appellantur; cf. P. 181, Barocius' translation of Proclus 7, 18 f.)[270] "A few men of outstanding gifts" (quidam ingeniosissimi viri) have, says Descartes, already "tried to bring it to life" (suscitare conati sunt). This "general discipline" is no other than the "true science" of which vestiges can be traced in Pappus and Diophantus: "for that art was seen to be nothing else than that which is called by the barbaric name of *algebra*." (Nam nihil aliud esse videtur ars illa, quam barbaro nomine Algebram vocant....) Descartes is, no doubt, thinking primarily of Cardano, Tartaglia, Bombelli, furthermore of Clavius (whose *Algebra* [1608] he probably[271] became acquainted with in his school days), and possibly also of Ghetaldi,[272] Cataldi,[273] and Stevin.[274] About his relationship to Vieta violent controversy had arisen already during Descartes' lifetime.[275] This much, at least, is clear: Even if Descartes in no way consciously continues Vieta's work, yet the "general algebra" he has in mind[276] is precisely that "new" and "pure" algebra *which Vieta first established as the "general analytic art."* The universal character of Vieta's analysis was therefore quite rightly emphasized by van Schooten in his notes to the *Isagoge* (cf. his Vieta edition of 1646, pp. 545 f.). He understands it as the general "doctrina quantitatis," the "universa Mathesis" in Descartes' sense:

"Everything which comes within the scope of 'mathesis' always enjoys the name of *quantity* and is precisely that which becomes apparent only through *equations and proportions*. Thus also *Vieta's* Analysis must come under this name as *being of the greatest possible universality*."[277] (Id omne, quod sub contemplationem Matheseos cadit, quantitatis nomine semper gaudet, illudque demum per aequalitatem aut proportionem elucescit. Ita ut hoc ipso nomine Vietaea Analysis habenda sit quam maxime universalis.) But this universal science of Vieta, which is in his own eyes, as in those of his contemporaries, the complete realization of the ancient *katholou pragmateia*, of the "general theory of proportion," has, as we have seen, inherently numerical characteristics. Its object is, its generality notwithstanding, "arithmetically" determined. As Plato had once tried to grasp the "highest" science "arithmologically" and therewith exceeded the bounds set for the *logos* (cf. Part I, Section 7C), so here the "arithmetical" interpretation of "general magnitudes" leads to a special — an "algebraic" — mode of cognition or, more exactly, to the conception of a *symbolic* mathematics, although success in fixing the frame of reference and in establishing the internal completeness which is still lacking even in Descartes' "Mathesis universalis" is yet to come. For Descartes too sees the "Mathesis universalis" first and last as an "art of finding" (ars inveniendi) and thus, above all, as a "*practical*" art.[278] He does, however, already ascribe to it an internal and an external *independence* not yet possessed by Vieta's "Analytic" (cf. P. 168). But above all — and it is this which gives him his tremendous role in the history of the origin of modern science — he was the first to assign to "algebra," to this "ars magna," *a fundamental place in the system of knowledge in general*.[279] From now on the fundamental *ontological* science of the ancients is replaced by a *symbolic* discipline whose ontological presuppositions are left unclarified. This science, which aims from the first at a comprehension of the totality of the world, slowly broadens into the system of modern mathematical physics. Within this discipline the things in this world are no longer understood as countable beings, nor the world itself as a *taxis* determined by the order of numbers; it is rather the *structure* of the world which is grasped by means of a symbolic calculus and understood as a "*lawfully*" *ordered course of* "*events.*" The very nature of man's understanding of the world is henceforth governed by the symbolic "number" concept, a concept which determines the modern idea of science in general.

In Vieta's "general analytic" this symbolic concept of "number" appears for the first time, namely in the form of the *species*. It lies at the origin of that direct route which leads, *via* the "characteristica universalis" of Leibniz,[280] straight to modern theories of "logistic" (i.e., that branch of symbolic logic dealing with the foundations of mathematics). The condition for this whole development is the transformation of the ancient concept of *arithmos* and its transfer into a new conceptual dimension. The thoroughgoing modification of the means and aims of ancient science which this involves is best characterized by a phrase, previously quoted, from the end of the *Isagoge*, in which Vieta expresses the ultimate problem, *the* problem proper, of his "analytical art": "Analytical art appropriates to itself by right the proud *problem of problems*, which is: TO LEAVE NO PROBLEM UNSOLVED" (fastuosum problema problematum ars Analytice . . . iure sibi adrogat, Quod est, NULLUM NON PROBLEMA SOLVERE).

12

The concept of "number" in Stevin, Descartes, and Wallis

A
The concept of "number" in Stevin

IN CONTRAST to Vieta, who is on principle conservative, Simon Stevin (1548–1620)[281] decidedly prefers novel approaches and unusual theses. While Vieta retains the tradition-bound and involved style of the jurists throughout his mathematical and astronomical writings, labors to preserve the spirit of the newly resurrected ancient world, and publishes his works in Latin, Stevin consciously breaks with the traditional forms of science and puts his "practical" commercial, financial, and engineering experience[282] into the service of his "theoretical" preoccupation — as, conversely, his "theory" is put to use in his "practical activity" — and as a Fleming, publishes his works first in Flemish and thereupon parts also in his own French translation.[283] This contrast in their personal ways, in the "style" of their activity, attaches also to their appraisal of their ancient models, although Stevin too, is possessed by the idea of a "renewal." He casts it into the special image of a "wise age," the "siècle sage," which once existed and which must be brought back. In the first book of the *Geographie*,[284] Def. VI (ed. Girard, 1634, II, pp. 106–128), Stevin treats of this second "wise age" at great length. He "defines": "We call the wise age that in which men had a wonderful knowledge of science which we recognize without fail by certain signs, although without knowing who they were, or in what place, or when." (Nous appellons siecle sage, celuy auquel les hommes ont eu une cognoissance admirable des sciences, ce que nous remarquons infailliblement par certains signes, toutesfois sans sçavoir qui ce sont esté, ou à quel lieu, ny quand — p. 106, col. 2.) In the observations which follow Stevin sketches out the horizon within which the science of the seventeenth century develops and sets down the pre-suppositions which form the basis of modern science in general: ". . . It has become a matter of common usage to call the barbarous age that time which extends from about 900 or a thousand years up to about 150 years past, since men were for 700 or 800 years in the condition of imbeciles without the practice of letters and sciences — which condition had its origin in the burning of the books through troubles, wars, and destructions; afterwards affairs could, with a great deal of labor, be restored, or almost restored, to their former state; but although the afore-mentioned *preceding* times could call themselves a wise age in respect to the barbarous age just mentioned, nevertheless we have not consented to this definition of such a wise age, *since both taken together are nothing but the true barbarous age* in comparison to that unknown time at which we state that it [i.e., the wise age] was, without any doubt, in existence." (. . . C'est une chose venue en usage d'appeller siecle barbare ce temps là [qui] depuis 900 ou mille ans en çà jusques à environ 150 ans passez, pource que les hommes avoient esté 7 ou 8 cens ans comme idiots, sans exercice des lettres et sciences: ce qui a pris son origine alors que les livres ont esté bruslez, par les troubles, guerres, et ruines; ce qui puis apres non sans grand travail, a esté remis en premier estat ou peu s'en faut; or ledit temps auparavant se pouvant nommer siecle sage au regard du susmentionné siecle barbare, toutesfois nous n'avons entendu à la definition d'un tel siecle sage, car les deux compris ensemble ne sont autres, que le vray siecle barbare, à comparaison de ce temps incogneu auquel nous remarquons iceluy [sc. le siècle sage], avoir esté sans aucune doute ... — *ibid*.) The "barbarous age," accordingly, extends "from the beginning of the Greeks to the present" (du commencement des Grecs jusques à present — p. 108, col. 2). For Stevin the "signs" that in earlier times a "golden age" (aurea aetas) of science actually existed are these:

1. The traces of a perfected astronomical knowledge found in Hipparchus and Ptolemy, whose writings he understands as mere "vestiges" of primeval knowledge, as remains of that which came before" (reliques de ce qui avoit esté auparavant — p. 107, col. 1). For in their time that "great experience and understanding of the course of the heavens" (grande experience et cognoissance du cours du ciel) "came near to extinction" (est presque venue à s'esteindre — p. 106, col. 2). So also "certain writings in the *Arabic* tongue" (quelques escrits en langue Arabique — p. 107, col. 1), which Stevin considers to have been current already *before* Ptolemy, point in this direction.[285] The heliocentric theory in particular is said to have had an origin of great antiquity; even Aristarchus of Samos, he thinks, hardly understood it anymore.

2. *Algebra*, as we have become acquainted with it through *Arabic* books and which represents one of the strangest "vestiges" of the "wise age." No trace of it is to be found in the writings of the Chaldaeans, the Hebrews, the Romans, and even the Greeks, for, as Stevin expressly adds, "*Diophantus is modern*" (car Diophante est moderne — p. 108, col. 1). Stevin has a very definite opinion about the reasons for this want of algebra, especially among the Greeks, which will be more closely dealt with later.

3. Evidence of the *foreign* origin of Greek geometry. In this branch of study tradition is, according to him, indeed most reliable. The books of Euclid pass on to us "something admirable and very necessary to see and to read, namely *the order in the method of writing on mathematics* in that aforementioned time of the wise age" (quelque chose d'admirable et fort necessaire à voir, et à lire, nommément l'ordre en methode d'escrire les Mathematiques en ce susdit temps du siecle sage — p. 109, col. 2).

4. Information concerning the height of clouds, which appears in an *Arabic* work and which Stevin does not hesitate to trace back to the science of the "wise age."[286]

5. "Alchemy," which was unknown to the Greeks and whose most expert representative Stevin sees in Hermes Trismegistos![287, 288]

On the presupposition that the human ability to know has not changed since that time, Stevin projects a general plan for the gradual recovery of the knowledge of the "wise age." This plan comprises four articles and represents the first project for "organized research" (II, p. 110 ff.):

1. Many *observations* (especially in astronomy, "alchemy," and medicine) must be made, and this must be done by *many* people living at many different points of the earth and belonging to nations as different as possible (cf. Bacon, *The Great Instauration*, Part 3, *Preparative toward a Natural and Experimental History*, beginning).

2. The last is possible only if for the communication and the (primarily mathematical) exploitation of observations there is used not learned Latin, which is accessible to few, but each man's own mother tongue — something which has not occurred since the days of the Greeks.

3. However, not all languages are fit for this purpose.[289] Therefore we must first of all determine in what the virtue of a language consists. Greek was very appropriate for the mathematical sciences, which is why even now so many

mathematical terms of Greek origin are in use.[290] The reason for this is that in Greek words are very easily *compounded*. In this respect, however, Greek is far surpassed by *Flemish*, since the number of words which contain one syllable and are consequently especially fit for making compounds is greatest in this tongue; Stevin undertakes to supply statistical proof of this assertion.[291]

4. In every scientific presentation and in all teaching activity the *right order* must be preserved (le bon ordre en la description et instruction des arts). Here the procedure of the mathematical disciplines is exemplary: "I have not noted any better [order] for the matter of mathematics than that of the wise age" (je n'en remarque de meilleur pour le faict des Mathematiques, que celui du siecle sage — p. 110, col. 2); Stevin means the mode of presentation of the Euclidean *Elements* (see P. 189, the third "sign"), which he calls the "ordre naturel" (p. 125, col. 2).

Stevin lightly pushes aside the science of the schools and has little respect even for the authority of the Greeks, while, on the other hand, he is forever drawing on his own practical experience; thus he is just the man to see in the notion of a wise age, once a fact and now to be reestablished, the basis for the necessity of a thorough investigation of traditional views both for the sake of the *primeval* truths which might be contained in them and also to test the conventional *concepts* for their reliability and usefulness. Now it is characteristic of his bias that he always draws invidious comparisons between Arabic and Greek science. The basic reason for this is the Arabic *digital and positional system*, which he is inclined to view as the heritage of the "wise age,"[292] since it appears to him to be immeasurably superior to the Greek notation. On the basis of this Arabic digital system he undertakes a fundamental critique of the traditional concept of "numbered assemblage" (*arithmos*), beginning with the concept, crucial indeed, of the "one," the "monad," the "unitas."

Definition I of his *Arithmetique* (p. 1ᵛ) states: "Arithmetic is the science of numbers." (Arithmetique est la science des nombres.) Definition II: "Number is that by which *the quantity* of each thing is revealed." (Nombre est cela par lequel s'explique la quantité de chacune chose — *ibid.*). This definition is followed by a precise explication in which the symbolic character of the new "number" concept fully appears. Stevin's main thesis is: "*that the unit is a number*"[293] (que l'unité est nombre). He declares (p. 2ʳ) that he has read all the "old and new philosophers" who have pronounced on this question and has spoken with many scholars, not because he had any doubts with respect to this assertion — "no, definitely not, since I was as assured of this as if nature herself had told me with her own mouth" (non certes, car i'en estois ainsi asseuré, comme si la Nature mesme me l'eust dict de sa propre bouche) — but in order to be armed against all objections. It is generally claimed that the unit is not "nombre," but only "its principle or beginning" (son principe, ou commencement). This is completely false, for, as Stevin argues: the part is "of the same material" (de mesme matiere) as the whole; the unit is a part of a multitude of units; consequently the unit is "of the same material" as a multitude of units; "but the 'material' of a multitude of units is 'number'" (mais la matiere de multitude d'unitez est nombre); therefore the "material" of the unit, and thus the unit itself, is "number." He who denies this last step can be compared to someone who denies "that a *piece* of bread *is* bread" (qu'une piece de pain soit du pain).

What is happening in this syllogism? How must the concepts used in it be interpreted, of what kind must they be, if the argument is to be sound? The decisive premise is the one in which the "material of a multitude of units" is equated with "number." Stevin here simply accepts the classical definition of *numerus* as "a multitude consisting of units" (cf. P. 51), but he understands this conceptual determination itself as the "material" of the thing to be defined,[294] in the same sense in which one speaks of the material (materia) of water or of bread. [For just as that which is the "stuff" of bread is understood by Stevin as identical with "what it is," namely bread, so the object intended by a concept is understood by him as a "piece" of the conceptual content taken as a "stuff," a material, which is, in turn, identical with the concept itself, namely with "what the object is."] Only on the basis of such an interpretation is the first premise of the syllogism, according to which the "part" is of the same material as the "whole," relevant. This does *not* mean that Stevin commits a paralogism. The fundamental *presupposition* which underlies his understanding — although he hardly recognizes it as such — is precisely the identification of the mode of being of the *object* with the mode of being of the *concept* related to the object. This means that the one immense difficulty within ancient ontology, namely, to determine the relation between the "being" of the object itself and the "being" of the object in thought, is here (and elsewhere) accorded a "matter-of-course" solution whose presuppositions and the extent of whose significance are simply bypassed in the discussion. The *consequence* of this solution is the *symbolic* understanding of the object intended, an understanding in which its *actual* objectivity is posited as identical with the mode of being of a "general object," or, in other words, in which the object of an "*intentio secunda*" (second intention), namely the concept as such, is turned into the object of an "*intentio prima*" (first intention). Such a symbolic understanding of *numerus* or *quantitas* is precisely what is *presupposed* in this syllogism. Again Stevin himself confirms this, namely in the first book of his *Geographie* (II, p. 108).

The true reason for the assertion that the unit is not "number" but the *principium*, the *arche* of "number" (cf. Pp. 53 and 108)[295] — an assertion in his opinion deeply false and fatal in its consequences — was the absence of a real *symbolic notation* among the Greeks, the "absence of the necessary equipment, namely of *ciphers*" (faute d'appareil necessaire, nommément des chiffres), which explains immediately why they were not arithmeticians. Misinterpreting a sign *called* a "point" (".") which they had inherited from the "wise age" and which was identical with the present day sign "o," they had understood the "points" as — "units" and consequently represented the latter with the aid of such points.[296] Now as the *geometric* point is indeed the *principium* of the line and not itself "line," they, misled by their false interpretation of the *arithmetic sign* "." (the "poinct Arabique"), were led to conceive of "one" as the *principium* of "number" *and not itself* "number," thereby giving the whole "barbarous age" its characteristic "unarithmetical" stamp. In truth, not the unit but the nought is the *principium* of "number" — it, and not the unit, is the true analogue to the geometric point. The sign "o," nowadays used for *nought* (in place of the primeval "Arabic" sign "."), was, he says, chosen only in order to avoid all confusion with the period sign at the conclusion of a sentence. But in order to retain at least a nomenclature consonant with the "wise age," Stevin calls the nought "*poinct de nombre*" (in analogy to and in distinction from the "poinct Geometrien").[297]

That the unit itself must be "number" is, then, for Stevin so essential because he transfers the role of *arche*, which had hitherto been assigned to the unit, to the *nought*: "zero is the true and *natural* beginning" (le o est le vrai et naturel commencement — *Arithmetic*, p. 4^r). What determines him to do this is his regard for the *sign notation*, for he completely identifies the *nought* with the sign "o," whose full significance can be conceived *only within the sign-system as a whole*. This, however, holds not only for the nought, but for all "quantities" represented by "ciphers."[298] As "arithmetician" Stevin no longer deals with numbers of units which are determinate in each case but with the *unlimited possibility of combining ciphers* according to definite "rules of calculation."

But this means that he conceives the "quantities" with which he is dealing in a symbolic way; he no longer knows "numbers of units" but only "numbers" directly expressed in ciphers.

Now the symbolic understanding makes "number," as we have seen, appear as a "material" comparable to the material of bread or water. In the traditional conception, by which Stevin is here governed, such "material" is characterized by ever-continuing divisibility, by "continuity." This immediately leads to a far more complete assimilation of "numbers" to geometric formations than was ever possible for the "numbered assemblages" and "magnitudes" of antiquity: "The community and similarity of magnitude and number is so universal that it almost resembles identity" (la communauté et similitude de grandeur et nombre, est si universelle qu'il resemble quasi identité — *Arithmetic*, p. 3ʳ).[299] Thus one of Stevin's supports for his thesis that not the unit but the *nought* is the principle of "number" is the following characteristic argument: Just as a line is not lengthened by the addition of a point, so also "number," e.g., 6, is not increased by the addition of zero, for $6 + 0 = 6$; therefore neither can infinitely many points "taken together" form a line, nor can infinitely many zeros "together" form a "number," while as few as *two* units can do this. If, however, one admits that a line $\overline{AB}$ is continuously lengthened by the addition of a point C in such a way that a new line $\overline{AC}$ results, then, by the same right, one may say *that the number 6, by the "addition" of 0, continuously increases to the number 60*![300] Here, then, the continuous extension of a line is compared to an arbitrarily continuable lining up of ciphers which yield continually new "numbers." This means precisely that numbers too are continuous structures and not "discrete," as the "barbarous age" asserted. Stevin formulates this thesis explicitly: "*that number is not at all discontinuous quantity*" (que nombre n'est poinct quantité discontinue — *Arithmetic*, p. 4ᵛ). The ancient view of the "discreteness" of numbers was based on an insight into the indivisibility of the "unit" (cf. Pp. 53 and 108). According to Stevin the unit is divisible as a matter of course, an opinion for which he, incidentally, appeals to the authority of Diophantus, the "prince des Arithmeticiens," who demanded just such a division of the unit in certain problems (cf. P. 132) contained in his work (*Arithmetic* p. 3^r f.). The unit is as much "part" of a "number," and thus itself "number" (see P. 191), as a smaller line is part of a larger and is certainly itself "line," while neither the nought nor a point are "parts" of a "number" or a line. The "parts" of the unit are, in turn, also "numbers," namely "fractional numbers,"[301] which decrease infinitely. Thus arises the complete correspondence of geometric magnitudes and "numbers": "As to a continuous body of *water* corresponds a continuous *wetness*, so to a continuous *magnitude* corresponds a continuous *number*. Likewise, as the continuous wetness of the entire body of water is subject to the same division and separation as the water, so the continuous number is subject to the same division and separation as its magnitude, in such a way that *these two quantities cannot be distinguished by continuity or discontinuity*." (Comme à une continue eau correspond une continue humidité, ainsi à une continue grandeur correspond un continue nombre: Item comme la continue humidité de l'entiere eau, souffre la mesme division et disioinction que son eau, Ainsi le continue nombre souffre la mesme division et disioinction que sa grandeur; De sorte que ces deux quantitez ne se peuvent distinguer par continue et discontinue . . . — *Arithmetic*, p. 5^r.) The "units" of a "number" are not "disioinctes" but "conioinctes" (p. 5^v).

Basing himself on the new "number" concept, Stevin now attacks the traditional appellations of "absurd" or "surd" or "irrational" (i.e., un-speakable) numbers, a mode of speaking which resulted from the schism between the actual understanding of "number" and the traditional concept of "number" which is nevertheless retained. His thesis is "that there are no absurd, irrational, irregular, inexplicable, or surd numbers" (qu'il n'y a aucuns nombres absurdes, irrationels, irreguliers, inexplicables, ou sourds — *Arithmetic,* p. 33 and p. 202, Thesis IV).[302] For "incommensurability does not cause the incommensurable terms to be surds" (l'incommensurance ne cause pas absurdité des termes incommensurables), which is immediately obvious for incommensurable lines and planes. For example, $\sqrt{8}$ is a "root" (racine). "Every 'root' is a 'number'" (racine quelconque est nombre — p. 30; cf. p. 25).[303] He therefore rejects the Diophantine (or Anatolian) terms "side," "square," "cube" (latus, quadratum, cubus), etc., but still retains the distinction between "arithmetical" and "geometric" numbers; Definition IV states: "An arithmetical number is one expressed without an adjective of size." (Nombre Arithmetique est celuy qu'on explique sans adjectif de grandeur — p. 6^r; cf. Pp. 172 f.) In contrast, "roots," "quadratic" numbers, "cubic" numbers, etc., are called "nombres Geometriques" (p. 6^v; 9 ff.), although it should be added "that any arithmetic *numbers whatsoever* can be squared or cubed numbers, etc." (que nombres quelconques [sc. arithmétiques] peuvent estre Nombres quarrez, cubiques, etc. — p. 30). But insofar as their "absolute," i.e., numerical, value is not known, the "geometric numbers" enter algebraic computations as indeterminate "quantities" and are designated in the following way: ① ② ③ ④ etc. (corresponding to our symbols x, x^2, x^3, x^4, etc.). Now just as 0 is the "beginning" of "arithmetic" numbers, so any "arithmetic" number you please is the "beginning" of these algebraic "quantities" — Def. XIV: "The beginning of quantity is every arithmetic number or any *radical* whatsoever." (Commencement de quantité, est tout nombre Arithmetique ou radical quelconque.) It is thus designated by ⓪, insofar as its "absolute" value is not known (p. 15).[304] Stevin is also the first mathematician who understands the subtraction of a "number" as the addition of a "negative number" (cf. Bosmans, p. 899). In every one of these theses Stevin is merely assimilating the *concept* of "number" to *operations* on "numbers" already long established (cf. Pp. 147 f.; P. 178; Note 260; Note 302). He thus once and for all fixes the ordinary understanding of the nature of number, for which being able "to count" is tantamount to knowing how to handle "ciphers."[305] It is an open question how much particularly Descartes owes to him.[306] In any case, Descartes' conception of "numbers" is far more tradition-bound than Stevin's, although, on the other hand, he sees through their conceptual structure with much more clarity. In the following section we shall therefore try to confront the "number" concept of Descartes, using as a guide the early *Regulae ad directionem ingenii* (*Rules for the Direction of the Mind*, ca. 1628), because it is apparently here that the original intentions of Descartes and the specific characteristics of his conceptual mode receive their clearest expression.

B
The concept of "number" in Descartes

Descartes' thinking, as he himself points out in the *Regulae*, *presupposes* the fact of symbolic calculation, namely in the form of contemporary "algebra."[307] As we have seen, he understands this "new" discipline not as "some kind of arithmetic" (genus quoddam Arithmeticae — Rule IV, 373, 16) but as *the general "art"* simply, an art of which he finds indications in Proclus (i.e., Barocius) and which had, albeit without his knowledge, already been realized by Vieta. Descartes' great idea now consists of identifying, by means of "methodological" considerations, the "general" object of this *mathesis universalis* — which can be represented and conceived only *symbolically* — with the "substance" of the world, with corporeality as "extensio."[308] Only by virtue of this identification did symbolic mathematics gain that fundamental position in the system of knowledge which it has never since lost (cf. Pp. 184 f.), even though Descartes may have failed subsequently in working out completely his original conception. It was, at any rate, on the basis of this beginning that Descartes chose a symbolism *of figures* for the *Regulae*, which he restricts only later (cf. *Discours*, Ad.-Tann., VI, 20) to lines.[309] For Descartes this conception of a figure-symbolism connects two different trains of thought: (1) the conception of algebra as a "general" theory of proportions, whose object, only symbolically comprehensible, acquires its specific characteristics from the *numerical* realm (cf. Pp. 171 ff. and also Pp. 204 f.), and (2) the identification of this "symbolic" mathematical object with the object of the "*true physics*." The connection of these two trains of thought is made possible by the "methodical" concept of "certain and evident knowledge" (cognitio certa et evidens).[310] This Cartesian concept of knowledge is unmistakably *Stoic* in origin: to it corresponds "apprehension" (κατάληψις), that is to say, the "assent" (συγκατάθεσις) given to an "*apprehensional image*" (φαντασία καταληπτική).[311] For Descartes this concept is so essential because, and only because, it allows him to assign a fundamental role to the *imaginatio* (= phantasia). Thus the *imaginatio* is, in the *Regulae*, always in the foreground.

In Rule XIV (444 ff.), while treating of the role of the imagination, Descartes includes a discussion of the equivocality of certain concepts. In reference to the assertion that "extension is not body" (extensio non est corpus) he claims that *in this case* "no special idea corresponds to this word 'extension' in the imagination" (nulla illi [sc. extensionis vocabulo] peculiaris idea in phantasia correspondet), that rather "this whole assertion is effected by the *pure intellect* which alone has the ability of *separating abstract beings* of that sort" (tota haec enuntiatio ab intellectu puro perficitur, qui solus habet facultatem ejusmodi entia abstracta separandi).

This holds also for the assertion "number is not the thing enumerated" (numerus non est res numerata)[312] and "a unit is not a quantity" (unitas non est quantitas)! "All of these and similar propositions must be altogether divorced from the imagination in order to be *true*." (Quae omnes et similes propositiones ab imaginatione omnino removendae sunt, ut sint verae.) If one were to try to "represent" them to oneself by means of the imagination, one would necessarily arrive at contradictions — for in the imagination the "idea" of extension cannot be separated from the "idea" of body, nor the "idea" of number from the "idea" of the thing enumerated, nor the "idea" of unity from the "idea" of quantity. There are propositions, however, in which the words mentioned are used with the same meaning and in the same way (namely as referring to structures which have been "abstracted" by the "naked intellect" from the "ideas" accessible to the imagination), but in which they are *not* explicitly separated from the actual content of these "ideas" — for it is this very content which, in accordance with the actual thing (i.e., the thing as it appears in the "imagination"), is intended by them. In such propositions it is allowed *and even necessary to call to aid the imagination*: "We must carefully note that in all other propositions in which these terms — although they retain the same meaning and are asserted in the same way as when abstracted from their subjects — do not [explicitly] exclude or deny anything from which they are not really distinct, we both can and *ought to call our imaginations to aid*." (Notandum est diligenter, in omnibus alijs propositionibus, in quibus haec nomina, quamvis eamdem significationem retineant, dicanturque eodem modo a subjectis abstracta, nihil tamen excludunt vel negant, a quo non realiter distinguantur, imaginationis adjumento nos uti posse et debere. . . .) When we speak of "*numerus*," for instance, we shall have to "represent" to ourselves an object which can be measured by "a multitude of units" (per multas unitates). And even if the ("naked")

intellect intended the "mere multitude" (solam multitudinem), namely multitudinousness as such, yet we should
not make the error of believing that the concept (conceptus)
of the *numerus* (to which corresponds the "idea" of the
numerus in the "imagination") excludes the *res numerata*, the
"enumerated thing" itself. For if one does this, one eventually begins to ascribe deep secrets to "numbers" and to hunt
after phantoms.[313] Even proponents of the genuine arithmetic and calculational art commonly cling to this erroneous
view: "For what master-calculator does not think that the
numbers with which he calculates are separated [abstracted]
not only by the [naked] intellect from any content-bearing
subject but can also be actually distinguished by the imagination?" (Quis enim Logista numeros suos ab omni subjecto,
non modo per intellectum abstractos, sed per imaginationem
etiam vere distinguendos esse non putat?) But this is exactly
what is not, and indeed cannot be, the case, in Descartes'
opinion.[314] In these formulations Descartes, therefore,
postulates — with an explicitness perhaps novel in the history
of science — a new mode of "abstraction" and a new
possibility of "understanding." The "abstract beings" (entia
abstracta) of which he speaks here are the products of the
"naked" or "pure intellect" (intellectus purus), which is
called "pure" only insofar as the "cognitive power" (vis
cognoscens) which it represents is free of all admixture of
"images" or "representations," is "divorced from the aid of
any bodily image" (absque ullius imaginis corporeae
adjumento) and "acts alone" (sola agit — Rule XII, 416, 4
and 419, 10). For under those circumstances the *intellectus* or
"mind" (mens) is dealing only *with itself*, and only in this
case can one speak of an "*intellectio*" and "*intellegere*," "in
that the mind, when it thinks, in a way turns *itself toward
itself*" (. . . quod mens, dum intelligit, se ad seipsam quodammodo convertat; cf. *Meditations* VI, Ad.-Tann., VII, 73, 15
ff.; see also the *Passions de l'âme* I, art. XX) and thereupon
"beholds some one of the ideas which are *within itself*"

(respiciatque aliquam ex ideis, quae illi ipsi insunt — *ibid.*).
These are: "simple things purely intellectual" (res simplices
pure intellectuales — Rule XII, 419, 8 ff.), such as "cognitio,"
"dubium," "ignorantia," "volitio," etc. (cf. Rule III, 368,
21 ff.), and furthermore the "res simplices" which belong to
the "spiritual" as well as to the "bodily" realm and must
therefore be called "communes," such as "existentia,"
"*unitas*," "duratio," etc. (According to Descartes, the
traditional "communes notiones" — κοινὰ ἀξιώματα or
ἔννοιαι — such as, for instance, the assertion that "two
magnitudes which are equal to a third are equal to each
other," Euclid I, Common Notion I, also belong among
these).³¹⁵ Now this "pure" intellect, which refers only to
itself, is also able to turn (or "apply") itself to the "ideas"
which the *imagination* offers it, and can even "separate" single
constituents of these ideas. In this "turning toward" the
imagination, the intellect has, strictly speaking, already
ceased to be "pure," but it retains the ability *proper* to it —
and *foreign* to the imagination — of carrying out this kind of
a "separating" operation. However, within the "realm" of
this "alien" imagination, *it must make use of this very imagination.* When, for instance, the "naked" or "pure" intellect
separates from a multitude of units "represented" in the
imagination (a number of units, that is), their "multitudinousness" as such, i.e., the "mere multitude" (sola multitudo), the "naked" indeterminate manyness to which simply
nothing "true," nothing truly in "being," and hence no
"true idea" of a being corresponds, it must employ the
imagination in order to be at all able to get hold of the thing
separated. Thus the *imaginative* power, which ordinarily
allows us to envisage, for instance, "five units" (perhaps as
"points"), here enters the service of a faculty directed
precisely toward something not "perceptually clear,"
namely the "pure intellect," which, *being bare of any
immediate reference to the world*, comprehends "fiveness" as
"something separated" from "five" counted points or other

arbitrary objects — as mere "multiplicity in general," as "naked" multiplicity. *Thus the imaginative power makes possible a symbolic representation of the indeterminate content which has been "separated" by the "naked" intellect* — this is what Descartes must emphasize. The "abstraction" which is here intended we may therefore call a "*symbol-generating abstraction*." It alone gives rise to the possibility of contrasting "intuition" and "conception," and of positing "intuition" as a *separate* source of cognition alongside reason. The ancient mode of separation *(ἀφαίρεσις)* appears from this point of view as a "direct" or "imaginative" abstraction: The insights of the ancient mathematicians "pertain more to the eyes and the imagination than to the intellect" (magis ad oculos et imaginationem pertinent, quam ad intellectum — Rule IV, 375, 18 f.); this is exactly why the ancient concept of number can now be described as having "intuitive" character ("Anschaulichkeit," cf. Pp. 62 f.). Here the constant presupposition is that *the "pure" intellect in itself has no relation at all to the being of the world and the things in the world*. What characterizes it is not so much its "incorporeality" as just this unrelatedness. *But this implies immediately that the separation between it and the "corporeal" is itself conceived completely on the analogy of corporeal separateness*. Descartes' examples are characteristic of this. "We must comprehend," he says, "that that power through which we properly know things is a *purely spiritual* one and no less distinct [separate] from all body *than blood from bone or hand from eye*." (. . . Concipiendum est, vim illam, per quam res proprie cognoscimus, esse pure spiritualem, atque a toto corpore non minus distinctam, quam sit sanguis ab osse, vel manus ab oculo . . . — Rule XII, 415, 13 ff.) The "pure" intellect, in order to be at all able to come into "contact" with the objects of the corporeal world "which are without us and *very much foreign*" (quae extra nos sunt, et valde aliena — Rule VIII, 398, 13) — that is to say, in order to come in contact with the "world" in general — needs the mediation of a special faculty, namely precisely

that of the imagination: "If the pure intellect places anything
before itself to be examined, such as can be referred to a
body, its idea must be formed as distinctly as possible in the
imagination." (Si . . . intellectus [sc. purus] examinandum
aliquid sibi proponat, quod referri possit ad corpus, ejus idea,
quam distinctissime poterit, in imaginatione est formanda —
Rule XII, 416 f.) Furthermore: "Nor, in general, do we
recognize those beings of the philosophers which really
cannot come before the imagination." (. . . Neque in
universum nos agnoscere ejusmodi entia philosophica, quae
revera sub imaginationem non cadunt — Rule XIV, 442,
26–28.) How this mediation, that is to say, the relation
between "body" and "soul," is to be understood is, as is
well known, *the* insoluble problem of Cartesian doctrine.[316]
For Descartes himself this difficulty is not a crucial one only
because he meets it originally in the realm of mathematics, namely
at that moment when it becomes important to reconcile the
traditional determinateness of the *numerus*, which Descartes,
in contrast to Stevin, perfectly appreciates, with the indeterminacy of the new "algebraic" quantities.[317]

This reconciliation is achieved by means of symbolic
figural representation. Everything depends on understanding
that the "figures" with which the "mathesis universalis"
deals, namely "rectilinear and rectangular planes" as well as
"straight lines" (cf. Note 309), have, as far as their mode of
being is concerned, no longer anything to do with the
"figures" of what had up till then been the ordinary
"geometry": "We easily conclude: here propositions must
be just as much *abstracted from those very figures with which
geometers deal*, if the inquiry involves these, as from any other
subject matter; and for this purpose none need be retained
besides rectilinear and rectangular plane surfaces, or straight
lines, which we also call figures because by means of them we
can imagine a subject which is in truth extended [namely in
three "dimensions"][318] just as well as by means of a plane
surface." (. . . Facile colligitur: hic non minus abstrahendas

esse propositiones ab ipsis figuris, de quibus Geometrae tractant, si de illis sit quaestio, quam ab alia quavis materia; nullasque ad hunc usum esse retinendas praeter superficies rectilineas et rectangulas, vel lineas rectas, quas figuras quoque appellamus, quia per illas non minus imaginamur subjectum vere extensum quam per superficies . . . — Rule XIV, 452, 14 ff.) As "*continuous* and undivided magnitudes" (magnitudines continuae et indivisae), "lines" and "planes" permit us to set up proportions (and thus equations) only insofar as their "common measure" (communis mensura), namely their particular "*unit*" (unitas), is known for each case. But if that condition is met, these continuous magnitudes can, in turn, immediately be understood as "numbers": "We must also understand that continuous magnitudes can, by the help of the unit that has been assumed, sometimes be reduced entirely to multitude." (Sciendum etiam, magnitudines continuas beneficio unitatis assumptitiae posse totas interdum ad multitudinem reduci . . . — Rule XIV, 451 f.) "Multitude" here means one of those numbers with which "algebra" deals in setting up proportions among "A," "B," and "C" (cf. Rule VII, beginning); these need no longer be referred to the "common measure" because measuring (mensura) is no longer our concern, but only "arrangement" (ordo): "[We must understand], too, that we are able afterwards to *arrange the multitude of units in such an order* that the difficulty which before was one of solving a problem of measurement now depends only on observing an order, and that the aid which our art gives us in this process is very great." (Atque multitudinem unitatum posse postea tali ordine disponi, ut difficultas, quae ad mensurae cognitionem pertine⟨b⟩at, tandem a solius ordinis inspectione dependeat, maximumque in hoc progressu esse artis adjumentum — Rule XIV, 452, 2–6.) Thus a plane or linear "figure" represents no less and *no differently* a "multitude or number" (multitudinem sive numerum) than a "continuous magnitude." What is more, it is the primary task and the *truly proper function* of such "figures" to "image" the "true idea" (vera idea) of a number. Confronted with an indeterminate multitude, i.e., with any algebraic quantity, which we, misled by the pure intellect, want to understand as a "mere multitude" (sola multitudo), that is, as a structure separated from all "enumerated things" (res numeratae), we must, in order not to make mistakes, indeed "represent" to ourselves "some subject measurable by many units" (subjectum aliquod per multas unitates mensurabile) — we must keep before our eyes a *picture* of the "countable *in general.*" Each "figure" is therefore not a representation of a *determinate number* of units of measurement (as are the straight lines of Euclid's arithmetical books) but the "*symbol*" of an *indeterminate multitude* obtained by "symbol-generating abstraction"; in other words, it is exactly the same as the *letter* (together with its "degree" designation) which occurs in "algebra," especially in Vieta's analytic (cf. Pp. 174 f.).[319] This "symbolic" character of Cartesian "figures" first makes possible that mutual correspondence of "lines" with letters or "ciphers" which obtains in Cartesian mathematics (cf. *Discours*, Ad.-Tann., VI, 20). Descartes himself says explicitly: "It should be noted . . . that we . . . here *abstract no less from numbers themselves than* we did just before [cf. P. 203] *from geometric figures* or from anything else you like." (. . . Advertendum est, . . . nos . . . hoc in loco non minus abstrahere ab ipsis numeris, quam paulo ante a figuris Geometricis, vel quavis alia re — Rule XVI, 455 f.) The symbol-generating abstraction which leads from the ordinary numbers to "letter-signs" (notae) or "termini generales" (457, 20), also called "puri et nudi" (455, 21),[320] needs the imagination in exactly the same way as when it leads to figures. In the former case, however, it is the retentive ability of the imagination, the "memoria," which is being aimed at (cf. Rule XII, 414, 23 f. and 416, 1 f.);[321] this is why the place of "whole figures" (integrae figurae) is taken by "signs as short as possible" (brevissimae notae — Rule XVI,

454, 10 ff.).[322] The mode of being of these "figures" is therefore, to repeat, none other than that of the algebraic "numbers" — of Vieta's "species." Precisely the same reinterpretation which the traditional concept of *arithmos* undergoes at the hands of Vieta, Stevin, and the other contemporary algebraists, is effected by Descartes — and this is his original achievement — in the realm of traditional *geometria*. The essential difference between Descartes and Vieta is not in the least that Descartes unites "arithmetic" and "geometry" into a single science while Vieta retains their separation. As we have seen, both have in mind a *universal* science: Descartes' "*mathesis universalis*" corresponds completely to Vieta's "zetetic," by means of which is realized, with the aid of "logistice speciosa," the "new" and "pure" algebra, interpreted as a general "analytic art" (cf. Pp. 157 f.; 168 f.; 183 f.). But whereas Vieta sees the most important part of analytic in "rhetic" or "exegetic" (cf. Pp. 168 f.), in which the numerical computations and the geometric constructions indeed represent two different possibilities of application (so that the traditional conception of geometry as such is here preserved), Descartes *begins* by understanding geometric "figures" as structures whose "being" is determined *solely* by their "symbolic" character. The truth is that Descartes does not, as is often thoughtlessly said, identify "arithmetic" and "geometry" — rather he identifies "algebra" understood as symbolic logistic with geometry *interpreted by him for the first time as a symbolic science.*[323]

We are now able, by using Descartes' assertions as a basis and taking into account the contemporary literature of the schools, to fix yet more exactly that conceptual character of algebraic symbols which has already been variously outlined. We saw that Descartes designates the "sola multitudo" which the intellect "separates" from the "idea" of number it finds available in the imagination as an "abstract being" (ens abstractum), also called an *ens rationis* in the language of the schools. In the *Summa* of Eustachius a Sancto Paulo, IV,

17–19 (quoted after Gilson, *Index*, p. 107) three kinds of "entia rationis" are enumerated: "Beings of reason are either negations, or privations, or *second intentions*." (. . . Entia rationis aut sunt negationes, aut privationes, aut secundae intentiones. . . .) "The two first kinds appertain to things in their own mode before any operation of the intellect" (duo priora [sc. genera] rebus suo modo conveniunt citra omnem operationem intellectus); only the third kind owes its "being" to the operation of the intellect alone: "But the last kind does not belong to things *unless a certain operation of the intellect is presupposed*, wherefore these beings of reason are said to depend on the intellect for their existence and connection; . . . this is why a 'being of reason' in its proper and strict sense is agreed to be only that last kind." (Postremum vero genus non nisi praesupposita aliqua intellectus operatione rebus competit, unde illa [sc. entia rationis] dicuntur pendere ab intellectu quoad existentiam et convenientiam . . . quo fit ut ens rationis proprie et presse sumptum pro isto postremo duntaxat genere accipiatur.) Thus "mere manyness" (sola multitudo), multitudinousness as such, which has its "being" by grace of the "pure intellect" is truly an *ens abstractum* or *ens rationis* in the sense of a "*second intention*" (cf., Pp. 174 and 192, furthermore, Note 319). Now Eustachius, appealing to established usage, defines "second intention" more narrowly as an *ens rationis* "which is conceived as belonging to a thing known by virtue of its being known, and which cannot exist except *as something present in the intellect*, since it is conceived [not originally but] *secondarily and by a reflexive operation of the mind*" (quod concipitur accidere rei cognitae, ex eo quod cognita est, quodque non aliter potest existere quam objective in intellectu, cum secundario et per reflexam mentis operationem concipitur). That process of "separating" in the course of which the "pure intellect" ventures into the "alien realm" of the "imagination" (cf. P. 201) could thus be described more exactly as follows: The intellect, when directed to the "idea" of a number as a "multitude of units" (multitudo unitatum — πλῆθος μονάδων) offered to it by the imagination (cf. P. 51), turns, as its nature requires, toward its own "directedness," its own knowing (cf. Pp. 200 f.). Consequently it sees the multitude of units no longer "directly," no longer in the "performed act" (actus exercitus; cf. P. 49), but "indirectly," "secondarily" (secundario), or, in terms of another scholastic expression, in the "signified act" (actus signatus [i.e., an act the object of whose intention is already an expressly signified concept, as opposed to the object of the first intention which is a being on which the actus exercitus is immediately exercised]). Its immediate "object" is now its own conceiving of that "multitude of units," that is, the "concept" (conceptus) of the number as such; nevertheless this multitude itself appears as a "something," namely as *one* and therefore as an "ens," a "*being*." This is precisely what the abstraction which the intellect undertakes consists in: It transforms the multitude of the number into an apparently "independent" being, into an "*ens*," if only an "*ens rationis*." When now—and this is of crucial importance—the *ens rationis* as a "second intention"[324] is grasped *with the aid of the imagination* in such a way that the intellect can, in turn, take it up as an object in the mode of a "first intention," we are dealing with a *symbol*, either with an "algebraic" letter-sign or with a "geometric" figure as understood by Descartes. This is the sense in which we spoke earlier of "symbol-generating abstraction."

But how is the imagination a bridge to the world of "bodies" for Descartes? How can the *mathesis universalis* turn, with its aid, into the "true physics" as well? In respect to the intellect, the imagination is defined by its "service" function, which insures the possibility of symbolic knowledge in general and, in particular, of the *mathesis universalis* as a general theory of proportions and equations. Only that kind of thing can be made the subject of proportions and equations, Descartes argues in Rule XIV, "which accepts the more or less" (quod recipit majus et minus), that is, all such things and none but such things as can be called "magnitudes" (cf. the concept of magnitude — μέγεθος in Euclid V). In this sense the *mathesis universalis* deals only with "magnitudes *in general*" (magnitudines in genere), which, we may add, are conceived by the intellect as "*entia abstracta.*" But since we *need*, as we already know, the "aid" of the imagination, we shall have to deal with "special" magnitudes, which we may at least so choose that they are as easily handled as possible. This demand is fulfilled by the *real* "*extension*" of bodies, i.e., by their *corporeality as such*, of which we disregard everything but its "figurality." *But this very same corporeal nature belongs, according to Descartes, to the imagination* (or *phantasia*) *with all the ideas present* "*in*" *it*: "In order to *imagine* something even then [i.e., in the treatment of general magnitudes] and so as not to use the intellect purely but *with the aid of specific* [i.e., particular] *forms depicted in the image-making organ*, we must finally note that nothing can be said of magnitudes in general which cannot also be ascribed to some specific form or other. From this we easily conclude that there will be no little profit in transferring that *which the intellect allows us to say about magnitudes in general* to that specific form of magnitude which is depicted most easily and distinctly of all in our imagination; this is indeed the *real extension of a body* abstracted from everything else, except that it *has figure*, as follows from what was said about it in the twelfth rule, where we conceived of the *image-making organ itself*, with the ideas existing in it, as being nothing but *a true body, really extended and having figure.*" (Ut vero aliquid etiam tunc imaginemur, nec intellectu puro utamur, sed speciebus in phantasia depictis adjuto: notandum est denique, nihil dici de magnitudinibus in genere, quod non etiam ad quamlibet in specie possit referri. Ex quibus facile concluditur, non parum profuturum, si transferamus illa, quae de magnitudinibus in genere dici intelligemus, ad illam magnitudinis

speciem, quae omnium facillime et distinctissime in imaginatione nostra pingetur: hanc vero esse extensionem realem
corporis abstractam ab omni alio, quam quod sit figurata,
sequitur ex dictis ad regulam duodecimam, ubi phantasiam
ipsam cum ideis in illa existentibus nihil aliud esse concepi-
mus, quam verum corpus reale extensum et figuratum —
Rule XIV, 440 f.)[325] *Everything* which we receive from the
"world" with the aid of the "outside senses" (sensus externi),
is transferred by way of the *sensus communis* to the imagination by a series of "seal impressions" (impressiones), and this
occurs in such a way that the parts of the body affected by
these "impressions," *including the imagination*, take on the
shape, the "figure," which belongs to that part of the world
which is making the impression (cf. Rule XII, 412 ff.). "This
must not be thought to be said metaphorically" (neque hoc
per analogiam dici putandum est), as Descartes explicitly
emphasizes. Exactly as happens in a seal impression, the
contours of the part of the world in question are "im-
printed" "realiter" on the receptive parts of the body, *up to
and including the imagination*. (Here "part of the world" must
be understood to mean *everything* which comes into *immediate* "touch" with the "organs" of our body by means of
locomotion, "motus localis".)[326] The "service" rendered by
the imagination depends, therefore, on a "real," and not a
"figurative," "rendering" of the corporeal world; the
imagination always represents precisely that within the
corporeal world which *really* constitutes its true nature, its
"substance," its "corporeality" — namely "figurate"
extension as such. (Although here the intellect in effect
makes use only of its planar and linear extension.)[327] This
and nothing else explains why the imagination can guarantee
the ability of the *mathesis universalis* to grasp the structure of
the "true world" and so to prove itself a "wonderful
science" (scientia mirabilis) indeed.

Extension has, accordingly, a *twofold* character for Descartes: It is "symbolic" — as the object of "general algebra,"

and it is "real" — as the "substance" of the corporeal world. More exactly, in Descartes' thinking, the dignity of representing the substantial "being" of the corporeal world accrues to extension precisely by reason of its *symbolic* objectivity within the framework of the *mathesis universalis*. Only at this point has the conceptual basis of "classical" physics, which has since been called "Euclidean space," been created. This is the foundation on which Newton will raise the structure of his mathematical science of nature.[328]

C
The concept of "number" in Wallis

The final act in the introduction of the new "number" concept is due to Wallis (1616–1703). On the one hand, Wallis belongs to the tradition founded by Vieta and mediated by Harriot's *Artis analyticae praxis* (cf. Note 275) and Oughtred's *Clavis mathematicae*,[329] while, on the other, he succumbs to the influence chiefly of Descartes, but also of Stevin. Since he is in the habit of combining his mathematical presentations with thorough (although not always reliable) historical and philological discussions and is therefore much better able than his predecessors to do justice to the ancient conceptions, he presents to us, for the last time and with the greatest distinctness, a clear picture of the reinterpretation which these conceptions undergo within the framework of symbolic intentionality. He makes a continual effort to adhere to ancient terminology, and when forced to depart from it he always supplies a precise justification, introducing distinctions in the "scholastic" manner, which are intended to enable the traditional concepts to co-exist with those that are new.

In Chapter II of his *Mathesis universalis* (1657)[330] the "*unitas*" is still introduced as "*principium numeri*," corresponding to the "*numerus*" understood as "*unitatum multitudo*," but this is done explicitly only to save the "usual" definitions: "As far as principles are concerned, the point is that of magnitude, while the unit is that of number, a principle *commonly* proposed ... since number is *commonly* defined as a multitude of units." (Principia quod attinet: *Punctum* quidem magnitudinis, *Unitas* autem numeri, principium vulgo perhibetur...; cum numerus vulgo definiatur, Unitatum multitudo — p. 20.) Thereupon, in Chapter IV, the question "whether the unit is a number" (an Unitas sit Numerus — pp. 24–27) is discussed at length, and in conjunction with it the problem of the "nought" is treated. Wallis does not omit to point to the crucial significance of the question concerning the true "principle of number." Here the guiding notion, by now accepted altogether as a matter of course, is the complete parallelization of "arithmetical" with "geometrical" procedure: "The usefulness of this matter is remarkable, especially in analytic, i.e., algebraic investigations, since through it the coincidence of arithmetical operations with geometrical constructions is better seen." (Insignis enim est hujusce rei utilitas, praesertim in Analytices sive Algebrae speculationibus; quo melius operationis Arithmeticae cum Geometricis constructionibus congruentia percipiatur.) He refers in this connection to Descartes and Francis van Schooten,[331] although, as we have seen, the chief role here should be assigned to Stevin (see Pp. 191 ff.). The problem of the "principle of number" must, according to Wallis, be decided as follows: "One" is a number, for "one" answers the question "How many are there?" (Quot sunt). More precisely: "When I assert that 'four' has the same force as 'four units,' then 'units' [here] are neither number nor a [constituent] part of a number, but either the denomination [i.e., the number-name] of the number or the term giving the denomination, or the thing numbered itself. However, 'four' is indeed the *number* of these units. Thus also, when I assert that 'one' has the same force as 'one unit,' 'one monad,' then 'unit' is the denomination of the term giving the denomination of the number, but 'one' is a *number*, namely the multitude of units (the word *multitude* being taken in a loose sense, . . . [cf. P. 52]), for 'one' says 'how many' or 'how many units' are asserted to be present, namely a single one." (Dum . . . *Quatuor* dico tantundem valere ac *quatuor unitates*; τὸ *unitates* nec numerus est, nec pars numeri, sed vel numeri denominatio seu denominator, vel ipsum numeratum. Est autem τὸ *Quatuor* earum unitatum numerus. Ita cum *Unum* tantundem valere dico ac *unam unitatem*, μίαν μονάδα, *Unitas*, est numeri denominatio, seu denominator. *Una*, numerus est, seu unitatum multidudo [*Multitudinis* nomine laxius accepto . . .] dicit enim *Quot* vel *quam multae* unitates adesse dicuntur, nimirum unicam.) Therefore one might, he says, deny that the "unitas" or *monas* is "number," but the "unum" or *hen* cannot possibly be denied a numerical character.

Here we see in what the difference between "thinking" (cogitare) and "saying" (dicere), as Bachet formulates it (cf. P. 177), really consists. By "four" or "six," we do, indeed, mean four or six "units," but indirectly, by way of a detour through "number," which alone can be directly expressed and addressed, since it is understood precisely as such, namely as a "number" *apart from the things counted*, and consequently only as "symbol" (cf. above, Pp. 199 ff.). This "indirection" in apprehending the "numbered assemblage," which characterizes the modern understanding, becomes apparent, however, only if the conceptual mode of ancient arithmetic is kept in mind; for within the latter, the concept of the "four" or the "six," or even of the "unit," refers *immediately* to the particular unit or units, be they objects of sense or purely noetic structures of either the independent or the merely "abstracted" type. So much is clear: The whole complex of ontological problems which surrounds the ancient concept of number loses its object in the context of the symbolic conception, since there is no immediate occasion for questioning the mode of being of the "symbol" itself. This state of affairs is, of course, completely distorted if it is simply asserted that the modern number concept possesses a remarkably high degree of "abstractness," and no inquiry into the real nature of this "abstractness" is made.

The true "principle of number," for Wallis as for Stevin, is the "nought." It is the sole numerical analogue of the geometric point (just as the "instant" is its temporal analogue): "As can be ascertained with certainty by anyone who looks [into the matter] more deeply as soon as the arithmetic operations [i.e., calculations] are compared with the [accompanying] geometric constructions." (Ut ex operationibus Arithmeticis cum constructione Geometrica comparatis liquido constare possit penitius intuenti.) Wallis expressly rejects the accusation that he is relinquishing the unanimous opinion of the ancients and the moderns, who all saw the unit as the element of number.[332] What he does, he says, is not done lightly, and besides, the traditional opinion can be brought into accord with his own if the following distinction is taken account of: Something can be a "principle" of something (1) which is the "first which is such" (primum quod sic) as to be of the same nature as the thing itself and (2) which is the "last which is not" (ultimum quod non) such as to be of the same nature as the thing itself. In the first sense the unit may indeed be called the "principle of number," while the nought is a "principle" in the second sense. True, the point can be a "principle" only in the second sense, since in the case of "magnitude" (i.e., extended and infinitely divisible magnitude) a primum quod sic cannot be found.[333] The ancients happened to have overlooked the fact that the analogy which exists is not between the "point" and the "unit," but between the "point" and the "nought." For this reason, as is shown in Chapter XI (p. 53), they were able to develop their algebra only for "geometric magnitudes" (quantitates Geometricae)[334] and, worse yet, only for "heterogeneous" geometric magnitudes: "for lines, and planes compared with solids, and also other, imaginary, quantities of still more dimensions"[335] (nempe per lineas, et superficies, simul cum solidis, aliisque etiam plurium adhuc dimensionum imaginariis quantitatibus comparatas); these are the structures to which correspond the expressions "latus," "quadratus," "cubus" (side, square, cube), and furthermore the "quantitates imaginariae," i.e., magnitudes which are biquadratic, quadratocubic, etc.: "But why did they resort to geometric (rather than arithmetic), and also to heterogeneous (rather than homogeneous) quantities? I see no reason more likely than that they chose the 'one' (and not, as they ought to have, the 'nought') among arithmetical objects to equate with the geometric 'point'."[336] (Cur autem illi ad Quantitates Geometricas [potius quam Arithmeticas] et quidem Heterogeneas [potius quam Homogeneas] confugerint; nullam ego rationem video verisimiliorem, quam quod Arithmeticorum *Unum* [non vero, ut oportuit, *Nullum*] cum *Puncto* Geometrico comparabant.) But this, he says, is just the crucial point — to understand "Algebra or Analytic" as an "*Universal Art*" and yet to confine it within the bounds of the realm of arithmetic: "What they tried to explain by many geometric dimensions, we contain within the bounds of arithmetic." (Quod illi per plurium dimensionum Geometricarum suppositionem conantur explicare, nos inter Arithmeticae fines continemus.) But if we wish to exhibit algebraic facts in geometric structures, we should use only "homogeneous magnitudes," "either only lines, or only planes, or only solids"[337] (nempe vel per solas lineas, vel per solas superficies, vel per sola corpora). In any case, in Wallis' opinion, which he considers in agreement with that of the ancients (cf. Note 268), the purely "arithmetic" treatment of algebra deserves priority, "since, in truth, the objects of arithmetic are of a higher and more abstract nature than those of geometry" (cum ... revera res Arithmeticae altioris sint et magis abstractae naturae, quam Geometricae; cf. Note 314): "Universal *algebra is in truth arithmetic, not geometry*, and must therefore be explained rather on arithmetic than geometric principles." (Universa

Algebra est vere Arithmetica, non Geometrica; ideo potius Arithmeticis quam Geometricis principiis explicanda — p. 56.) The fact that within geometry many solutions can be found algebraically does not mean that algebra is fundamentally a "geometric" discipline; rather "this comes about because of the intimate relationship of arithmetic to geometry" (ob intimam Arithmeticae Geometriaeque affinitatem illud evenit); or, more exactly: "Since geometry is, as it were, subordinated to arithmetic, it makes to that extent a special application of the *universal assertions of arithmetic* to its [special] objects" (... quoniam Geometria sit Arithmeticae quasi subordinata, adeoque universalia Arithmetices effata rebus suis specialiter applicet); algebraic assertions in the narrower sense also fall under these universal assertions of arithmetic: "And the same general account holds for all operations, whether arithmetic *or specifically algebraic*." (Atque eadem omnino ratio est operationum omnium sive Arithmeticarum sive speciatim Algebricarum.)

We saw how in Vieta the twofold character of the art of calculating with species, the *logistice speciosa*, which produces a "*general* analytic," on the one hand, and an art immediately related to the Diophantine *Arithmetic* on the other, also lent a twofold character to its object: The species was a "general" magnitude, which, *at the same time*, displayed a nature essentially "arithmetic" (cf. Pp. 172 ff.). In Vieta the emphasis was undoubtedly on the universality of the species. Now the same twofold character is displayed by the algebraic magnitudes, the "symbols or species" (symbola seu species) in Wallis (p. 58), only that here their "arithmetic" nature comes out more strongly — they are unambiguously "*numeri*" — "numbers." Therefore the requirement of "homogeneity," which is that "all comparisons [i.e., all relations expressed in equations] of quantity with a view to their equality must be made only among quantities of the same degree" (omnes ... quantitatum comparationes quoad aequalitatem, inter Homogeneas tantum sunt faciendae), is fulfilled, as it were, automatically (pp. 56 f.); since algebraic equations are in themselves homogeneous, "it often happens that equations are made among various powers not of the same *height* [of degree]; *this, although acceptable enough for arithmetic degrees, is by no means acceptable for geometric dimensions*" (saepissime ... contingit aequationes fieri inter varias potestates non ejusdem altitudinis; quod quamvis satis conveniat Gradibus Arithmeticis, Geometricis tamen dimensionibus neutiquam convenit). For equations between powers of various "height" Wallis gives as examples: $2a^2 = 6a$ and $2a^3 = 6a^2$. The "arithmetic degrees" (Gradus Arithmetici) are nothing other than "*numbers* in continuous proportions" (numeri continue proportionales; cf., P. 172, and especially Notes 248 and 249);[338] *all* numbers, however, Wallis argues, are "homogeneous": "Since all numbers [*properly so-called*] are constituted out of units, they are in fact homogeneous quantities"[339] (cum ... numeri omnes [proprie dicti] ex unitatibus constituantur [...] sunt vere Homogeneae [sc. quantitates] ...); consequently the algebraic magnitudes are, as *numbers*, "homogeneous" with each other. This argumentation, however, apparently fails to do justice precisely to the universal character of the "algebraic numbers," for it holds only if the algebraic "numbers" are conceived also as "numbers properly so-called" (numeri proprie dicti). In other words, it makes the algebraic "numbers" appear as "numbers of units," entirely in the Diophantine mode (if, in this context, we disregard "fractions," and "numeri surdi," and "irrational numbers"). But actually the argumentation itself already presupposes a "symbolic" understanding *even* of the ordinary "numbers of units," i.e., of the "numbers properly so-called" (cf. Pp. 212 f. and Pp. 175 f.).[340] The homogeneity among all "ordinary" as well as all algebraic "numbers" is for Wallis, in effect, the result of their membership in one and the same "genus," namely the genus of "number" as such. The unity of this "genus" is immediately manifested in the fact that all its members *have their character as signs in common*; this and *only* this makes them appear homogeneous. Homogeneity is here no longer, as it was in Vieta, a requisite *condition* for operating with "algebraic magnitudes," but characterizes their very nature; they are *essentially* homogeneous *and just for this reason* "*universal.*" Thus Wallis can say, in speaking of algebraic nomenclature (p. 57): "Let it suffice at any rate to point out that the various algebraic powers, by whatever name they may be called, *are nothing else* than numbers *or* lines *or also* other *mutually homogeneous* quantities in continuous proportion." (Sufficiat saltem monuisse, varias Potestates Algebricas, quocunque appellentur nomine, nil aliud esse quam Numeros sive Lineas sive alias etiam quantitates invicem homogeneas, continue proportionales.) On the other hand, this universality of the "potestates algebricae" excludes them neither from having a "numerical" character attributed to them nor from having an "arithmetic" or "logistic" interpretation attached to them. In other words, they are not seen as mere (reference) signs but as *symbols* — they do not only "represent," they "are" in themselves, mathematical objects.

As symbols they are both "*general magnitudes*" and precisely also — "*numbers.*" Therefore Wallis no longer needs special "rung" (i.e., degree) designations for each single algebraic "number." The dimension, more exactly, the "altitude" or height of an algebraic magnitude no longer changes its "genus" (cf. Pp. 172 f.). All possible "numbers" now belong to one and the same, dimensionless, "genus" — *their homogeneity is identical with their symbolic character as such.*[341]

This may be confirmed in the following way. The crucial difference between the "methodus symbolica" and the ancient procedure lies, as we have seen, in the conception of the "unit": For the ancients it was (as a "pure" unit) the "principle of number" and, as such, simply indivisible (cf. Pp. 53, 39 ff. and 109); for the "moderns" it is, "*as something continuous*" (ut quid continuum), divisible into as many (equal) parts as you please: "When arithmetic wishes to imitate in some way the infinite divisibility of geometry, it supposes a unit or a one which is something whole, as it were, but divisible into as many parts as you please." (... Infinitam Geometriae divisibilitatem, cum quodammodo imitari velit Arithmetica; supponit Unitatem sive Unum, quasi jam quid integrum, in quotvis Partes divisibile —Chap. XII, p. 60; cf. also Chap. XLI, p. 210.) Accordingly, the moderns now have to understand "number" otherwise than did the ancients. Of the ancients it is true that "they allowed almost no numbers other than integers; nor did they allow the material of arithmetic to be infinitely divisible like that of geometry, but they required [division] to stop at the unit" (vix alios numeros quam Integros admiserint [nec enim materiam Arithmeticam sicut Geometricam, in infinitum divisibilem admiserunt, sed in unitate sistendum voluere] — Chap. XXXV, p. 183, cf. P. 42). Therefore, "for Euclid and others none but integers were designated by the term number" (apud quos [sc. Euclidem aliosque] numerorum apellatione non nisi Integri insigniuntur — Chap. XLI, p. 210). "Euclid speaks only of *true numbers and properly so-called* (meaning *integers*) which are composed of units; correspondingly he considers the unit as always indivisible, that is, as among numbers 'the last which is of the same sort' [as they are]." ([Euclides] de veris tantum numeris et proprie dictis [integros intellige] verba facit, qui ex Unitatibus componuntur; adeoque Unitatem semper habet pro indivisibili, quodque est, in numeris, minimum quod sic... — Chap. XIX, p. 93; cf. Note 333). The moderns, on the other hand, are entitled, on the basis of the internal "continuity" of the unit as well as of the whole numerical realm, to speak of fractional numbers (numeri fracti), of irrational numbers (numeri surdi), and also of "algebraic numbers." But we should not forget that numbers, and especially the unit itself, can, in turn, be characterized as continuous only by reason of being symbolic or, more exactly, by reason of being symbols produced by abstraction (cf. Pp. 194 ff. and 201 f.). Consequently, Wallis, who makes a continual effort to remain as true as possible to tradition and to retain the ancient terminology, has, on the one hand, great doubts whether fractions, for instance, can be understood "as numbers": "But 'fractions,' or 'broken numbers' *are not so much numbers as* 'fragments of the unit'." (Sunt autem *Fractiones* seu *Numeri fracti*, non tam numeri, quam *Unitatis fragmenta* — Chap. XII, p. 60.) "I admit that the numbers which are called 'broken' or 'fractions' are, as it were, in between 'one' and 'nought' [cf. above, P. 137]. But I add that we have now passed 'into another genus' . . . so that the 'broken number' is *not so much a number as an index of the ratio numbers have to one another*." (Concedo etiam numeros quos *Fractos* vocant, sive *Fractiones*, esse quidem *Uni* et *Nulli* quasi intermedios. Sed addo, quod jam transitur εἰς ἄλλο γένος . . . adeoque numerus Fractus non tam Numerus est, quam numerorum ad invicem Rationis indicium — Chap. IV, p. 27.) But, on the other hand, this last remark itself shows that "fractions" *are*, in fact, nothing but "numbers." For, according to Wallis, a "ratio," a "relation," underlies every "number" as such. In his discussion of the fifth book of Euclid's *Elements* (that is, of the "general theory of proportions"), whose propositions Wallis undertakes to prove "arithmetically," i.e., algebraically (Chap. XXXV),[342] he says: "For this fifth book of the Elements is, like the whole theory of proportions, arithmetical rather than geometric. *And so also the whole of arithmetic itself seems, on closer inspection, to be nothing other than a theory of ratios*, and the numbers themselves nothing but the 'indices' of all the possible ratios whose common consequent[343] is 1, the unit. For when 1 or the unit is taken as the [unique] *reference quantum, all the rest of the numbers (be they whole, or broken or even irrational) are the* 'indices' *or* 'exponents'[344] *of all the different ratios possible in* relation to the reference quantum." (Est autem illud Elementum quintum, ut et tota rationum Doctrina, Arithmetica potius quam Geometrica ... Quid quod et ipsa Arithmetica tota, si strictius spectetur, vix aliud videatur quam Rationum doctrina. Ipsique Numeri rationum totidem indicia quarum communis consequens est 1, Unitas. Ubi enim 1, sive Unitas, habetur pro *quantitate exposita*; reliqui omnes numeri [sive integri, sive fracti, sive etiam surdi] sunt rationum totidem aliarum ad expositam quantitatem indices sive exponentes — p. 183.) Here "number" no longer means a "number of ..."; rather a number now *indicates* a certain "ratio," a *logos* in the sense of Euclid V, Defs. 3–5, and can be designated as a "whole" or a "broken" or an "irrational" — more briefly: as a "rational" or "irrational" — *number* only with reference to this ratio. However, the real reason for this interpretation of "number" becomes evident only in the following argument.

In Chapter XXV of the *Mathesis universalis*, "comparatio," the relating of magnitudes to one another, is discussed for the first time. According to the requirement of homogeneity (Euclid V, Def. 3), only magnitudes of "the same kind" can be "compared" with one another, can have a relation to one another: "Quantities cannot be related unless 'homogeneous,' and only in so far as they are homogeneous." (Quantitates non nisi *homogeneas* comparandas esse, et quidem prout sunt homogeneae.) But there are two possible ways of "comparing" such magnitudes: We may ask either whether, or by what "part," or by "how much," one exceeds the other, i.e., what the "difference" between them is; or we may ask how many times one magnitude contains the other, or how many parts of one the other represents, i.e., what their "relation," in the narrower sense of *ratio*, is (p. 134). The essential distinction between "difference" (differentia) and "ratio" (ratio) is the following (pp. 135–136): The "difference," which is found by "subtraction" (subductio), is always *of the same kind* as the magnitudes compared, from which it follows that the *differences themselves cannot simply be "compared"*, i.e., related to one another. On the other hand, the "ratio" is always the result of a division of one magnitude by another; in other words, it can be read off the "quotient" which is the result of this division; every such "quotient" is, however, itself a "number." For *either* the dividend and the divisor are "of the same kind" (homogeneous), in which case they have the same number of "dimensions,"[345] so that the quotient indicates *how often or how many times* one magnitude is contained in the other; consequently the quotient itself has *no* dimension, since "where any species is divided by another of exactly the same dimension, *a quantity of no dimension arises*" (ubi species aliqua per aliam totidem praecise dimensionum dividitur, quantitas nullius dimensionis oritur — cf. p. 103). *Or* the dividend and the divisor are "of a different kind" (heterogeneous), i.e., do not have the same number of dimensions, in which case division proper is not involved but an "application" (applicatio),[346] although it is carried out *as if* the "quantities were considered *as good as* numbers" (quantitates illae singulae ad instar numerorum considerantur); consequently they too produce quotients which are a dimensionless number.[347] Thus all "quotients" (and this means *all "ratios," all relations, no matter of what kind of magnitude*) are, insofar as they are "dimensionless" structures, quotients "of the same kind" (homogeneous), so that they *can all be compared with one another*: "And hence it is clear that *all ratios of whatever quantities taken in turn are homogeneous among one another*." (Atque hinc patet, *Rationes omnes, quarumcunque ad invicem quantitatum, esse inter se homogeneas*.) "When a comparison in terms of ratio is made, the resultant ratio often [namely with the exception only of the 'numerical genus' itself] leaves the *genus of the quantities compared, and passes into the numerical genus*, whatever the genus of the quantities compared may have been." (Ubi . . . comparatio fit quoad Rationem, quae emergit ratio comparatorum genus non raro deserit, et transit in genus numerosum, cujuscunque sint generis quae comparantur.) But this can only mean that *all numbers are homogeneous. Their homogeneity is identical with their dimensionlessness.* This dimensionlessness, again, is identical with the *"exponential"* or "indexical" role of numbers as "*indices* of ratios," a role immediately apparent in the sign language, the *"notatio."* The dimensionlessness of numbers is therefore actually identical with their *symbolic character.* This is why Wallis can say: "And since indeed 'double' and 'half' and 'triple' and 'third,' etc., are to be taken just as names of ratios, while the symbols of the half and the third, i.e., $\frac{1}{2}$, $\frac{1}{3}$, are reckoned among the numbers (namely those that are fractioned), *why not so reckon the symbols of the double and the triple, i.e., $\frac{2}{1}$, $\frac{3}{1}$ or 2, 3 ?"* (Et quidem cum duplum et dimidium, triplum et triens, etc. perinde pro rationum nominibus habenda sint; dimidii autem et trientis notae $\frac{1}{2}$, $\frac{1}{3}$, numeris (fractis) accenseantur; quidni et dupli, triplive notae $\frac{2}{1}$, $\frac{3}{1}$, vel 2, 3 — *ibid.* cf. p. 103.) Thus, the "notation" itself leads Wallis to call the unit a "*denominator*" (see Pp. 212 f.). And in this very passage he adds explicitly: "*And this is the chief reason* why I assert that the whole theory of ratios belongs more to arithmetical than to geometric investigations" (Atque hac potissimum de causa, ego totam Rationum doctrinam Arithmeticae potius quam Geometricae speculationis autumo.)[348] In other words, the universality of arithmetic as a "general theory of ratios," which depends on the homogeneity of all "numbers," can be understood only in terms of a symbolic reinterpretation of the ancient "numbered assemblage," of the *arithmos.* The object of arithmetic and logistic in their algebraic expansion is now defined as "number," and this means as a symbolically conceived ratio — a conception consonant with that of algebra as a general theory of proportions and ratios (see Pp. 179 ff.). The "material" of this universal and fundamental science is no longer furnished by "pure" units whose mode of being may be subject to dispute, since they can be conceived *either* as independent beings *or* as obtained by "abstraction" (ἀφαίρεσις); the "material" is now rather constituted by — "numbers," whose being no longer presents any problem since, as the products of symbol-generating abstraction, they can be immediately grasped in the notation.

NOTES

Notes to Part I

1. See J. Ruska, "Zur ältesten arabischen Algebra und Rechenkunst," *Sitzungsberichte der Heidelberger Akademie der Wissenschaften, Philosophisch-historische Klasse,* 1917, 2. Abhandlung, especially pp. 35, 49, 60 f., 69 f., 104, 109 f., 113 f. It remains an open question whether the Indian sources in turn go back to Greek or to earlier oriental sources.

2. Cf. Tannery, *Zeitschrift für Mathematik und Physik, Historisch-LiterarischeAbteilung,* XXXVII (1892), pp. 41 ff., and *Mémoires scientifiques,* II, 428 ff.; IV, 275 ff.

3. Cf. Ruska, *Sitzungsberichte,* 1917, pp. 28 and 69. The lost writings of Anatolius (third century A.D.) seem to have constituted yet another Greek source.

4. The *scholium* as well as the Geminus fragment are traced by Tannery to Anatolius, who, in turn, must have used Geminus as a source (*La géométrie grecque* [Paris, 1887], pp. 42–49).

5. Cf. *Theon,* Hiller, 19, 13–22.

6. Cf. Archimedes, *Opera* (Heiberg), II^1, 450 f.=II^2, 528 ff.

7. μῆλον means apple as well as sheep — here too the reference is to the passage in Plato's *Laws* cited previously.

8. Cf. Tannery, *La géométrie grecque,* pp. 49 f.; Hultsch in A. Pauly and G. Wissowa, *Real-Encyclopaedie der klassischen Altertumswissenschaft* (Stuttgart, 1894—), see "Arithmetica," paras. 8, 9 (pp. 1070 ff.); Heath, *A History of Greek Mathematics* (Cambridge, 1921), I, pp. 52–54. Neugebauer has collected the available documents on Egyptian "logistic" in "Arithmetik und Rechentechnik der Ägypter," *Quellen und Studien,* Section B, I (1930), 301 ff., cf. especially pp. 377 f.

9. Tannery (*La géométrie grecque,* p. 50) thinks that this may refer to problems in Diophantus. Probably numerical calculations involving triangles and polygons, like those in Book I of Heron's *Metrics* (cf. Tannery, *ibid.,* p. 47; but see also the text of Diophantus, ed. Tannery, I, 14, 25–27), are intended as well.

10. The numerical value 10 is here substituted for the value 8 which permitted an integral solution in the problem of Olympiodorus. The same problem (with the value 10) is, incidentally, to be found in Diophantus I, 21, where it has been reduced to its "abstract" expression.

11. In the text edited by Jahn the words are πρὸς ἄλλα, but cf. the following *scholium* to *Gorgias* 451 C (Hermann, VI, 301): πρὸς αὐτά, ὡς ὅταν ἄρτιος [sc. ἀριθμός] πρὸς ἄρτιον ἢ περιττὸς πρὸς περιττὸν πολλαπλασιασθῇ· πρὸς ἄλληλα δέ, ὅταν περιττὸς πρὸς ἄρτιον ἢ ἀνάπαλιν— "*with themselves*, as when an even number has been multiplied with an even, or an odd with an odd; *with one another*, as when an odd [number is multiplied] with an even and vice versa"; cf. also the Platonic text itself (see P. 17).

12. Cf. also Proclus: [ἡ ἀριθμητικὴ] τὰ εἴδη τοῦ ἀριθμοῦ καθ᾽ αὑτὰ σκοπεῖ — "[Arithmetic] examines the *eide* of number by themselves" (Friedlein, 39, 17); furthermore, the *scholium* to *Gorgias* 451 B (Hermann, VI, 300): ἐπίστησον ὅτι ἡ ἀριθμητικὴ κατὰ τὴν ἑαυτῆς φύσιν τὰ εἴδη τοῦ ἀριθμοῦ σκοπεῖ, τὰ δὲ σχήματα τὰ ἐπ᾽ αὐτοῖς καὶ τοὺς λόγους ὡς ἀρχὴν γεωμετρίας καὶ μουσικῆς . . . — "Know that arithmetic, *according to its own nature*, studies *the kinds of number*, while it sees the associated figures and ratios as the basis of geometry and music" (cf. P. 31 and Pp. 43 f.).

13. Cf. Nicomachus (Hoche), 19, 9 ff.

14. Nor is it likely that Olympiodorus is here thinking of Euclid IX, 21–23 and 28–29.

15. Cf. *Republic* 525 A: Ἀλλὰ μὴν λογιστική τε καὶ ἀριθμητικὴ περὶ ἀριθμὸν πᾶσα — "But *all logistic* as well as arithmetic is *about number*"; cf. also *Gorgias* 453 E 2 f.

16. Cf. also *Epinomis* 978 E – 979 A.

17. E.g., G. H. F. Nesselmann, *Die Algebra der Griechen* (Berlin, 1842), p. 40; M. Cantor, *Vorlesungen über Geschichte der Mathematik*, I³, p. 157; Hultsch, Pauly-Wissowa, see "Arithmetica," para. 2, p. 1067; Heath, *A History of Greek Mathematics*, I, pp. 13 f. Exceptions: Ruska, *Sitzungsberichte*, 1917, p. 93, and Heidel, "Πέρας and Ἄπειρον in the Pythagorean philosophy," *Archiv für Geschichte der Philosophie*, XIV (1901), p. 398, note 44.

18. Cf. *Laws* 818 C.

19. Cf. also *Ion* 537 E: . . . ὅτι πέντε εἰσὶν οὗτοι οἱ δάκτυλοι . . . τῇ αὐτῇ τέχνῃ γιγνώσκομεν τῇ ἀριθμητικῇ . . . — ". . . that these fingers amount to five . . . we know by this same arithmetical art. . . ."

20. Cf. also Archytus, Diels, I³, p. 337, fr. 4., where "logistic" is used in a sense which apparently comprehends arithmetic and logistic.

21. Cf. [Aristotle], *Problemata* Λ 6, 956 a 11 ff., the comment, ascribed to Plato, that the superiority of man over other life consists in this: ὅτι ἀριθμεῖν μόνον ἐπίσταται τῶν ἄλλων ζῴων — "that he alone of all animals, knows how to *count*"; cf. also *Republic* 522 E 4, *Laws* 818 C.

22. Cf. also *Theaetetus* 198 A: arithmetike as θήρα ἐπιστημῶν ἀρτίου τε καὶ περιττοῦ παντός — "pursuit of the sciences of *all of even and odd*"; see *Euthydemus* 290 C; also *Protagoras* 357 A; *Epinomis* 981 C, 990 C; furthermore, *Laws* 818 C.

23. E.g., *Republic* VIII, 546 A ff.; cf. *Laws* V, 746 E ff.

24. Cf. Archytas, Diels, I³, p. 331, fr. 1.

25. This was pointed out by Nesselmann, *Algebra der Griechen*, p. 195.

26. Cf. Iamblichus' *Commentary on Nicomachus*, Pistelli, 56, 18 ff., where this theory is described as ἴδιος τόπος τοῦ καθ' αὐτὸ πόσου — "the place proper of absolute quantity"; see also 35, 11 ff.

27. *Ibid.* 8, 12 f.: . . . τὸ δὲ πρὸς ἕτερόν πως ἔχον [sc. ποσόν] (ὃ δὴ πρός τι ποσὸν ἰδίως λέγεται) . . . — ". . . that quantity which has some relation toward another (which is properly called relational quantity). . . ."

28. Tannery (*Mém. Scient.*, III, 27), however, thinks that this is a later interpolation, the source of which he sees in the lost *Theologoumena* of Nicomachus.

29. But which is probably to be traced back to later Byzantine interpolations; see Tannery, *ibid.*, II, 459 f.

30. Cf. Pappus, Hultsch, II, 636, 24 f.

31. Cf. a similar appeal to εὐμάρεια in Iamblichus, Pistelli, 35, 16 ff.

32. *Mém. scient.*, III, 70. It is characteristic of this ("Pythagorean") tradition that it apprehends the numbers themselves directly in the visible world (cf. Section 7A) but their ratios in the audible world (cf. P. 38). See also Plato, *Timaeus* 36 E–37 A; Proclus, *in Euclid.* 35, 28 ff.

33. Cf. also the *Gorgias* scholium cited in Note 12.

34. *Anecdota Graeca* edited by Boissonade (1832), IV, 413–429. Emendations of the text by Tannery, *Mém. Scient.*, II, 213–220, and Hultsch, *Neue Jahrbücher für Philologie und Pädagogik*, 1897, pp. 507–511.

35. Cf. Tannery, *Mém. Scient.*, II, pp. 105 ff., and III, 266 ff.

36. Tannery (*ibid.*, II, p. 214) and, independently, Hultsch (*Neue Jahrbücher*, p. 508; Pauly-Wissowa, see "Domninos") assume the reading τῆς λογιστικῆς . . . θεωρίας. This is plausible especially

because the numerical and calculational system of Apollonius, which is preserved in Book II of Pappus (ed. Hultsch, I, pp. 2 ff.), was probably expounded in the books of Apollonius called "*logistic*" (cf. *Heron*, Heiberg-Schmidt, V, p. 114, 11 f.: Ἀπολλώνιος ἐν τῷ γʹ τῶν Λογιστικῶν — "Apollonius in the third book of the *Logistics*"; cf. Pappus I, p. 20, 16: ... ἐκ τοῦ ... λογιστικοῦ θεωρήματος ιβʹ ... — "from the twelfth logistic theorem"), but hardly in the book mentioned by Eutocius (*in Archim.*, Heiberg, III¹, p. 300, 17=III², p. 258, 16 f.), called ὠκυτόκιον — "good for inducing speedy birth," i.e., for speedy results of calculation (cf. Pappus, Hultsch, III, p. 1212, and Nesselmann, *Algebra der Griechen*, pp. 126, 132). On the other hand, the term λογικὴ θεωρία should probably not be translated simply as "logic," as Tannery does. It is worth recalling what Hultsch (Pauly-Wissowa, see "Apollonius," p. 159) says about this work of Apollonius, namely that "he [Apollonius] proved that numbers whose size far surpasses the human imagination can be expressed by such Greek words as are borrowed from common language and not coined anew." Archimedes too attempted to show in his *Psammites* "that even the largest numbers can be expressed *in words*," though he had to add some expressions which diverged from those current in common language. (He is concerned with the κατονόμαξις τῶν ἀριθμῶν — "the nomenclature of numbers"; cf. *Archimedes*, Heiberg, II¹, 266, 10; 246, 11; 242, 17 f.= II², 236, 18; 220, 4; 216, 17 f.) In this connection it should be remembered that in ancient times an incomparably greater importance was attached to the *oral* teaching of mathematics than nowadays (cf. Cantor, *Vorlesungen*, I³, p. 157; Tannery, *Mém. scient.*, I, p. 83 f.; also Nesselmann, *Algebra der Griechen*, p. 302, on "rhetorical algebra").

37. On the other hand, the words *pros auta* in the Platonic definition of logistic (see P. 17) can refer only to the *inner* composition of numbers. But since their configuration (which is achieved by summation with the aid of the respective gnomon or by multiplication of two or three factors) has to be disregarded, precisely because it is concerned with their *eidetic* character (see Note 26 and Pp. 55 f.), only the "perfect" or "superabundant" or "deficient" character of numbers remains for the *pros auta* (see P. 27). Against this is the fact that *before* Euclid (IX, prop. 36) "perfect" numbers are never mentioned *in this sense*, see Cantor, I³, p. 168. But cf. Aristotle, *Metaphysics* Γ 2, 1004 b 10 ff., where of the ἴδια πάθη, the "characteristics proper" of the *arithmos*, the following are enumerated: περιττότης, ἀρτιότης, συμμετρία, ἰσότης, ὑπεροχή, ἔλλειψις

(oddness, evenness, commensurability, *equality*, *excess*, *defect*), and Aristotle says expressly: καὶ ταῦτα καὶ καθ' αὐτοὺς καὶ πρὸς ἀλλήλους ὑπάρχει τοῖς ἀριθμοῖς — "and these belong to numbers *both by themselves and in relation to one another*." Cf. also Cantor, I³, p. 225. It should also be noted that Schleiermacher in his translation of *Gorgias* 451 A–C (as also elsewhere) renders "logistic" as "Rechenkunst," i.e., "art of calculation," while in the definition he translates it as "Verhältnislehre," i.e., "theory of proportions."

38. Cf. Tannery, *Mém. scient.*, I, 106 ff., Cantor, I³, 158 f.; also Nesselmann, *Algebra der Griechen*, pp. 232 ff., Heath, *Diophantus of Alexandria*, I, 94 ff. In Iamblichus ἐπάνθημα is, incidentally, throughout a *terminus technicus* designating a rule for solving logistic problems (cf. Pistelli, Index).

39. See Rudolf Fecht, "De Theodosii vita et scriptis," *Abhandlungen der Gesellschaft der Wissenschaften zu Göttingen, Philologisch-historische Klasse*, new series, Vol. XIX, 4 (Berlin, 1927), especially p. 9; furthermore Heath, *A History of Greek Mathematics*, II, 245 f. "Spheric," the study of celestial spheres, probably goes back to Eudoxus (cf. Fecht, p. 7).

40. Cf. E. Frank, *Plato und die sogenannten Pythagoreer* (Halle, 1923), especially pp. 163 ff. and 65 ff. I cannot, however, follow Frank's general interpretation.

41. Tannery (*Mém. scient.*, III, 246 ff.; cf. p. 69) has called attention to a fragment of Archytas in Boethius, *De institutione musica* III, 11, pp. 285 f. Friedlein=Diels, I³, 329, A 19, which treats a musical problem as a purely arithmetical, or, as Plato would say, as a purely "logistic" theorem.

42. Cf. Stenzel, *Zahl und Gestalt*¹ (Leipzig, 1924), pp. 36 f.

43. Cf. the Aristotle passage quoted herein (Pp. 109 f.): *Metaphysics* M 3, 1078 a 23 ff.

44. Cf. Nicomachus, Hoche, 5, 1–5; Iamblichus, Pistelli, 7, 18–22; Proclus, *in Euclid.*, Friedlein, 184, 24 f., etc.

45. It is a ὕλη νοητή or φανταστή — "thought material" or "imagined material," since within the Neoplatonic framework, which on this point goes back to Aristotelian and perhaps to Stoic notions, the corresponding faculty of the soul is φαντασία — "imagination" (cf., e.g., Proclus, *in Euclid.*, Friedlein, 53, 1 and 21 f.; 51, 13 ff.; 55, 5).

46. We here disregard the fact that in Plato's time a debate arose over the question whether "problems," since they deal with the "construction" of definite figures and the "computation" of definite numbers, in short with the *genesis* of mathematical structures, can

indeed belong to the realm of those sciences which are turned toward what *always* is (cf. Plato, *Republic* 527 A f.). To be exact, they should be called "porisms" (cf. Proclus, *in Euclid.*, Friedlein, 77 ff., also 178 f., 201 and 212; also Pappus, Hultsch, II, 650, and Heiberg, *Literarisch-geschichtliche Studien über Euklid* [Leipzig, 1882], pp. 62 ff.). In this context the original meaning of *problema*, i.e., something thrown forward, should be noted; see Plato, *Republic* 530 B, 531 C; cf. *Sophist* 261 A–B, *Statesman* 279 D1.

47. "Sur la constitution des livres arithmétiques d'Euclide et leur rapport à la question de l'irrationalité," *Oversigt over det Kongelige Danske Videnskabernes Selskabs Forhandlinger*, 1910, pp. 405 ff., especially 419–421. Cf. also Hultsch, Pauly-Wissowa, see "Eukleides," p. 1014, and Tannery, *La géométrie grecque*, p. 102, note.

48. Cf. Proclus, *in Euclid.*, Friedlein, 68, 20–23: Καὶ τῇ προαιρέσει δὲ Πλατωνικός ἐστι [sc. Εὐκλείδης] καὶ τῇ φιλοσοφίᾳ ταύτῃ οἰκεῖος, ὅθεν δὴ καὶ τῆς συμπάσης στοιχειώσεως τέλος προεστήσατο τὴν τῶν καλουμένων Πλατωνικῶν σχημάτων σύστασιν — "By inclination Euclid is a Platonist and at home in this philosophy, whence also the end which governs his whole *Elements* is the construction of the so-called Platonic figures." In spite of this testimony Euclid should not simply be classified among the Platonists. The way in which he fits Theaetetus' work into his *Elements* by no means proves that he belongs to a definite philosophical school. He seems to have been influenced no less by Peripatetic notions (cf. Pp. 111 f.; furthermore Heiberg, *Studien über Euklid*, p. 27).

49. Cf. the *scholium* to Book XIII, Heiberg-Menge, V, p. 654, 1–10; in Suidas, see Θεαίτητος, with the comments of Tannery, *La géométrie grecque*, p. 101; also Eva Sachs, "Die fünf platonischen Körper: Zur Geschichte der Mathematik und der Elementenlehre Platons und der Pythagoreer," *Philologische Untersuchungen*, eds. A. Kiessling and U. v. Wilamowitz-Moellendorff, fasc. 24, 1917, pp. 76 ff.

50. Hence, mainly, arises the representation of numbers by means of straight lines (cf. P. 111; also Heiberg, *Studien über Euklid.*, pp. 30 f.).

51. This must, incidentally, lead to a discussion concerning the relationship between διάστημα (interval) and λόγος (ratio); cf. Theon, Hiller, 81 f.; furthermore Aristotle, *Physics* Γ 3, 202 a 18 f.; also B 3, 194 b 27 f. and 195 a 31.

52. This book, as is well known, goes back to Eudoxus. Cf. Aristotle, *Metaphysics* Δ 15, 1020 b 26–28 and b 32–1021 a 14, where one of the possible meanings of *pros ti* is exemplified precisely by the arith-

metical theory of proportions, which already constitutes a reference to its *generalization* (cf. also Plato, *Parmenides* 140 B–C).

53. Cf. Notes 8 and 36; also Friedlein, *Die Zahlzeichen und das elementare Rechnen der Griechen und Römer und des christlichen Abendlandes vom 7. bis 13. Jahrhundert* (Erlangen, 1869), pp. 73 ff.; Tannery, *Mém. scient.*, III, p. 331; a new detailed survey can be found in Heath, *A History of Greek Mathematics*, I, pp. 39–64. Friedlein (*in Euclid.*, p. 74) gives a quotation from Lucian which shows that in common language "arithmetic" and "logistic," both being understood in their practical sense, may easily coincide (see P. 20).

54. Cf. furthermore the title of Archimedes' Ψαμμίτης, by which ψαμμίτης ἀριθμός — "the *counted* sand-heap," is to be understood (Hultsch in Pauly-Wissowa, see "Archimedes," p. 515). So too in Archimedes, Heiberg, II², 244, 18: ὁ τοῦ ψάμμου ἀριθμός — "the heap of sand."

55. Cf. Aristotle, *Metaphysics* N 5, 1092 b 19 f.: Καὶ ἀεὶ ὁ ἀριθμὸς ὃς ἂν ᾖ τινῶν ἐστίν . . . — "And a number, whichever it may be, is always of something. . . ."

56. Cf. Pp. 107 f.

57. Cf. Plato, *Theaetetus* 204 E: ΣΩ. Ὁ δὲ ἑκάστων ἀριθμὸς μῶν ἄλλο τι ἢ μέρη ἐστίν; ΘΕΑΙ. Οὐδέν — "So.: And the number of each thing is nothing other than its parts? Theae.: Nothing other"; Aristotle, *Metaphysics* B 4, 1001 a 26: ὁ μὲν γὰρ ἀριθμὸς μονάδες . . . — "for number is units . . ."; Alexander, *in Metaph.*, Hayduck, 55, 25 f.: αἱ δὲ μονάδες ἀριθμοί.

58. Cf. the discussion in Plato, *Theaetetus* 203–206. Its true meaning becomes clear only with reference to the *Sophist*; cf. Section 7 C.

59. There is, of course, an especially close connection between the nature of *episteme* and the nature of the object to which Theaetetus' researches are devoted; cf. Section 7 C, especially Pp. 96 f.

60. For a possible emendation concerning the term δύναμις, cf. Tannery, *Mém. scient.*, II, pp. 91 ff., and also, on the other side, Stenzel, *Zahl und Gestalt*¹, p. 94 (see Note 66). The conclusive formulation is probably that of H. Vogt, *Bibliotheca mathematica*, 3rd series, X, 113 f.

61. Cf. Domninus, *Anecdota Graeca*, 413, 13 f.: The odd numbers cannot be divided into two equal parts διὰ τὸ τὴν μονάδα ἀδιαίρετον εἶναι τῇ αὐτῆς φύσει — "because the unit is indivisible by its very nature" (Nicomachus, Hoche, 15, 9: . . . τὴν φύσει ἄτομον μονάδα . . . — "the unit [is] by nature . . . uncuttable"); cf. Nicomachus, Hoche,

13, 13: διὰ τὴν ... τῆς μονάδος μεσιτείαν — "because of the intervention of the unit." See also Pp. 39 f. and Aristotle, *Metaph.* M 8, 1083 b 29 f.; *Topics* Z 4, 142 b 8; Plato, *Phaedo* 105 C; finally, Euclid VII, Def. 7.

62. Cf. Aristotle, *Physics* Γ 7, 207 b, 10 f.: ἄπειροι γὰρ αἱ διχοτομίαι τοῦ μεγέθους — "For the bisections of a magnitude are unlimited."

63. Cf. Heidel, "Πέρας and Ἄπειρον in the Pythagorean philosophy," *Archiv für Geschichte der Philosophie,* XIV (1901), especially pp. 395 f. for the passages from Plutarch and Aristoxenus. See Aristotle, *Metaphysics* A 5, 986 a 17: τοῦ δ'ἀριθμοῦ στοιχεῖα τὸ τ'ἄρτιον καὶ τὸ περιττόν, τούτων δὲ τὸ μὲν πεπερασμένον, τὸ δὲ ἄπειρον — "The elements of number are the even and the odd; of these, the former is unlimited and the latter limited." See, above all, *Physics* Γ 4, 203 a 10 ff., and the definition of "one" as the περαίνουσα ποσότης — "the limiting quantity" — by Thymaridas (Iamblichus, Pistelli, 11, 3, and Theon, Hiller, 18, 5).

64. Cf. Plato, *Critias* 119 D, concerning the possible neglect of one of these two *eide.* The appearance of the περιττόν, the uneven, on the "positive" side of the Pythagorean table of opposites marks a characteristic reversal of the "natural" sense of values (cf. especially the English word "odd").

65. Cf. Stenzel, *Studien*[1], pp. 59, 62 f.; *Zahl und Gestalt*[1], p. 21 (see Note 66).

66. J. Stenzel, "Arete und Diairesis," *Studien zur Entwicklung der platonischen Dialektik von Sokrates zu Aristoteles* (Leipzig, 1917; 2d enlarged ed. 1931); *Zahl und Gestalt bei Platon und Aristoteles* (Leipzig, 1st ed., 1924; 2d ed., 1933); for the present study only the first edition was available.

67. A list of these studies is to be found in the second edition of *Zahl und Gestalt,* pp. vii f. [This review includes no scholarly literature published after 1933.]

68. Especially the results of Toeplitz' work overlap the present study. But the difference in the points of departure makes it difficult to harmonize the results completely.

69. *Die Lehre von den Kegelschnitten im Altertum* (Copenhagen, 1886); "Hvorledes Mathematiken i Tiden fra Platon til Euklid blev rationel Videnskab (Avec un résumé en français)," *Det Kongelige Danske Videnskabernes Selskabs Skrifter,* 8 Raekke, *Naturvidenskabelig og mathematisk Afdeling,* I, 5, 1917; "Sur l'origine de l'Algèbre," *Det Kgl. Danske Videnskabernes Selskab, mathem.-fys. Medd.,* II, 4, 1919.

70. Cf. Zeller, *Die Philosophie der Griechen* (7th ed.; Leipzig, 1923), p. 574; Burnet, *Die Anfänge der Griechischen Philosophie*, German ed. 1913, pp. 262 ff. [*Early Greek Philosophy* (Cleveland, 1961), Ch. VII]; J. Stenzel, *Metaphysik des Altertums*, 1931, p. 46.

71. Cf. the two first of the three kinds of Pythagorean Ἀκούσματα — "oral teachings" (Diels, I³, 358, 16 f.): τὰ μὲν γὰρ αὐτῶν τί ἐστι σημαίνει, τὰ δὲ τί μάλιστα . . . — "for some of them signify *what* something *is*, others what it is *above all*."

72. Namely the relations of numbers (cf. P. 31).

73. Cf. the *scholium* to *Gorgias* 451 A (Hermann, VI, 300): πάντα γὰρ ἐστι πρώτως ἐν τοῖς ἀριθμοῖς — "for everything is *primarily* in numbers." (This sentence follows immediately on the text quoted in Note 12.)

74. Cf. Aristotle, *Metaphysics* A 5, 986 a 8: ἐπειδὴ τέλειον ἡ δεκὰς εἶναι δοκεῖ καὶ πᾶσαν περιειληφέναι τὴν τῶν ἀριθμῶν φύσιν — "since the decad is considered to be perfect and to comprise the total nature of number."

75. See A. Delatte, *Études sur la littérature pythagoricienne* (Paris, 1915), p. 139. The term "arithmology" arose in the seventeenth century (cf. Note 313).

76. Cf. Note 32, and P. 38.

77. *Philolaos des Pythagoreers Lehren* (Berlin, 1819), p. 60. Heeren had already interpreted the *morphai* in the sense indicated above (cf. also Frank, *Plato und Pythagoreer*, p. 307, note 1).

78. We may add that the method of Eurytos as described by Aristotle (*Metaphysics* N 5, 1092 b 8 ff.) and by Theophrastus (*Metaphysics* 6 a 19 ff.) is meaningful only with reference to a number *eidos* which is determinate for every case.

79. Cf. also Theon, Hiller, 106, 7 ff. = Diels, I³ 337, 15 ff.

80. Cf. J. Cook-Wilson, *Classical Review*, XVIII, p. 258.

81. The close relation of *dianoia* to *logos* is, for instance, shown by the following passage in the *Sophist* (263 E): ΞΕ. Οὐκοῦν διάνοια μὲν καὶ λόγος ταὐτόν· πλὴν ὁ μὲν ἐντὸς τῆς ψυχῆς πρὸς αὐτὴν διάλογος ἄνευ φωνῆς γιγνόμενος τοῦτ᾽ αὐτὸ ἡμῖν ἐπωνομάσθη, διάνοια; ΘΕΑΙ. Πάνυ μὲν οὖν. — "Stranger: Are not thinking and reasonable speech the same, except that the former, which takes place inside as a voiceless dialogue of the soul with itself is called by us thinking ? Theaetetus: Quite so." Cf. furthermore *Theaetetus* 206 C, D, where the *logos* is described as τὸ τὴν αὑτοῦ διάνοιαν ἐμφανῆ ποιεῖν διὰ φωνῆς μετὰ ῥημάτων τε καὶ ὀνομάτων — "the making apparent of one's thinking through

the voice using verbs and nouns," and 208 C: διανοίας ἐν φωνῇ ὥσπερ εἴδωλον — "the image in voice, as it were, of thinking." Cf. also 189 E f. and *Philebus* 38 E, 39 A.

82. Cf. in reference to this H. G. Gadamer, *Platos dialektische Ethik, Phänomenologische Interpretationen zum "Philebos"* (Leipzig, 1931), pp. 56 ff.

83. Cf. Socrates' role as arbiter in the *Laches* 184 D.

84. Cf. Stenzel, *Zahl und Gestalt*, pp. 22, 44 f. See also *Republic* 602 D–E. (Hence also the connection between the περὶ τοὺς ἀριθμοὺς διατριβή — "occupation with numbers" — and the "sharpening" of the understanding, cf. *Laws* V, 747 B; *Republic* VI, 526 B.)

85. The opposition which is mentioned here (as also in the *Phaedo* 102 B, where Simmias appears at the same time as big and as small) is, to be sure, easily resolved. For the finger is called "one" and "many" in totally different respects. This is the most elementary form of the "one and many" problem (cf. *Philebus* 14 C, D) which reaches its full weight and range only in the realm of the *noeta* themselves (*Philebus* 14 E ff.). Nevertheless in this elementary and easily penetrable form, which immediately reminds Glaucon (as it does Protarchus in the *Philebus*) of the familiar line of questioning, the contradiction is palpable enough to spur the *dianoia* on to activity — which is here the sole point (cf. also *Parmenides* 129 f. and *Sophist* 251 B, C).

86. It is to be noted that here also, as in the *Philebus* (16 C–E), the *arithmos* stands "between" one (ἕν) and infinity (ἀπειρία); see also *Sophist* 256 E 5 f.

87. Theaetetus explicitly adds: δῆλον δὲ ὅτι καὶ ἄρτιόν τε καὶ περιττὸν ἐρωτᾷς καὶ τ'ἄλλα ὅσα τούτοις ἕπεται — "clearly you are asking also about *even and odd* and whatever else follows on these." (Cf. P. 69.) The passage at 185 B echoes the formula of *Republic* 524 B (see P. 75): [διανοῇ] καὶ ὅτι ἀμφοτέρω δύο, ἑκάτερον δὲ ἕν — "you think also that *both are two and each is one.*"

88. On the connection of the μᾶλλον καὶ ἧττον and ἐναντίωσις (contrariety, obstacle) cf. Aristotle, *Physics* E 2, 226 b 7 f.; furthermore Plato, *Republic* 479 B and 602 D ff.

89. Cf. *Theaetetus* 177 C ff. and 186 A–B.

90. See *Republic* 509 D and 534 A. Cf. furthermore H. G. Gadamer, *Platos Ethik*, p. 77, note 1.

91. Cf. Heath, *History of Greek Mathematics*, I, 304 f.

92. As also in the *Republic* (476 A), where the ontological *methexis* problem is expressly mentioned alongside the dianoetic one.

93. See *Euthydemus* 293 C ff. The fact that Euthydemus and Dionysiodorus form a "pair" here points toward the essential nature of the sophist in general (cf. 271 A; 296 D.).

94. It should not be forgotten that "irrational" magnitudes are explicitly mentioned in *Hippias major* 303 B (cf. Note 91). And Theaetetus is (along with Theodorus, who is likewise present in the *Sophist*) the classical master of the "irrational."

95. Cf. the three possibilities in Aristotle, *Metaphysics* M 7: 1081 a 5 ff; a 17 ff.; b 35 ff. The problem of "mixing" must always, as in the case of letters (252 E–253 A) and tones (253 B), be understood from the point of view of the genesis of meaningful formations "capable of being" (cf. 261 D–E).

96. See W. D. Ross, *Aristotle's Metaphysics* (Oxford, 1924), II, 427, the note on the term ἀσύμβλητος.

97. The meaning of the term *arithmos eidetikos* was clearly recognized by O. Becker ("Die diairetische Erzeugung der platonischen Idealzahlen," *Quellen und Studien*, I, pp. 483 ff.), who did not, however, draw the full consequences of his insight (see P. 62).

98. Cf. Aristotle, *Metaphysics* I 4, 1055 a 6 f: τὰ μὲν γὰρ γένει διαφέροντα οὐκ ἔχει ὁδὸν εἰς ἄλληλα, ἀλλ᾽ ἀπέχει πλέον καὶ ἀσύμβλητα — "For there is no way to one another for things different *in genus* but they are too far apart and *not to be thrown together*."

99. A list of the sources concerning this lecture is given by Toeplitz in his article "Das Verhältnis von Mathematik und Ideenlehre bei Plato," Section 5, *Quellen und Studien*, Section B, 1 (1930), p. 18. [See the collection of *testimonia* in K. Gaiser, *Platons Ungeschriebene Lehre* (Stuttgart, 1963), pp. 443–557.]

100. Aristotle, *Metaphysics* B 3, 999 a 6 f.: ἐν οἷς τὸ πρότερον καὶ ὕστερόν ἐστιν, οὐχ οἷόν τε τὸ ἐπὶ τούτων εἶναί τι παρὰ ταῦτα — "In respect to things in which there is a prior and a posterior it is not possible for that which is [said] of them to be something apart from them"; furthermore, *Nicomachean Ethics* A 4, 1096 a 17–19 and *Metaphysics* M 6, 1080 b 11 ff. See also J. Cook-Wilson, *Classical Review*, XVIII, pp. 247 f. and 253 ff.

101. The dialogues *Theaetetus*, *Parmenides*, *Sophist*, and *Statesmen*, when taken in this order, do, without doubt, form a unity. Their *external* connection is immediately given by the fact that the *Sophist* is expressly tied in with the *Theaetetus* and that the *Theaetetus*, in turn, refers just as clearly to the *Parmenides* (*Theaetetus* 183 E), while the *Statesman* immediately continues the *Sophist*. Their *inner* connection is beyond the framework of this study, except to note that in the

Theaetetus and the *Parmenides* the *aporiai* of *episteme* are developed on very different levels; in the *Sophist* and the *Statesman* these are then brought as close to resolution as the dianoetic-dialectic procedure allows. The true solution should not and cannot be given in this way (cf. *Phaedrus* 277 E ff.; *Seventh Letter* 342 E ff.; also *Cratylus* 438 D ff.).

102. This state of affairs is obscured for us by the "symbolic" concept of number (see Part II, Sections 11 and 12), but it poses a nearly insurmountable difficulty for Eleatic, Pythagorean, and Platonic philosophy.

103. Cf. *Seventh Letter* 342 E: τὸ τῶν λόγων ἀσθενές — "the weakness of speech."

104. On the linguistic "duplicity" of the ἕτερον, cf. Stenzel, "Zur Theorie der Logik bei Aristoteles," *Quellen und Studien*, Section B, I (1930) p. 39.

105. Cf. Aristotle, *Physics* Γ 2, 201 b 20 f., where the Platonists are mentioned as ἑτερότητα καὶ ἀνισότητα καὶ τὸ μὴ ὂν φάσκοντες εἶναι τὴν κίνησιν — "ever saying that motion is otherness and inequality and not-being."

106. The connection between the diairetic procedure and the theory of eidetic numbers in Plato furnishes Stenzel's main theme. His interpretation of *arithmos* as a *logos*, which follows Toeplitz' thesis (*Quellen und Studien*, Section B, I (1930), pp. 34 ff. and pp. 3 ff.), is, however, inadequate for the clarification of this context. In any case, this interpretation fails to recognize that "arithmetic" has priority over "logistic," however great a role the latter may, especially in the form of the theory of proportions, play for Plato.

107. Cf. *Parmenides* 139 C 3–5; E 3; 164 C 1–2.

108 Cf. on this also Nicomachus, Hoche, II, 20, p. 117 f.

109. At most, the *aoristos dyas* may be understood as a "twofold infinite." It was probably with this in mind that Plato called it — as the *arche* of the "more and less" — the "great and small" (*Philebus* 24 A ff.; cf., e.g., Aristotle, *Physics* Γ 4, 203 a 15 f.: Πλάτων δὲ [φησὶν εἶναι] δύο τὰ ἄπειρα, τὸ μέγα καὶ τὸ μικρόν — "But Plato says that there are two infinites, the great and the small"). Or it may also be understood, more loosely, as ὑπεροχὴ καὶ ἔλλειψις — "excess and defect" (cf. *Statesman* 283 C ff.; Aristotle, *Physics* A 4, 187 a 16 f.). From the *dyas aoristos* as the *arche* of the "not" (and thus of all manyness) a direct road leads to the Aristotelian doctrine of the ὄνομα ἀόριστον — the "indefinite name" (*On Interpretation* 2, 16 a 30 ff. and 10, 19 b 8 f., where, incidentally, he explicitly states: ἐν γάρ πως σημαίνει καὶ τὸ ἀόριστον

[e.g., οὐκ ἄνθρωπος] — "for the indefinite [name] somehow signifies a single thing [e.g., not-man]"; cf. P. 85), and further to the "infinite judgment."

110. Incidentally, only the *even* eidetic numbers can be generated in this way (cf. Aristotle, *Metaphysics* N 4, 1091 a 23 f.), while the corresponding *odd* numbers arise through a "delimitation" on the part of the "one." (See Ross, I, 173 ff., and II, 484; cf. also the definition of Thymaridas: the "one" as περαίνουσα ποσότης — "limiting quantity," and Aristotle, *Physics* Γ 4, 203 a 10 ff; also Aristotle, *Metaphysics* M 8, 1083 b 29 f.)

111. Cf. Sextus Empiricus, *Against the Mathematicians*, IV, the section Πρὸς ἀριθμητικούς (*Against the Arithmeticians*), ed. I. Bekker (Berlin, 1842), p. 724, 6–21. Sextus' discussion, which follows on pp. 724–728, is, incidentally, very enlightening on the Greek concept of number and confirms the interpretation here given.

112. But, on the other hand, cf. *On the Soul*, B 2, 413 a 13 ff.

113. For the text, cf. Alexander on this passage; also *Metaphysics* N 2, 1089 a 23; *Posterior Analytics* A 10, 76 b 41 f; see also *Physics* B 2, 193 b 31–35.

114. Cf. *Metaphysics* M 6, 1080 a 22 f.; M 7, 1081 a 19 f.

115. In contrast, for instance, to a συλλαβή, a syllable (*Metaphysics* Z 17, 1041 b 11–13).

116. Aristotle's critique of Plato's generic concept of unity is not confined to the realm of numbers but pervades his whole teaching. Especially instructive in this connection is his criticism of the fifth book of the *Republic*, where Plato supports more thoroughly the demand for the communal possession of children, women, and goods. Here, too, Plato is interested in creating a single *genos*, as it were, out of the warrior and ruler class of the *polis*, a *genos* designed to take the place of the multitude of clans and families. Nothing but this could insure that greatest possible *unity* of the *polis* (462 A, B; cf. Book II, 375 B ff.), which is described as its "greatest good" (μέγιστον ἀγαθόν). Aristotle's argument against this is, once again, that the "community" (κοινωνία) of the *polis* is not to be understood as a "single thing" (ἕν). To do so would abolish the *polis* as *polis*: πλῆθος γάρ τι τὴν φύσιν ἐστὶν ἡ πόλις — "For the city is by nature a certain *multitude*" (*Politics* B 2, 1261 a 18, cf. a 6 ff.).

117. These facts form the basis of the analysis of χρόνος (time) in *Physics* Δ 10–14.

118. *Metaphysics* M 8, 1083 b 16 f.: ἀλλὰ μὴν ὅ γ᾽ἀριθμητικὸς ἀριθμὸς μονα-δικός ἐστιν — "but the arithmetical *arithmos* is in fact monadic."

119. This ποιότης (quality) is defined as διαφορὰ οὐσίας (difference of being), for: ταύτης δέ τι καὶ ἡ ἐν τοῖς ἀριθμοῖς ποιότης μέρος· διαφορὰ γάρ τις οὐσιῶν, ἀλλ᾽ ἢ οὐ κινουμένων ἢ οὐχ ᾗ κινούμενα — "the quality of numbers also falls within this [definition], for it is a difference in the being of things, although these are either non-moving, or not considered insofar as they move," *Metaphysics* Δ 14, 1020 b 15-17. (cf. also *Topics* Z 4, 142 b 9 f.).

120. Cf. Plato, *Theaetetus* 204 B, C, *Republic* 337 B.

121. In view of the way in which Plato has him explicate his classification of numbers (*Theaetetus* 147 C ff.; cf. P. 55) this description could, to be sure, come from Theaetetus (or Theodorus) himself (cf. P. 43). It seems that the transformation of the ontological conception of *mathematika* which leads by way of Eudoxus to the Aristotelian doctrine is incipient in Theaetetus. The result of this development is the reversal of the "Pythagorean" thesis that the *mensurability* of things is grounded in their numerability (cf. P. 67): now *numerability* is, conversely, understood as a — not even always complete — expression of mensurability. Plato's position is determined by his attempt to *come to terms* with both of these extremes, except that he remains fundamentally oriented toward the "Pythagorean" thesis, as the content and the structure especially of the *Timaeus* show. On the relations of Euclid, Aristotle, and Theudios, see Heiberg, "Mathematisches zu Aristoteles," *Abhandlungen zur Geschichte der Mathematischen Wissenschaften*, fasc. XVIII (1904); especially note the relation between κοινὰ ἀξιώματα (common axioms) in Aristotle and the κοιναὶ ἔννοιαι (common notions) in Euclid, on which see also Proclus, *in Euclid.*, Friedlein, 194, 7-9.

122. In Theon, on the other hand, the influence of the peripatetic Adrastus can be felt in this respect also (cf. P. 31).

123. See Heiberg, *Studien über Euklid*, pp. 197 ff.

124. With the exception of Domninus, who here follows Euclid (cf. P. 32).

125. Compare, for instance, *Metrica* B 11 (Schöne, 120, 27 ff.), with *Stereometrica* 1 (Schmidt-Heiberg, *Opera*, V 2, 3 ff.), where for the same problem the former uses monads and the latter feet (πόδες) as the basic unit of measurement. Cf. also *Geometrica* 12 (*Opera*, IV, 236 ff.), where the monads which are the final result of the calculation are immediately replaced by the corresponding measurements of length. The writings which have come down to us under the

titles *Geometrica* and *Stereometrica* do not represent genuine works of Heron but are textbooks composed of various writings, based on works indeed originally by Heron but constantly revised and augmented in the course of time; among the works so utilized was also the *Metrica* of Heron (cf. *Opera* V, p. xxi, xxiv f., xxix, xxxii).

Notes to Part II

126. This has been shown very forcefully by Leonardo Olschki in his *Geschichte der neusprachlichen wissenschaftlichen Literatur* I–III, Heidelberg, 1919–1927)

127. See Zeuthen, *Die Lehre von den Kegelschnitten im Altertum* (1886), pp. 6 f.; cf. also Part I, P. 62, and Part II, P. 157; Note 238 and Note 323 (P. 303); P. 214.

128. *Ibid.*, p. ix, pp. 3 ff.; cf. also Zeuthen, "Sur la constitution des livres arithmétiques d'Euclide ...," *Oversigt over det Kongelige Danske Videnskabernes Selskabs Forhandlinger*, 1910, p. 404.

129. This holds true in exactly the same way for the relation of the *one* apodeixis to the *indeterminately many* solutions in Diophantus (see P. 132). The *apodeixis* as such is *general* but it furnishes only *particular* numbers in each case. (Cf. Ptolemy, *Syntaxis* (Heiberg), I, 9, 16: διὰ τῶν ἐν ταῖς γραμμικαῖς ἐφόδοις ἀποδείξεων — "by proofs using *the linear approach.*")

130. This holds for the use of lettered magnitudes up into the sixteenth century (cf. J. Tropfke, *Geschichte der Elementarmathematik*[3] [Berlin, 1930–1940], II, 48 f.).

131. In particular, this fails to be the case for the μυριάδες ὁμώνυμοι τῷ κ — "myriads synonymous with κ," i.e., the κ-fold myriads, of Apollonius and Pappus (Hultsch, I, prop. 23, pp. 14, 27 f.–15, 1 f.; prop. 25, pp. 18, 9 f.–18, 20 ff.; cf. also P. 131), which were considered by Cantor (I[3], 347) to represent a "highly significant generality." The tension between method and object is directly expressed in a fundamental difference between "analysis" and "synthesis" which we shall discuss in Section 11 C, 2. Additional proof will there be given for what has been said earlier.

132. This identification is certainly also suggested by the fact that the mathematical disciplines traditionally belong among the "artes liberales." In the Middle Ages the "artes liberales," and especially arithmetic and geometry, continue to be considered "theoretical" disciplines, but precisely in explicit counterdistinction to the

"practical" "artes mechanicae" (cf., for instance, Thomas Aquinas, *Summa Theologica* II, 1, Qu. 57, Art. 3, Ob. 3: "... sicut artes mechanicae sunt practicae, ita artes liberales sunt speculativae" (just as the mechanical arts are practical so the liberal arts are theoretical). Now it is indeed the case that this traditional "arithmetic" contains from the very beginning considerable "logistic" elements, which provide the theoretical foundation for "practical" calculations (cf. Part I, Pp. 20; 38 f.; 43) — thus it is very hard to draw the boundary between arithmetic, on the one hand, and logistic as the art of calculation, on the other (cf. also Note 46). But the crucial point is that *in general* the distinction between the *artes liberales* and the *artes mechanicae* is slowly obliterated, i.e., that the "artful" character of *all* mathematics (cf. Part I, Pp. 73 f.), which has its roots in the original kinship of *techne* and *episteme*, is identified with "practical application" in the sense of the application of a skillful *method*.

133. Editions: Bachet de Meziriac, 1621; 2d. ed. (with notes by Fermat), 1670; Tannery, 1893–1895 — this is the edition here referred to. Some codices, incidentally, divide the same material into seven books. [For a sample problem see Appendix, Note 22.]

134. Especially by H. Hankel, *Zur Geschichte der Mathematik im Altertum und Mittelalter* (Leipzig, 1874), p. 158; and by Hultsch, Pauly-Wissowa, see "Diophantos," paras. 6 and 9 (pp. 1055; 1059 f.). Montucla, *Histoire des mathématiques* (1758), I, 315, had already, very cautiously, observed that "it is not possible to determine whether Diophantus was the inventor of algebra," but that "one can from that work form an idea of what algebra would have been in Diophantus' time."

135. *Die Algebra der Griechen* (1842), pp. 284 ff.

136. For example, *Mém. scient.*, III, 158; 357; *La géométrie grecque*, pp. 50–52; Vol. II of the Diophantus edition, p. xxi.

137. *Diophantus of Alexandria*² (1910), pp. 111 ff.

138. Cf. Heath, *Diophantus of Alexandria*², pp. 118–121; *A History of Greek Mathematics*, II, 444–447.

139. Cf., for example, Nesselmann, *Algebra der Griechen*, p. 285, note 51; Heath, *Diophantus of Alexandria*², p. 124; also see Note 146. The words in the proemium (2, 8 f.) — Ἴσως μὲν οὖν δοκεῖ τὸ πρᾶγμα δυσχερέστερον, ἐπειδὴ μήπω γνώριμόν ἐστιν ... — "perhaps, indeed, the undertaking seems very hard since it is not yet familiar" — should certainly not be understood to imply that Diophantus intended to teach things totally unknown before; they are uttered

with students in mind (cf. Nesselmann, *Algebra der Griechen*, pp. 286 f.; Cantor, I³, 469). The expression "ἐδοκιμάσθη" — "it is a confirmed opinion" (4, 12), which occurs in the same sense also in the work on polygonal numbers (451, 11), "indicates that our author is reproducing a hallowed tradition" (Tannery, *Mém. scient.*, II, 68).

140. Cf. Cantor, I³, 466, also 74 ff.; Heath, *Diophantus of Alexandria*², pp. 112 f.; *A History of Greek Mathematics*, II, 440 f.; Hultsch, Pauly-Wissowa, see "Diophantos," para. 7, pp. 1056 ff. (See also Note 8.)

141. *Die Naturwissenschaften*, fasc. 30, p. 564 (review of the third edition of Vol. II of J. Tropfke, *Geschichte der Elemtarmathematik*). [Cf. O. Neugebauer, *The Exact Sciences in Antiquity* (2d. ed., New York, 1962), pp. 146 ff.]

142. For the term "algebra" or "al-muqābala", cf. Neugebauer, "Studien zur Geschichte der antiken Algebra I," *Quellen und Studien*," Section B, II, 1 f., note 1.

143. This is especially clear, for instance, in "Def. IX" (12, 19 f.) where are given, without further elucidation, the rules for the multiplication of "defect" (λεῖψις) and "presence" (ὕπαρξις) which correspond to the modern *signs* "minus" and "plus," while the concept of "negative magnitude" is quite unknown to Diophantus. (On calculation with fractions in Diophantus, cf. also Tannery, *Mém. scient.*, II, 155 f.)

144. "Apollonius-Studien," *Quellen und Studien*, Section II, 216 f.

145. Neugebauer, "Studien zur Geschichte der antiken Algebra," *Quellen und Studien*, Section B, II, 2.

146. Cf. also the expressions: προβλήματα ἀριθμητικά (4, 10), ἡ ἀριθμητικὴ θεωρία (4, 13 f.) — "arithmetical problems," "the *arithmetical* science." Traditionally the Diophantine work is also called ἀριθμητικὴ στοιχείωσις — "elementary exposition of arithmetic," and even simply Στοιχεῖα — "Elements" (second volume of the Tannery edition, pp. 72, 18 f.; 62, 24 f.; cf. also the quotation later in the text.

147. *La géométrie grecque*, pp. 50–52. What is said there forms the general basis for Tannery's later publications on Diophantus. Cf. also Nesselmann, *Algebra der Griechen*, p. 44.

148. *Mém. scient.*, II, 535–537.

149. The corresponding passage in Psellus runs: Περὶ δὲ τῆς αἰγυπτιακῆς μεθόδου ταύτης Διόφαντος μὲν διέλαβεν ἀκριβέστερον, ὁ δὲ λογιώτατος Ἀνατόλιος τὰ συνεκτικώτατα μέρη τῆς κατ᾽ ἐκεῖνον ἐπιστήμης ἀπολεξάμενος

ἑτέρως [in the MS: ἑτέρω; Tannery reads ἑταίρῳ or τῷ ἑταίρῳ] Διοφάντῳ συνοπτικώτατα [another reading: συνεκτικώτατα as earlier, whence Tannery deletes the word] προσεφώνησε (cf. second volume of the Tannery edition p. xlvii). The end of this sentence is obviously corrupt. That ἑταίρῳ is hardly possible follows, as Hultsch (Pauly-Wissowa, see "Diophantus" pp. 1052 f.) has already pointed out, from the word order, which would have to be: Δ. τῷ ἑταίρῳ. Hultsch therefore reads ἑτέρως, namely ἀπολεξάμενος ἑτέρως. Heath (*Diophantus of Alexandria*², p. 2) likewise reads ἑτέρως and translates: "Diophantus dealt with it more accurately, but the very learned Anatolius collected the most essential parts of the doctrine, as stated by Diophantus, in a different way and in a most succinct form, dedicating his work to Diophantus." This does not give a sufficiently clear sense. The assertion that Diophantus was a "friend" of Anatolius is hardly supported by the text; neither does the whole context of the passage seem in the least to constitute a claim that Anatolius "dedicated" his work to Diophantus, the sense in which the word προσεφώνησε — admittedly in accordance with general usage — is usually understood. Tannery himself traces the nomenclature for powers of unknowns given by Psellus, which diverges from that of Diophantus, back to Anatolius (cf. *Mém. scient.*, II, 430 f.; IV, 276 ff.). Apparently our passage refers to the *differences of nomenclature* in Anatolius and Diophantus; compare the expression προσεφώνησε with the remarks in the proemium of Diophantus at 4, 12 ff. and 6, 22–25. Προσφωνεῖν would accordingly have to mean "calling a thing something, naming." Then we should read: ἑτέρως Διοφάντου — "differently *from* Diophantus."

Aside from the highly questionable identity of the Dionysios named by Diophantus (1, 4) with the Bishop of Alexandria, it seems that only this much emerges as certain from Psellus' words: (a) that Diophantus *cannot be later* than Anatolius; as a result of this the possible leeway for Diophantus' date is, at any rate, reduced by nearly a century; and (b) that there is no proof at all that Diophantus was an (older) contemporary of Anatolius. For a more exact dating Hippolytos and Heron should perhaps be drawn on more than hithertofore. As far as Hippolytos is concerned, the dubiously genuine section in the *Philosophoumena* (H. Diels, *Doxographi graeci* [Berlin, 1879], pp. 556 f.) probably represents a conflation of the Pythagorean *tetraktys*-theory with an " algebraic" nomenclature corresponding exactly to that of Diophantus but completely misunderstood by the author. Compare, in Hippolytos (*Doxographi graeci*, ed. Diels, p. 557, 3 f.), the series ἀριθμός μονάς δύναμις κύβος δυναμοδύναμις δυναμόκυβος κυβόκυβος — "number, monad, square [power], cube, squared square, squared cube, cubed cube" — with Diophantus (ed. Tannery) pp. 4, 14–6, 8; Hippolytos (cf. also *ibid.*, p. 556, 23 f.) places the terms ἀριθμός and μονάς, mentioned last by Diophantus, at the beginning of the series. Compare further-more Hippolytos, *ibid.*, p. 556, 8 f.: Ἀριθμὸς γέγονε πρῶτος ἀρχή, ὅπερ ἐστὶν ἀόριστον ἀκατάληπτον, ἔχων ἐν ἑαυτῷ πάντας τοὺς ἐπ' ἄπειρον δυναμένους ἐλθεῖν ἀριθμοὺς κατὰ τὸ πλῆθος — "the first *number* is the origin, which is *indeterminate* and not [finitely] conceivable, having in itself *all the numbers capable of going to infinity* in respect to multitude" — with Diophantus (ed. Tannery), pp. 2, 14–16 and 6, 3–5. On the other hand, there exist striking similarities of "style" between the *Metrics* of Heron and the *Arithmetic* of Diophantus, namely (1) in the calculation with "pure" monads (cf. Part I, P. 112), (2) in the far-reaching detachment of the calculations from geometric notions (cf. Note 169), (3) in the correspondence of Heron's *synthesis* with Diophantus' concluding *apodeixis* (cf. Note 161), (4) in the nomenclature for the fractions. Furthermore, the expression δυναμοδύναμις (Heron, ed., Schöne, p. 48, 11 ff.), and even the Diophantine subtraction sign (Diophantus, ed. Tannery, p. 156, 8 and 10), occur in Heron. (Concerning this whole matter, see Tannery, *Mém. scient.*, III, 147 f. and 208 ff.) Let us follow Ingeborg Hammer-Jensen (*Hermes*, XLVIII [1913], pp. 224–235) in dating Heron *after* Ptolemy but reject her and Heath's (cf. *A History of Greek Mathematics*, II, 306) specific dating at the end of the third century, accepting rather A. Stein's (*Hermes*, IL [1914] pp. 154–156) and Heiberg's date (Heron, *Opera*, V, p. ix; *Geschichte der Mathematik und Naturwissenschaft im Altertum* [1925], p. 37, note 4) of the *end of the second century* A.D., and let us, in addition, regard Diophantus' *Arithmetic* as the direct source of the passages quoted previously from the *Philosophoumena* of Hippolytos, which were composed in Rome in about A.D. 230 (cf. O. Bardenhewer, *Geschichte der altkirchlichen Literatur* ([Freiburg, II², 1914], pp. 554, 555, 602); then the possibility that *Diophantus was a contemporary of Heron* presents itself. The end of the second century A.D. was indeed a time distinguished by a special flowering of the mathematical and mechanical sciences in Alexandria, so that under Alexander Severus (A.D. 222–235) several Alexandrian scholars were called to Rome (cf. Hammer-Jensen, *Hermes*, XLVIII, p. 233). Furthermore, Heron composed an introduction to a στοιχείωσις ἀριθμητική (cf. Note 146), which he also dedicated (as he later did the corresponding "geometric" work, the so-called *Definitions*) to a Dionysios (cf.

Opera, IV, p. 14, 1 ff.; p. 76, 23; p. 84, 18). (The ἐγχειρίδιον, i.e., "handbook," of Domninus is, incidentally, of the same type; cf. Domninus, *Anecdota graeca*, p. 428, 16 f., where he refers to a projected ἀριθμητικὴ στοιχείωσις.) Thus it would indeed be possible that the Dionysios of Diophantus and Heron's Dionysios are one and the same person, as Heath conjectures (*A History of Greek Mathematics*, II, 306, note 1; cf. also Tannery, *Mém. scient.*, II, 538; *La géométrie grecque*, p. 180, note 2). There would, of course, remain the question whether this Dionysios was, as Stein thinks, M. Aurelius Papirius Dionysius, who was prefect (ἡγημών) of Egypt in A.D. 187–188. In this context we might refer to a salutation in a document from the second or early third century: Διονυσίῳ τῷ τιμωτάτῳ χαίρειν (B. P. Grenfell and A. S. Hunt, *The Oxyrhynchus-Papyri* [London, 1925], X, pp. 1 f.). Here the title causes a difficulty, since Diophantus addresses Dionysios as τιμιώτατος — "most honorable," while Heron calls him λαμπρότατος — "most illustrious."

We might note in addition:

a. Heiberg on the one hand places Heron at the end of the second century and on the other supports Hammer-Jensen's opinion that the Dionysios addressed by Heron is identical with L. Aelius Helvius Dionysius, *praeceptus urbi* (city prefect) in the year 301 (Heron, *Opera* V, p. xi, note 1, and *Geschichte der Mathematik und Naturwissenschaft im Altertum*, p. 38, note 4). This must be by mistake.

b. Tannery regards Hippolytos as an author of the second century (cf. *Mém. Scient.*, II, 68; 90; I, 186, note) and disregards him in the dating of Diophantus for this very reason. The passage mentioned in Hippolytos is for him merely evidence that the "algebraic" nomenclature was in use already before Diophantus.

Now as it happens, the date of Heron, Hammer-Jensen's argument notwithstanding, is by no means settled. And it is somewhat hard to see why the epigrams of Lucillus (or Lucillius) and of Nicarchos, which refer to the "astronomer" Diophantus (and to a physician Hermogenes), could not refer to the "arithmetician" Diophantus. Ever since the German translation of Diophantus by Otto Schulz (1822) the identity of the "astronomer" and the "arithmetician" is generally denied because (1) an "astronomer" simply cannot be an "arithmetician"; (2) these epigrams make fun of the tiny stature of Diophantus, and this cannot be reconciled with the "high genius of the arithmetician" (cf. Nesselmann, *Algebra der Griechen*, pp. 497 f.); (3) the epigrams speak of the sudden and

violent death of Diophantus, whereas according to the well-known
epigram (second volume of the Diophantus edition of Tannery, pp.
60 f.) he attained the age of eighty-four and "thus presumably died
a natural death." (Cf. Schulz, "Vorrede," p. xii; see also pp. ix ff.)
It will be admitted that this kind of argument is notably naïve. It
would probably be better to return to the opinion already voiced by
Bachet in the introduction to his Diophantus edition, according to
which the above *testimonia*, which come from the time of Nero
(W. v. Christ-Schmid, *Geschichte der griechischen Literatur* [Munich,
1920], pp. 329 f.; 416; cf. *Handbuch der klassischen Altertumswissenschaft*, VII, 2, 1) permit us to date Diophantus *at the first century* A.D.
This might then, reversely, help with the dating of Hero. [Cf. O.
Neugebauer, *The Exact Sciences in Antiquity*, 2d ed. (New York,
1962), 178–179.]

150. *Algebra der Griechen*, p. 296.

151. *Mém. scient.*, III, 160.

152. *Diophantus of Alexandria²*, p. 39.

153. Cf. Tropfke, *Geschichte der Elementarmathematik*, II³, 22 f. It occurs
for the first time in the second half of the fifteenth century (*ibid.*,
pp. 15 ff.).

154. Thus, corresponding to the ordinary usage, the sign $\overset{\circ}{M}$ sometimes
follows the number sign, e.g., p. 238, 11: αἱ δὲ θ $\overset{\circ}{M}$ ... — "and 9
monads"; or p. 334, 12 f.: ζητῶ ἄρα μόριον τετραγωνικὸν προσ-
θεῖναι ταῖς $\overline{κς}$ $\overset{\circ}{M}$ ... κτλ. — "I therefore seek a square fraction [i.e.,
the reciprocal of a square] to add to the 26 monads." This is natural
since the sign $\overset{\circ}{M}$ in Diophantus is — as are *all* his signs — nothing
but an abbreviation of a word (see P. 146); and it is still in accord
with common language that the $\overset{\circ}{M}$ sign occasionally drops out
entirely, e.g., p. 244, 14: Ἔστω ὁ δοθεὶς [sc. ἀριθμός] ὁ $\overline{δ}$ — "let
there be given 4"; or pp. 252, 22 f.: ἀλλὰ ὁ $\overline{κη}$ ἥμισύ ἐστι τῶν $\overline{νς}$
[sc. μονάδων], ὥστε τὰ $\overline{ιδ}$, $δ^{ον}$ ἐστι τοῦ $\overline{νς}$ — "but 28 is half of 56, so that
14 is $\frac{1}{4}$ of 56."

155. Heron, ed. Schöne, p. vii.

156. *Opera*, ed. Heiberg, II¹, 242, 17–19; 246, 11; 266, 11–14=II², 216,
17–19; 220, 3 f.; 236, 19–22. It is not quite certain whether this
work had the title *Archai*; cf. Hultsch, Pauly-Wissowa, see "Archimedes," pp. 511 f. At any rate, its content is noted in the third
chapter of the *Psammites* (II¹, 266 ff.=II², 236 ff.).

157. In Pappus-Apollonius (Hultsch, I, 28, 9) the number which we
would nowadays render as 780,337,152,000,000 is thus written as μ$^{\gamma}$

ψπʹ καὶ μᵝ͵γτοά καὶ μᵅ͵εσʹ — "seven hundred and eighty threefold [i.e., cubed] myriads, and three thousand three hundred and seventy-one twofold [i.e., squared] myriads, and five thousand two hundred simple myriads," or 780.3371.5200.0000.

158. Cf. Hultsch, Pauly-Wissowa, see "Arithmetica," p. 1069.

159. The problems begin with the words: Τὴν μονάδα διελεῖν [or τεμεῖν] εἰς δύο μόρια [or ἀριθμούς] ... — "to divide [cut] the unit into two parts [numbers]"; or: Μονάδα διελεῖν εἰς τρεῖς ἀριθμούς ... — "to divide a unit into three numbers."

160. Here παρομοίως undoubtedly has the sense of παρωνύμως, which is why Xylander, the first translator of Diophantus, as well as Bachet, replace it by the latter. (Bachet writes, παρονύμως, which must be simply a mistake, since he explicitly states that he is following Xylander's conjecture, i.e., παρωνύμως.)

161. Diophantus knows yet another kind of apodeixis, namely the subsequent "test," which will be more thoroughly discussed later (see Pp. 156 and 162).

162. See Nesselmann, Algebra der Griechen, pp. 309, 335 ff. and 352 ff.; Heath, Diophantus of Alexandria, pp. 73 ff. The mere fact of the generality of these rules does not, indeed, by itself suffice to prove the "theoretical" character of the work (cf. Pp. 122 ff.). In any case, these rules are to be understood as "rules of the 'art'," a fact which was to become of the greatest importance for the modern development and transformation of the Diophantine Arithmetic (cf. P. 168). Note also that an approximation procedure (παρισότητος ἀγωγή) is occasionally used by Diophantus; cf. Heath, Diophantus of Alexandria, pp. 95 ff.; 207 ff.

163. We are not contending that Diophantus was a follower of Peripatetic philosophy or even that he occupied himself with philosophical questions at all. So also the "theoretical" character of his work has nothing to do with the question whether he was himself a "scientist" or a "banausic," a mere technician; this holds true also for Heron. Whatever the position of Diophantus and the degree of his "scientific" consciousness may have been, in our context all that matters is that the grounds for the possibility of the Diophantine mode of presentation be understood, although this mode may well have had its actual origins in a "scientific" tradition already in existence (see P. 136).

164. Cf. Nesselmann, Algebra der Griechen, pp. 413 f.; 419 ff.; Tannery, Mém. scient., II, 371.

165. Cf. Nesselmann, *Algebra der Griechen*, pp. 314 ff.; Hankel, *Zur Geschichte der Mathematik* . . . pp. 164 f., Tannery, *Mém. scient.*, II, pp. 367 ff.; Heath, *Diophantus of Alexandria*, pp. 54 ff.

166. Cf. also problems III, 19 and V, 7 (with its second lemma), 8, 21 and 27 (see also Part I, P. 12).

167. Cf. Heath, *Diophantus of Alexandria*, pp. 8 ff. Possibly this does not even refer to a work of Diophantus. (The porisms at the end of I, 34 and 38 are of a different sort, but as "corollaries," they also correspond to the Euclidean terminology.)

168. Pappus, Hultsch, II, 636, 21; 648, 18 f. and 866 ff.; Proclus, *in Euclid.*, Friedlein, pp. 212, 13; 302, 11–13; cf. Heiberg, *Literargeschichtliche Studien über Euklid*, pp. 78 f.

169. As is shown especially by the fact that he, like Heron, adds or subtracts "planes" and "lengths" from one another without compunction; cf. Hankel, *Zur Geschichte der Mathematik* . . ., p. 159; Hultsch, Pauly-Wissowa, see "Diophantos," p. 1055.

170. See the second vol. of the Tannery edition of Diophantus, pp. 36, 20 ff. On the text cf. *ibid.*, pp. vii f.

171. This is, incidentally, sometimes also expressed in the written signs themselves, e.g., $\overset{x}{M}L'$ or $\overset{x}{M}\delta^x$, i.e., a half or a quarter monad. (The sign ^{x}, when added to a number sign, changes the latter into the corresponding reciprocal, for instance, since δ is 4, δ^x is $\frac{1}{4}$; see Pp. 141 f.).

172. It is quite possible that the work *On Fractional Parts* (Μοριαστικά) here mentioned was a separate work by Diophantus, the subject of which was calculation with fractions (cf. Hultsch, Pauly-Wissowa, see "Diophantos," p. 1071), and not merely a collection of *scholia* to the corresponding *Definitions* in the proemium of the *Arithmetic*, as Tannery conjectures (second volume of the Diophantus edition, p. 72, note 2), or even the *Definitions* themselves (cf. Heath, *Diophantus of Alexandria*, pp. 3 f.). Cf. on this also Diophantus, ed. Tannery, I, p. 288, 1 ff.

173. Iamblichus ascribes this definition to certain "Pythagoreans" which, coming from him, does not mean much. According to the wording of the *scholium* Diophantus could easily be its originator.

174. "Def. VI," p. 8, 13 f. (cf. p. 6, 6 f.): Τῆς οὖν μονάδος ἀμεταθέτου οὔσης καὶ ἑστώσης ἀεί . . .; Theon of Alexandria repeats precisely this passage in his commentary on Ptolemy: τῆς γὰρ μονάδος ἀμεταθέτου οὔσης καὶ ἑστώσης πάντοτε . . . — "for since the monad is invariable and *always constant* . . ." (second volume of the Tannery edition of Diophantus, p. 35).

175. In these problems the πυθμένες (bases, i.e., root ratios) are involved, that is, the *smallest* numbers which fulfill the given ratios. This must have been what was in Commandino's mind when he mistakenly identified the πυθμένες with the ὑποστάσεις, the hypothetical expressions, in Diophantus (Pappus, Hultsch, I, p. 81 note).

176. This understanding sets the tone — *the changed concept of number, notwithstanding* — up into the sixteenth century. Frater Fridericus (*ca.* 1460): "surdus numerus non est numerus. Nam numerus est, quem unitas mensurat" (a surd [i.e., an irrational] number is not a number. For a number is that which the unit measures; cf. Part I, P. 109); Michael Stifel in his *Arithmetica integra* (1544): "irrationalis numerus non est uerus numerus" (an irrational number is not a true number); see Tropfke, *Geschichte der Elementarmathematik*, II³, 91 and 92. Cf. furthermore the opinion of *Peletier* (1560) in Note 302.

177. We agree with Ruska, "Zur ältesten arabischen Algebra und Rechenkunst," *Sitzungsberichte der Heidelberger Akademie der Wissenschaften, Philosophisch-historische Klasse*, 1917, 2. Abhandlung, pp. 68 f., that the words δυναμοδυνάμεων (p. 4, 1), δυναμοκύβων (p. 4, 3) and κυβοκύβων (p. 4, 6) must be deleted, since these συντομώτεραι ἐπωνυμίαι (abbreviated descriptive names, p. 4, 13) are not introduced before p. 4, 14. While δύναμις (the square power) and κύβος (the cube) result immediately (p. 4, 14–18) from the preceding text (p. 2, 18–22), in the case of δυναμοδύναμις, δυναμόκυβος and κυβόκυβος the definitions given before (p. 4, 1–7) are first exactly repeated (p. 4, 19–6, 2).

178. These terms already occur, as is well known, in Thymaridas (Iamblichus, Pistelli, p. 62, 19 f.); cf. P. 36.

179. This expression, numbered continuously, was chosen by Bachet only in order to obtain a clearer view of the single paragraphs of the proemium; apparently he was following canonical mathematical terminology.

180. This is true particularly of the ἀριθμὸς τετράγωνος (square number) and κύβος (cube), p. 2, 18 ff.; p. 4, 14–18; cf. Part I, Pp. 26 ff.; 54 ff.; cf. furthermore Euclid VII, Defs. 19 and 20; also Nesselmann, *Algebra der Griechen*, p. 201.

181. A concept indeed common and used by Diophantus, for instance, to signify a certain kind of equation; cf. p. 96, 9: καὶ τοῦτο τὸ εἶδος καλεῖται διπλοισότης — "and this class [of equations] is called a double equation"; cf. P. 133.

182. Something similar holds true of those equations in which ultimately only numbers of *two* classes remain on one side; Diophantus promises to give the procedure for solving these later: ὕστερον δέ

σοι δείξομεν καὶ πῶς δύο εἰδῶν ἴσων ἑνὶ καταλειφθέντων τὸ τοιοῦτον λύεται
— "later I will show you how such a problem is solved if two classes remain [whose sum is] equal to one" (p. 14, 23 f.). Diophantus must have fulfilled this promise in one of the parts of his work which are lost. In the books which have come down to us the procedure for solving such equations is simply presupposed, although here too we occasionally find references to the rule in question (cf. on this Nesselmann, *Algebra der Griechen*, pp. 264 ff., 317–324; Heath, *Diophantus of Alexandria*, pp. 6 ff., 59 ff.; also Tropfke, *Geschichte der Elementarmathematik*, III², 40 f.).

183. Cf. Aristotle, *Metaphysics* Z 6, *Categories* 5.

184. Cf. Part I, Pp. 28, 29, 33; also Diophantus' work on polygonal numbers (450).

185. I.e., in our notation: 1.8747.4560 (cf. Note 157).

186. Cf. on this Heron, *Opera* (Schmidt-Heiberg), IV, 175.

187. Cf. Note 143. It is noteworthy that Maximus Planudes, clearly referring to "Def. X," where the λείποντα εἴδη (lacking classes) are mentioned only in connection with the ὑπάρχοντα (present [classes]), attempts to interpret "Def. IX" in the following way: Οὐχ ἁπλῶς λεῖψιν λέγει, μὴ καὶ ὑπάρξεώς τινος οὔσης, ἀλλὰ ὕπαρξιν ἔχουσαν λεῖψιν — "He does not say simply 'lack,' as if there were not some presence [which lacked something] but 'a presence having a lack'" (second volume of the Tannery edition of Diophantus, p. 139).

188. The nomenclature of Anatolius (cf. Note 149), which goes up to the ninth power of the unknown, recurs in the Italian algebraists; cf. Tannery, *Mém. scient.*, IV, 280 f.

189. Cf. the preface to *L'algebra, opera di Rafael Bombelli da Bologna*, libri IV e V, comprendenti *La parte geometrica* inedita, tratta dal manoscritto B 1569 della Biblioteca dell'Archiginnasio di Bologna, pubblicata a cura di Ettore Bortolotti, Bologna, 1929 (*Per la storia e la filosofia delle mathematiche*. Collezione diretta da Federico Enriques, promossa dall'Istituto nazionale per la storia della scienze), pp. 9, 17–18. The reference to the studies of Bortolotti is drawn from the introduction to the French translation of Diophantus by Paul Ver Eecke (*Diophante d'Alexandrie* [Brussels, 1926]), pp. lxiii ff. This translation is remarkable for its literal rendition of the Diophantine text; it avoids all modern terminology and notation.

190. The 1579 edition of Bombelli's *Algebra* says (p. 414) "... ed'io solo habbia posta l'operatione delle dignità Arimetiche ... immitando gli antichi scrittori ..." — "and I alone have restored the effectiveness of the dignity belonging to arithmetic ... imitating the ancient

writers." See Ver Eecke, *Diophante d'Alexandrie*, p. lxvii, note 1; cf. Bortolotti, *Bombelli* p. 16.

191. Potentia (power) is already used for δύναμις by Gerard of Cremona (twelfth century) and Leonardo of Pisa. Regiomontanus on occasion even uses this word in the generalized sense (cf. Tropfke, *Geschichte der Elementarmathematik*, II³, 160).

192. Bortolotti, *Bombelli*, pp. 14, 16; also pp. 50 ff.

193. Cf. Frédéric Ritter, "François Viète, inventeur de l'algèbre moderne, 1540–1603. Essai sur sa vie et son oeuvre," *La revue occidentale philosophique, sociale et politique*, 2d series, Vol. X (1895), pp. 239–241; 246. Ritter's study is at present [1934] the only monograph on Vieta's life and works. Ritter also made a modern French translation of the *Isagoge* and the *Notae priores ad logisticen speciosam* (*Bullettino di bibliografia e di storia delle scienze matematiche e fisiche*, I, pubbl. da B. Boncampagni, I [1868], pp. 225–276).

194. Ritter, *Revue occidentale philosophique* ..., Vol. X, p. 242.

195. *Ibid.*, p. 269.

196. *Ibid.*, p. 241.

197. In the Dedicatory Letter to Catherine of Parthenay, which is prefixed to the *Isagoge* Vieta says: "... tibi autem [debeo], o diva Melusinis, omne praesertim Mathematices studium, ad quod me excitavit tum tuus in eam amor, tum summa artis illius, quam tenes, peritia, immo vero nunquam satis admiranda in tuo tamque regii et nobilis generis sexu Encyclopaedia." (... And now, O divine Melusine, I owe to you especially the whole study of mathematics, to which I have been spurred on both by your love for it and by the very great skill you have in that art, nay more, the comprehensive knowledge in all sciences [Encyclopaedia] which can never be sufficiently admired in one of your sex who is of so royal and noble a race.) The fact referred to, which is veiled by the baroque ornateness of the style, is Vieta's instruction of Catherine, from which he drew the inspiration for his own extensive studies. How much he felt indebted to Catherine of Parthenay in all his mathematical labors, is shown by the playful remark which he prefixes to the third theorem of the ninth chapter of his work *Ad problema, quod ...proposuit Adr. Romanus, Responsum*, containing the answer to a problem proposed by Adrianus Romanus: Cujus inventi laetitia adfectus, o Diva Melusinis, tibi oves centum pro una Pythagoraea immolavi." (Moved by joy over this discovery, O divine Melusine, I sacrificed to you a hundred sheep instead of Pythagoras' one.)

198. Cf. Ritter, *Revue occidentale philosophique* . . . , Vol. X, pp. 251; 354–373; Zeuthen, *Geschichte der Mathematik im 16. und 17. Jahrhundert* (Leipzig, 1903), p. 115.

199. Ritter, *Revue occidentale philosophique* . . . , Vol. X, p. 242; cf. p. 354.

200. Cf. *ibid.*, pp. 251, 255 f., 366 ff.

201. Cf. Jacobi Peletarii Cenomani, *De occulta parte numerorum, quam Algebram vocant, libri duo* (Paris, 1560); in the preface an opinion according to which Diophantus is to be considered the founder of algebra is mentioned. Ramus had known of Diophantus as the author of the *Arithmetic* as early as 1569: ". . . Diophantus cujus sex libros, cum tamen author ipse tredecim polliceatur, graecos habemus de arithmeticis admirandae subtilitatis artem complexis, quae vulgo Algebra arabico nomine appellatur . . . ." (Diophantus of whom we have six books in Greek, although the author himself promises thirteen, on arithmetical matters, involving an art of admirable subtlety which is popularly called by its Arabic name, algebra — *Scholarum mathematicarum libri unus et triginta*, Book I, quoted according to the edition of L. Schoner [Frankfurt, 1599], p. 35.) Incidentally, in *Universalium Inspectionum ad canonem mathematicum liber singularis*, pl. 71, Vieta refers to Petrus Ramus explicitly, calling him "homo λογικώτατος," a "most perspicacious man."

202. Vieta quotes Diophantus I 27 as early as 1579, namely in the aforementioned *Canon*, pl. 30, clearly following Xylander's edition, where I 27 is the thirtieth problem of the first book.

203. Cf. Cantor, II², 630, note 4. Vieta must have had access to the manuscript then in the possession of Catherine of Medici (now Parisinus 2379), which represents a copy of two older Vatican codices, and which was later used by Bachet for his Diophantus edition. (Regiomontanus had already referred to the existence of Diophantus manuscripts in 1463–1464; note that Constantinople fell in 1453.)

204. See Ritter, *Revue occidentale philosophique* . . . , Vol. X, pp. 246, 250, 258.

205. They are: (1) the polemical writings against Scaliger (Ritter, pp. 260, 262); (2) the solutions of problems posed by Adriaen van Roomen and Vieta himself (Ritter, pp. 264–267; Zeuthen, *Geschichte der Mathematik*, pp. 119 ff.); (3) studies toward a reform of the calendar; Vieta himself valued these most highly and toward the end of his life allowed himself to be carried away on their

account and to engage in unjustified polemics against Clavius (Ritter, *ibid.*, pp. 270–273).

206. See Ritter, *Revue occidentale philosophique* . . . , Vol. X, pp. 270, 374, 370. Compare especially the letter of Marino Ghetaldi to M. Coignet (*ibid.*, p. 384), which is printed at the end of Vieta's work *De numerosa potestatum ad Exegesin Resolutione* (1600), on solving equations having numerical solutions; further Anderson's preface to *De aequationum recognitione et emendatione tractatus duo*, ed. van Schooten, p. 83, a work on the discovery and transformation of equations (cf. Note 239). Vieta himself says in the introduction to his *Ad problema, quod . . . proposuit Adr. Romanus, Responsum* [1595]: "Ego qui me Mathematicum non profiteor, sed quem, si quando vacat, delectant Mathematices studia . . ." (I, who do not profess to be a mathematician, but who, whenever there is leisure, delight in mathematical studies . . .).

207. His younger contemporary, the historian *De Thou*, whose juristic and political career recalls Vieta's and who shares Vieta's predilection for mathematical studies, says of him: "In mathematicis . . . adeo excelluit, ut quicquid ab antiquis in eo genere inventum, et scriptis, quae temporis injuria aut perierunt aut obsoluerunt, proditum memoratur, ipse adsidua cogitatione invenerit et renovarit, et multa ex suo ad illorum ingeniosa reperta addiderit." (In mathematical matters he so excelled that he discovered and *renewed* by his own diligent thinking, whatever of that sort had been discovered by the ancients and had been handed down, related in writings which either perished or decayed through the ravages of time, and he *added* many things of his own to those which they had ingeniously discovered. — Vieta, *Opera mathematica*, ed. van Schooten [1646] p. *4).

208. Cf. Ritter, *Revue occidentale philosophique* . . . , Vol. X, pp. 241, 269. On Vieta's position on Copernicus see especially *Apollonius Gallus*, Appendicula II (ed. van Schooten, p. 343; first edition 1600, p. 11$^{\rm r}$), where there is also a reference to a work intended to correct the errors in Copernicus and the defects in Ptolemy. It was to have had the name *Francelinis* and was to have contained, among other things, a composition called *Epilogistice motuum coelestium Pruteniana*: "Sed ea supplebimus omissa et emendabimus commissa in 'Francelinide,' in qua etiam ex(h)ibebimus 'Epilogisticen motuum coelestium Prutenianam' per hypotheses, quas vocant Apollonianas, si minus placent Ptolemaicae a motu in alieno centro et hypocentris seu ἐπι-κύκλων προνεύσεσι liberatae." (But we will supply ommissions and correct errors in the *Francelinis*, in which we shall set forth the

Prussian Reinterpretation of Heavenly Motions based on hypotheses called Apollonian [i.e., the hypothesis of the moving excentric] if the Ptolemaic hypotheses, when freed from motion around a foreign center [i.e., the equant] and subcenters or inclinations of epicycles [a hypothesis to correct a lunar anomaly, see *Almagest* V, 5] be not quite acceptable.) "Pruteniana," i.e., "Prussian," from Prut(h)enia is an allusion to Copernicus and the *Prutenicae Tabulae coelestium motuum* of Erasmus Reinhold (or Reynoldus), which first appeared in 1551 and incorporated the Copernican hypothesis. In the publisher's preface to the van Schooten edition of 1646 there is mentioned, besides the *Harmonicum coeleste*, a "fragment which investigates the same theme" (fragmentumque eodum spectans). Cf. also the title of a work by Anderson mentioned there: Πρόχειρον *ad triangulorum sphaericorum epilogismum*, a "handbook" on spherical triangles with new calculations. (The *Harmonicum coeleste* was, incidentally, not wholly completed even in 1600, as emerges from a letter of Marino Ghetaldi to Coignet, cf. Note 206).

209. Vieta must have taken his direction from Ptolemy's words in the proemium of the *Syntaxis* (Heiberg, I, 5–7) which we will repeat here because they show most clearly that the ideal of "certain" and "exact" knowledge is by no means a distinguishing characteristic of the "new science." Modern and ancient cosmology are distinguishable from each other not by the *function* but by the *nature* of their mathematics. The character of the "new" kind of mathematics is itself conditioned by a changed ontological understanding of the world; this is directly evidenced by the extension of its "realm of applicability" to *all* of "nature." Ptolemy, reinterpreting Peripatetic doctrine stoically, says that of the three *gene* of the "theoretical" part of philosophy (cf. Aristotle, *Metaphysics* E 1) "theology," the science which is removed from all sense perception and all change and which is concerned with the "outermost" original source of original motion "up high somewhere with the loftiest things of the cosmos" (ἄνω που περὶ τὰ μετεωρότατα τοῦ κόσμου), and "physics," the science of material "qualities" of impermanent things within the sublunar sphere, have a largely conjectural character, i.e., depend on εἰκασία (→ "image-making" → "conjecture"); only "*mathematics*," the science of σχῆμα (→ figura), of ποσότης (→ quantitas, numerus), of πηλικότης (→ magnitude), of τόπος (→ locus→ spatium), of χρόνος (→ tempus), which belong "to all beings simply" (πᾶσιν ἁπλῶς τοῖς οὖσι), represents a certain and irrefutable science; therefore he, Ptolemy, will dedicate himself to the latter and especially to its astronomical part which has for its

object the eternal, unchangeable course of the divine, immutable heavenly bodies, an object which is, indeed, appropriate to the character of a "science," of a *knowledge*, which remains ever the same and does not fluctuate with opinion. (Heiberg, I, 6, 11–7, 4: ... διανοηθέντες, ὅτι τὰ μὲν ἄλλα δύο γένη τοῦ θεωρητικοῦ μᾶλλον ἄν τις εἰκασίαν ἢ κατάληψιν ἐπιστημονικὴν εἴποι, τὸ μὲν θεολογικὸν διὰ τὸ παντελῶς ἀφανὲς αὐτοῦ καὶ ἀνεπίληπτον, τὸ δὲ φυσικὸν διὰ τὸ τῆς ὕλης ἄστατον καὶ ἄδηλον, ὡς διὰ τοῦτο μηδέποτε ἂν ἐλπίσαι περὶ αὐτῶν ὁμονοῆσαι τοὺς φιλοσοφοῦντας, μόνον δὲ τὸ μαθηματικόν, εἴ τις ἐξεταστικῶς αὐτῷ προσέρχοιτο, βεβαίαν καὶ ἀμετάπιστον τοῖς μεταχειριζομένοις τὴν εἴδησιν παράσχοι ὡς ἂν τῆς ἀποδείξεως δι᾽ ἀναμφισβητήτων ὁδῶν γιγνομένης, ἀριθμητικῆς τε καὶ γεωμετρίας, προήχθημεν ἐπιμεληθῆναι μάλιστα πάσης μὲν κατὰ δύναμιν τῆς τοιαύτης θεωρίας, ἐξαιρέτως δὲ τῆς περὶ τὰ θεῖα καὶ οὐράνια κατανοουμένης, ὡς μόνης ταύτης περὶ τὴν τῶν αἰεὶ καὶ ὡσαύτως ἐχόντων ἐπίσκεψιν ἀναστρεφομένης διὰ τοῦτό τε δυνατῆς οὔσης καὶ αὐτῆς περὶ μὲν τὴν οἰκείαν κατάληψιν οὔτε ἄδηλον οὔτε ἄτακτον οὖσαν αἰεὶ καὶ ὡσαύτως ἔχειν, ὅπερ ἐστὶν ἴδιον ἐπιστήμης, πρὸς δὲ τὰς ἄλλας οὐχ ἧττον αὐτῶν ἐκείνων συνεργεῖν. — "... thinking that the other two genera of the theoretical would be described as *conjecture* rather than scientific conception, the theological because it is entirely without appearance and inapprehensible, the physical because its material is unstable and unclear, so that, because of this, the philosophers could never hope to think the same about them; *only the mathematical*, provided it is approached in the spirit of inquiry, furnishes to those who attempt it such *certain* and *trustworthy knowledge* as comes from demonstration both arithmetical and geometric made according to an indisputable procedure — by these thoughts we were led rather to study to our utmost ability *a theoretical discipline of such a sort, especially insofar as it concerns divine and heavenly things,* for it alone is involved in the inquiry concerning *things which are always as they are* and because of this is able, since the kind of apprehension proper to it is neither unclear nor disorderly, *to be itself always such as it is* — which is the peculiar property of science — and yet to cooperate on an equal basis with those other disciplines." Cf. Part I, P. 10.)

On "figure," "quantity," "place," and "time," see Descartes, *First Meditation*: "... quaedam adhuc magis simplicia et universalia vera esse fatendum est ... Cuius generis esse videntur ... figura rerum extensarum; item quantitas, sive earumdem magnitudo et numerus; item locus in quo existant, tempusque per quod durent, et similia."(We must say that there are certain things still more simple and more universal, which truly are; ... Of this kind seem to be the *shapes* of extended things; furthermore quantity or the *magnitude and number* of these; furthermore the *place* in which they exist and the *time* during which they are and similar things. — Adam-Tannery, VII, 20.) Compare, on the other hand, Aristotle, *De anima* B 6, 418 a 16–19 and Γ 1, 425 a 15 f., also 425 b 5 f., furthermore *De Sensu* 1, 437 a 8 f. (Cf. also Note 326.)

210. Under the same general title nine further projected writings are here mentioned, of which only the *Supplementum geometriae* (1593) and *De numerosa potestatum ad Exegesin Resolutione* (1600) appeared in the form announced. (For the *Zeteticorum libri quinque* [1593] I was unable to ascertain the subtitle.) Vieta says already in the *Liber inspectionum*, commenting on the *new* "proportions" given by him for the calculation of right-angled spherical triangles, that their compilation opens the way "ad reparandam et instaurandam pene collapsam nobilem scientiam" (to the *repair and restitution* of a noble science which is nearly in ruins — pl. 37).

211. In the introduction to his work *Ad problema, quod . . . proposuit Adr. Romanus, Responsum* (1595) Vieta says: "Neque vero placet barbarum idioma, id est, Algebricum." (Nor surely is that *barbaric term, namely* "algebraic," satisfactory.)

212 Luca Pacioli, in his *Summa* (1494) already speaks of Algebra as the "Arte magiore" (cf. Cantor, II², 321). Following him, Cardano calls his work: Artis magnae *sive de regulis algebraicis liber unus* (1545); Peletier: De occulta parte *numerorum quam Algebram vocant* (1560); Bombelli: *L'Algebra*, parte maggiore *dell' Arithmetica* (1572); Gosselin: De arte magna *seu de occulta parte numerorum, quae et* Algebra et Almucabala *vulgo dicitur* (1577). The Arabic expression "Almucabala" was translated as "liber de rebus occultis" (book on hidden things), as, for instance, still by Schoner in *Petri Rami Arithmetices libri duo, et Algebrae totidem: a Lazaro Schonero emendati et explicati*, etc. (Frankfurt, 1586), p. 322. Cf. also Descartes, *Regulae ad directionem ingenii*, Rule IX, Ad.-Tann., X, 402: "Unumque est quod omnium maxime hic monendum mihi videtur, nempe ut quisque firmiter sibi persuadeat, non ex magnis et obscuris rebus, sed ex facilibus tantum et magis obviis, scientias quantumlibet occultas esse deducendas." (One thing there is which seems to me to have to be kept in mind above all else, namely that everyone should firmly convince himself that any science, however hidden, must be deduced *not from great and obscure things* but only from easy and very obvious ones.)

213. ". . . mihi pauca admiranti . . ." (to me, who marvels at few things) he says of himself in that same *Dedicatory Epistle.*

214. The Latin edition of Pappus by Commandinus appeared in 1588-1589 (cf. Hultsch, I, p. xvii); however, Vieta had, without doubt, access to Pappus manuscripts before that time. Incidentally, Apollonius Gallus (i.e., the French Apollonius) also represents a "restitution," namely of the work Περὶ ἐπαφῶν (On Tangents) by Apollonius, of which Pappus speaks in that very same seventh book (Hultsch, II, pp. 636, 21; 644, 23 ff.; 820-852).

215. Cf. Descartes, Regulae, Rule IV, Ad.-Tann., X, 376: "Et quidem huius verae Matheseos vestigia quaedam adhuc apparere mihi videntur in Pappo et Diophanto. . . ." (Indeed, certain traces of this true mathematics seem to me still to appear in Pappus and in Diophantus.)

216. Proclus, In Euclid., 211, 19-22 (cf. Diogenes Laertius III, 24; see also Note 218).

217. In a scholium to Euclid XIII, 1-5 (Heiberg-Menge, IV, 364 f.) we read: 'Ανάλυσις μὲν οὖν ἐστι λῆψις τοῦ ζητουμένου ὡς ὁμολογουμένου διὰ τῶν ἀκολούθων ἐπί τι ἀληθὲς ὁμολογούμενον. Σύνθεσις δὲ λῆψις τοῦ ὁμολογουμένου διὰ τῶν ἀκολούθων ἐπί τι ἀληθὲς ὁμολογούμενον. — "Analysis, then, is the taking of what is sought as admitted and [going through] the consequences from this to something admitted to be true, while synthesis is the taking of something admitted and [going through] the consequences from this to something admitted to be true." In the sixteenth century an opinion, which had on occasion been expressed earlier (cf. Cantor, II^2, 102), was current, to the effect that Theon's edition of the Euclidean Elements represented a complete reworking of the original. It was even supposed that the proofs were throughout by Theon and not by Euclid. Xylander, Petrus Ramus (and his circle, cf. Cantor, II^2, 549, and Notes 225 and 233) and Candalla (François de Foix-Candalle), the most important of the Euclid editors and revisers of the time, and many others (cf. Heiberg, Studien über Euklid, pp. 175 and 168) held similar opinions. Vieta became personally acquainted with the two latter, as well as with Forcadel, Peletier, Gosselin, and probably also Bressieu, in Paris; cf. Ritter, Revue occidentale philosophique . . ., Vol. X, p. 245; furthermore C. Waddington, Ramus, Sa vie, ses écrits et ses opinions (Paris, 1855), pp. 109, 156, 353. It was therefore natural for Vieta to ascribe the above scholium as well to Theon; see also Note 233; furthermore Nesselmann, Algebra der Griechen, p. 59; on the modern discussion of this scholium see T. L. Heath, The Thirteen Books of Euclid's Elements^2 (Cambridge, 1926), I, 137 and III, 442. The preference for Theon over Euclid corresponds to the general humanistic tendency to derogate the authority of those writers who

were recognized as authorities in the schools, on the grounds of a "better" knowledge of the ancients. (In reference to Theon, Johannus Buteo and, characteristically, Clavius form an exception, cf. Cantor, II², 563 and 556.)

218. Pappus says: ἀνάλυσις τοίνυν ἐστὶν ὁδὸς ἀπὸ τοῦ ζητουμένου ὡς ὁμολογουμένου διὰ τῶν ἑξῆς ἀκολούθων ἐπί τι ὁμολογούμενον συνθέσει ... ἐν δὲ τῇ συνθέσει ἐξ ὑποστροφῆς τὸ ἐν τῇ ἀναλύσει καταληφθὲν ὕστατον ὑποστησάμενοι γεγονὸς ἤδη, καὶ ἑπόμενα τὰ ἐκεῖ προηγούμενα κατὰ φύσιν τάξαντες καὶ ἀλλήλοις ἐπισυνθέντες, εἰς τέλος ἀφικνούμεθα τῆς τοῦ ζητουμένου κατασκευῆς· καὶ τοῦτο καλοῦμεν σύνθεσιν — "Analysis, then, is the *way* from what is sought, taken as admitted by means of a [previous] synthesis ... but in synthesis, *going in reverse*, we suppose as admitted what was the last result of the analysis, and, arranging in their natural order as consequences what were formerly the antecedents, and connecting them with one another, we arrive at the *completion of the construction* of what was sought; and this we call synthesis." Cf., incidentally, on the concepts of the ζητούμενον (the sought) and the ὁμολογούμενον (the admitted) Plato, *Meno* 79 D, where they are probably employed with a view to the general "geometrical" background of the dialogue. The assertion that Plato was the discoverer of the "analytic" method (cf. above, Note 216) loses its strangeness when understood in the original context of teaching and learning characteristic of Platonic philosophy. The Socratic question-and-answer game of the Platonic dialogues stands in a certain contrast to the "mathematical" method which begins with definitions, axioms and postulates. The purity of this mathematical *synthetic* procedure is *not* to be found in dialectic (cf. *Theaetetus* 196 E: ... πάλαι ἐσμὲν ἀνάπλεῳ τοῦ μὴ καθαρῶς διαλέγεσθαι — "for a long time now we have been infected by *impure* conversation"; cf. also *Meno* 75 C–D). Rather, dialectic always begins with "opinions" (δόξαι) which presuppose what is "sought" as "known," in order to arrive, by means of the refutation of these opinions as "false opinions" (ψευδεῖς δόξαι), at the "*true*" or "*right*" opinion" (ἀληθὴς or ὀρθὴ δόξα) which is sleeping in the soul and which must, once it is found, be fixed by means of an exact "*account of the reason why*" (λογισμὸς αἰτίας) in order to become "*knowledge*" (ἐπιστήμη; cf. *Meno*, 98 A). In the course of *this way to the truth* — the truly Socratic way of "recollection" (ἀνάμνησις) which is the very subject of the *Meno* — the *word* which *designates* the unknown or the thing sought (ζητούμενον) is thus always used as if the thing designated were something already known and admitted (ὁμολογούμενον). This is

precisely the root of the "analytical" power of Socratic conversation (*Theaetetus*, 196 E – 197 A: ΘΕΑΙ. Ἀλλὰ τίνα τρόπον διαλέξει ὦ Σώκρατες, τούτων ἀπεχόμενος; ΣΩ. Οὐδένα ὧν γε ὅς εἰμί … — "Theae.: But in what manner will you converse if you refrain from these [i.e., words designating things sought]? So.: *In no manner, at least while I am who I am*"; cf. *Meno* 98 B). Cf. also Tannery, *La géométrie grecque*, pp. 112 f., who, speaking of the analytic method, refers to the sixth book of the Republic. The dialectic ascent to the ἀνυπόθετον (the unsupposed) must be characterized as "analytic" in the sense set forth here, although the *last* steps would lead beyond the realm of the *logos* (cf. Part I, Pp. 79 f.; 93 ff.). [Cf. J. Klein, *A Commentary on Plato's Meno* (Chapel Hill, 1965), pp. 83 f., 120 ff., 158 ff.]

219. Cf. on this Tannery, *Mém. scient.*, III, 162 f.; Hankel, *Zur Geschichte der Mathematik* . . ., p. 144.

220. Ammonius, *Commentaria in Aristotelis Analyticorum Priorum* (ed. M. Wallies; Berlin, 1899), I, Proemium, 5, 27–31, quotes the definition of *analysis* by Geminus: ἀνάλυσις ἐστιν ἀποδείξεως εὕρεσις — "*analysis* is the finding of the *demonstration*." This holds for "theoretical" as well as "problematical" analysis. Marinus, too, repeats this definition: . . . ἀποδείξεώς ἐστιν εὕρεσις ἡ ἀνάλυσις (*in Euclid. Data*, Heiberg-Menge, VI, 252–254, printed in the Pappus edition of Hultsch, III, p. 1275). Here we read also: . . . μεῖζόν ἐστι τὸ δύναμιν ἀναλυτικὴν κτήσασθαι τοῦ πολλὰς ἀποδείξεις τῶν ἐπὶ μέρους ἔχειν — "it is a greater thing to possess the analytic power than to have many demonstrations about particulars." Cf., furthermore, Heron, *Metrics* (Schöne), p. 16, 12–14: . . . ἐξῆς δὲ κατὰ ἀνάλυσιν διὰ τῆς τῶν ἀριθμῶν συνθέσεως τὰς μετρήσεις ποιησόμεθα — "we will make our measurements through the *synthesis* of numbers [which follows] right after the *analysis*" (cf. Tannery, *Mém. scient.*, III, 146 f.).

221. The expression "canonical" for a certain form of an equation is of late origin; its author is Th. Harriot; cf. Cantor, II², 791; Note 275; also Vieta, *Isagoge*, V, 12, who uses the expression "ordinata" (ordered).

222. In Vieta, incidentally, the judicial term "cautio" (security) is used.

223. *De triangulis omnimodis libri quinque* (1533). Cf. II, 12 (p. 51): "Hoc problema geometrico more absolvere non licuit hactenus, sed per artem rei et census id efficere conabimur." (This problem has to this day admitted no solution in the geometric way, but we will try to achieve it through the *art of the thing* [i.e., the unknown] *and its power*; see P. 148.) Cf. also II, 23 (p. 56).

224. Cf. on this linguistic usage the first chapter of the *Variorum de rebus mathematicis responsorum, liber octavus*, where Vieta, basing himself on Plutarch, *Life of Marcellus*, Chap. 14, calls the traditional problem of the construction of two mean proportionals, i.e., the doubling of the cube, a problem ἄλογον (irrational) and also ἄρρητον (unspeakable) "non quod numeris explicari non possit, ut γραμμαὶ ἄλογοι dicuntur, sed cujus fabrica non ratione, sed instrumento constituatur" (not because it cannot be explicated in numbers, as lines are called irrational, but because its structure is devised *not by reason but by an instrument*).

225. Gosselin in his algebraic work (cf. Note 212) already adds "algebra" to the ancient "Pythagorean" division of mathematics into geometry and astronomy, arithmetic and music (cf. P. 26, and Proclus, *in Euclid.*, Friedlein, 35 f.), assigning it, together with "music," to the realm of relational quantity: "Contempletur numeros secundum se Arithmeticus, numeros certe sub ratione ad aliud cognoscent Musicus, et Algebraeus." (The arithmetician sees numbers *in themselves*, the musician and the *algebraist* indeed know numbers, but in their relation *to something* else — p. 2^r.) This is, so to speak, the first "official" introduction of algebra into the system of sciences recognized by the schools (cf. Pp. 147 f.). Up to that point the "ars rei et census"(art of the thing and its power, i.e., the unknown and its square) was a more or less obscure curiosity (cf. P. 154). It was even considered suitable for public exhibitions in the form of contests and aroused the wonder of the crowd much as did acrobatic or magic tricks: Thus as late as 1548 there took place in Milan a turbulent public contest between Tartaglia and Ferrari, Cardano's pupil (cf. Cantor, II^2, 494), which followed on a dispute about the solution of cubic equations. The tradition of these contests was still continued by Vieta and Romanus and even by Descartes and Stampioen, albeit no longer in the form of public exhibitions; cf. Descartes, *Oeuvres*, Adam-Tannery, XII, 272 ff. In the beginning of the seventeenth century algebra was almost "in fashion" and the predilection of high society; cf. G. Cohen, *Écrivains français en Hollande dans la première moitié du XVIIe siècle* (Paris, 1920) p. 378. It was, above all, thanks to Petrus Ramus, to whose circle Gosselin also belongs, that algebra was admitted into the realm of official science; indeed Ramus generally defended mathematics as the *model* science, although he himself may have had no very great understanding of it (cf. Notes 201, 217, 233, 235, 241, 251, 269). In particular, he was the founder of a chair for mathematics at the Collège royal, which was later occupied by Roberval (from 1634

until 1675; cf. Waddington, *Ramus . . .*, pp. 188 and 326 ff.: also pp. 108 ff.). Ramus himself, incidentally, published an Algebra in 1560 (anonymously, in Paris, through Andr. Wéchel), which was republished in 1586 by Lazarus Schoner (cf. Note 212) and continued to undergo a series of reprintings.

In the work mentioned previously, Gosselin says furthermore (p. 3) "huius scientiae [sc. Algebrae] . . . tota ratio in proportione occupata est" (the whole object of this science of Algebra is comprised *in proportions*) and, surprisingly, sees the foundation of this science in Euclid IX, 8. On Gosselin, as the French translator of Tartaglia (*L'arithmetique de Nicolas Tartaglia Brescian, grand mathematicien, et prince des praticiens etc.*, 1st ed. 1578, 2d ed. 1613), and therefore a major contributor to the dispersion of Algebraic technique, cf. H. Bosmans, "Le 'De arte magna' de Guillaume Gosselin," *Bibl. math.*, 3d series, Vol. VII, pp. 44–66.

Cf. also Vieta's definition of "equation" in *Isagoge*, Chap. VIII, 2: "Aequatio est magnitudinis incertae cum certa comparatio." (An equation is a comparison of an unknown magnitude with a determinate one). Furtnermore see P. 151.

226. Symb. 2   = Euclid, Book I, Common Notion 1
Symb. 3   = Euclid, Book I, Common Notion 2
Symb. 4   = Euclid, Book I, Common Notion 3
Symb. 7   = Euclid, Book V, Def. 13, 12, Prop. 16
Symb. 10   = Euclid, Book VI, 23; VIII, 5
Symb. 13   = Euclid, Book II, 1
Symb. 15 ⎫
Symb. 16 ⎭   = Euclid, Book VI, 16 and 17; VII, 19 (significantly not in Book V!).

Symb. 5 is prefigured in Euclid, Book I, Common Notion 5
Symb. 6 is prefigured in Euclid, Book I, Common Notion 6
Symb. 8 is prefigured in Euclid, Book V, 12
Symb. 9 is prefigured in Euclid, Book V, 19
Symb. 12 is prefigured in Euclid, Book VII, 17
Symb. 14 is prefigured in Euclid, Book VII, 16
Symbol 1: "Totum suis partibus aequari" (the whole is equal to its parts) occurs in the much-read Euclid edition of Clavius as axiom XIX (edition of 1589, Vol. 1, p. 72). Symbol 11 is the converse of Symbol 10.

The enumeration in Chapter II is introduced by the following sentence: "Symbola aequalitatum et proportionum notiora quae habentur in Elementis adsumit Analytice ut demonstrata, qualia sunt fere . . . ." (The Analytic art assumes as demonstrated the better known stipulations governing equations and proportions found in the *Elements*, which are roughly these. . . .) In Chapter I it is said of the *symbola* that they are "tam ex communibus derivanda notionibus, quam ordinandis vi ipsius Analyseos theorematis" (as much derived from common notions as from theorems provided by the force of analysis itself).

By a "symbolum," i.e., a stipulation, Vieta here understands a "contractual stipulation" which corresponds to the judicial concept of the Greek *symbolon*. Perhaps the influence of theological usage also plays a part (cf. however, Note 257; see also Aristotle, *De Interpretatione*, Chap. 1).

227. The significance of the concept and the theory of proportion for the origin of modern science cannot be overrated. Since Thomas Bradwardine (first half of the fourteenth century) and Albert of Saxony (second half of the fourteenth century) there exists a fixed school tradition concerning these matters, which, through the mediation of intellects as different as Nicolaus of Cusa and Luca Pacioli, exerts a crucial influence on the science of the sixteenth and seventeenth century. This influence has two tendencies which, having a common origin, finally merge again; one of these leads, as the present study attempts to show, to *algebra*, the other to the foundation of modern *physics* and *astronomy*. Galileo in his theory of motion, as well as Kepler in his astronomical investigations, are — though this is usually overlooked — in search of certain "proportions." Kepler works under the immediate influence of ancient science, especially of the Platonic *Timaeus* and of Proclus, while, on the other hand, Galileo is stimulated in his mathematical-physical studies by the "practical" disciplines of statics, mechanics, hydraulics, military architecture, etc., and comes to Euclid and Archimedes through these; yet both continue the Pythagorean-Platonic tradition, albeit in different respects and with diverging tendencies. For them, as for Descartes (*Principia*, III, 47), the guiding principle of their investigations is signified by the words "proportio vel ordo," although the meaning of these words as used by them represents a major transformation of the meaning of the corresponding Greek words *analogia* and *taxis*.

228. Cf. *Metaphysics* K 4, 1061 b 17 ff. and *Posterior Analytics* A 10, 76 a 41; b 14 and 20 f.

229. In this respect Diophantine arithmetical "problems" are therefore closer to "geometric" theorems than to "geometric" problems. The synthetic *apodeixis*, i.e., the "test," is here simply the reversal of the analytic *apodeixis* (cf. P. 156 and P. 132). At any rate, here too

the geometric (Euclidean) background of the Diophantine terminology should be noted.

230. In Vieta the term "logistice" means "art of calculation," namely the "theoretical" art of calculation, without particular reference to the "relations" of numbers to one another. In any case, this term is a perfect, if roundabout, description of the "logistic" character of the Diophantine *Arithmetic* (cf. Pp. 134 f.).

231. Xylander translates *eidos* sometimes as "species" and sometimes as "forma."

232. "Et quanquam veteres duplicem tantum proposuerunt Analysin ζητητικὴν καὶ ποριστικήν, ad quas definitio Theonis maxime pertinet, constitui tamen etiam tertiam speciem, quae dicatur ῥητικὴ ἢ ἐξηγητική, consentaneum est, ut sit Zetetice qua invenitur aequalitas proportiove magnitudinis, de qua quaeritur, cum iis quae data sunt. Poristice, qua de aequalitate vel proportione ordinati Theorematis veritas examinatur. Exegetice, qua ex ordinata aequalitate vel proportione ipsa de qua quaeritur exhibetur magnitudo. Atque adeo tota ars Analytice triplex sibi vendicans officium definiatur, Doctrina bene inveniendi in Mathematicis." (And although the ancients proposed only a twofold analysis, the "seeking" [zetetic] and the "providing" [poristic] art, *to which Theon's definition mostly pertains*, I nevertheless established yet a third kind, which is called the "telling or exhibiting" [rhetic or exegetic] art: it is suitable that there might be [1] a *zetetic* art, by which is found the equation and proportion between the magnitude which is sought and those which are given, [2] a *poristic* art by which from the equation or proportion the truth of the theorem set up is examined, [and 3] an *exegetic* art by which from the equation or proportion set up there is exhibited the magnitude *itself which is being sought.* And the whole threefold Analytic Art which claims this office for itself may be defined as the *discipline of right finding in mathematics.*)

233. What this definition is intended to mean can be seen from the short sixth chapter: "De Theorematum per Poristicen examinatione," (On the investigation of theorems by the *poristic* art), which deals, however, more with "synthesis" and its relation to "analysis" than with "poristic." According to it the "via Poristices," the poristic way, is to be taken when a problem is given which does not fit immediately into the systematic context, i.e., which occurs by chance or incidentally; "Quod si alienum proponitur inventum, vel fortuito oblatum, cujus veritas expendenda et inquirenda est; tunc tendanda primum Poristices via est . . ." (But if something *unfamiliar*

is *discovered,* or some *chance finding,* the truth of which must be weighed and investigated, is proposed for proof, then the way of poristic must first be tried.) Examples of this, he says, were given by Theon in the *Elements* (by which Vieta means the *scholium* to Euclid XIII, 1–5, mentioned earlier, P. 155, where analysis and synthesis are defined), further by Apollonius in the *Conics* and by Archimedes in various books. Clearly Vieta is here thinking of the term *porisma* in its meaning of "corollary." Cf. Proclus, *in Euclid.,* Friedlein, p. 212, 12–17: τὸ δὲ πόρισμα λέγεται . . . ἰδίως, ὅταν ἐκ τῶν ἀποδεδειγμένων ἄλλο τι συναναφανῇ θεώρημα μὴ προθεμένων ἡμῶν, ὃ καὶ διὰ τοῦτο πόρισμα κεκλήκασιν, ὥσπερ τι κέρδος ὂν τῆς ἐπιστημονικῆς ἀποδείξεως πάρεργον. — "We refer specifically to a porism when, as a consequence of things demonstrated, there appears at the same time *some other* theorem which we did not ennunciate, and which because of this they call a 'porism' [i.e., a profit obtained], as being *a gain* which is a *by-product* of scientific demonstration." Cf. p. 301, 22–25; see also Note 235.

Vieta seems also to have composed a "poristic" work to parallel his "zetetic" (*Zeteticorum libri quinque,* 1593). In the fifth chapter of his *Variorum de rebus mathematicis responsorum, liber octavus,* containing various answers to mathematical problems, he treats the traditional problem of the doubling of the cube, i.e., of the construction of two mean proportionals, in the synthetic way, referring to the "*ex Poristicis* methodus" which he had presented in the *Supplementum Geometriae.* (Both of these works also appeared in 1593.) Now in the *Supplementum Geometriae* the problem is indeed solved in the "analytic way" (with the aid of the ancient *neusis,* i.e., "verging," procedure, cf. Note 238), namely as a consequence of Proposition VII. Here, as well as in Proposition VI, Vieta uses lemmas of which he says that they were proved "in Poristicis." This work is not preserved. Perhaps it belonged to the contents of the other seven books of "various responses" which, in the *Isagoge,* Vieta assigns to the *Opus restitutae mathematicae Analyseos* (*Opus of the restored mathematical Analysis*), and which Marino Ghetaldi claimed to have had in his possession (Ritter, *Revue occidentale philosophique . . .,* X, p. 408).

As far as the equating of "theorem" and "problem" is concerned, Ramus was already its forceful advocate. In the third book of his much-read work *Scholarum mathematicarum libri unus et triginta* (1569) he says in reference to the relevant definitions of Proclus: "Tota ista problematis et theorematis differentia scholastica et commentitia est, mathematica non est. . . ." (That whole difference between problems and theorems *belongs to the schools* and is *fictitious* but not mathematical.) In another passage he says that this difference "inanis est et plane sophistica" (is empty and clearly sophistical — quoted from L. Schoner's edition [Frankfurt, 1599], pp. 86; 84). The purpose of Ramus' polemic is to "popularize" mathematics by removing all "obscurities" and superfluous distinctions; the whole tenth book of Euclid, incidentally, becomes the victim of this tendency (*ibid.*, p. 252, cf. also Note 217). Ramus, who according to the testimony of his contemporary Estienne Pasquier was "grandement désireux de nouveautez" (Waddington, *Ramus* . . . p. 13), is one of those to whom the commanding position of "inventio" in the science of the second half of the sixteenth century may be traced. By this term, which was taken from rhetoric, Ramus designates the first and most important of the two parts of "dialectic," which for him takes the place of "logic" and which he understands as "dialectica" or "logica *naturalis*"; cf. *Dialecticae institutiones* (1575, reprinting of the first edition of 1543), pp. 1 ff. and 100; see also Waddington, *Ramus* . . ., pp. 367 ff. The ambiguity of the term "inventio," which means both "discovery" and "invention," should not be overlooked; cf., for instance, the letter of Johannes Sturm to Heinrich Schor, printed in the beginning of the *Commentarius doctissimus in Dialecticam Petri Rami* by Rudolph Snell (1595): "Si idem semper sentiendum et loquendum sit: nihil novi invenire liceat." (If the same thing were always to be thought and said — *it would be forbidden to come up with anything new*.) Sturm here explicitly includes the mathematicians among those to whom he appeals. See furthermore Note 332.

234. Cf. Chapter VII: "('De officio Rhetices'): Ordinata Aequatione magnitudinis de qua quaeritur, ῥητικὴ ἢ ἐξηγητική . . . . suum exercet officium; tam circa numeros, si de magnitudine numero explicanda quaestio est, quam circa longitudines, superficies, corporave, si magnitudinem re ipsa exhiberi oporteat." ("On the functions of the rhetic art": When the equation of the magnitude sought has been set [in canonical order] the rhetic or exegetic art exercises its function — on numbers, if the search is for a magnitude expressible *in a number*, as well as on lengths, planes, or solids if the *thing itself* must be shown; cf. P. 157, the quotation from *Apollonius Gallus*). But both designations are used by Vieta promiscuously; thus, in Chap. VIII, 23, he explicitly speaks of the "Exegetice in Arithmeticis" and calls the corresponding work *De numerosa potestatum ad Exegesin resolutione* (*On the Numerical Resolutions of Powers* [i.e. of the unknown] *in respect to Exegesis*).

235. "Synthesis" in Vieta generally takes second place to "analysis," although in geometric problems he frequently makes use of it and recognizes its traditional priority. In the sixth chapter of the *Isagoge* he says expressly that the results of analysis have to be brought "under the order of the art" (in artis ordinationem) according to the "laws" (leges) κατὰ παντός, καθ᾽ αὑτό, καθόλου πρῶτον (i.e., in school language: predicated "of every instance of its subject," "essentially," "commensurately with the universal"). These formulations, which go back to Aristotle (*Posterior Analytics* A 4), are here obviously interpreted by Vieta according to Ramus, who emphasizes them on every occasion, especially in the preface to the *Scholae physicae* ([1565]; cf. also *Scholae mathematicae*, pp. 78 ff.), as the "leges logicae" valid for all sciences. Through the law "of every instance," says Ramus, "non modo falsa sed fortuita tollentur" (not only what is false but also what is *accidental* is removed); cf. Vieta's definition of "poristic," which he seems, in the sixth chapter, purposely to contrast with this law. The law "*essentially*" demands that every "rule of the art" (artis decretum) be "homogeneum, et tanquam corporis ejusdem membrum" (of the same genus and a member of the same body as it were); cf. also the "lex homogeneorum" of Vieta to be discussed later. Therefore such results "as are demonstrated and firmly established by *zetetic*" (quanquam suam habent ex Zetetesi demonstrationem et firmitudinem) must be subjected to "synthesis," "which is commonly considered the logically tighter way of demonstration" (quae via demonstrandi censetur λογικωτέρη); this means that "the tracks of analysis are thus repeated" in reverse (atque idcirco repetuntur Analyseos vestigia). But, Vieta significantly adds, "*this is itself also analytical*" (quod et ipsum analyticum est) and "not troublesome, on account of the species calculation introduced" by him (neque propter inductam sub specie Logisticen iam negociosum). In the context of this preference for "analysis" the equivocality of the corresponding Latin term "resolutio" should be noted, which means (1) "reverse solution" in Pappus' sense (ἀνάπαλιν λύσις — Hultsch, II, 634, 18), (2) "resolution" into the fundamental elements (cf. *De numerosa potestatum ad Exegesin resolutione*, beginning: "Nihil tam naturale est, secundum Philosophos omnes, quam unumquodque resolvi eo genere quo compositum est." —"Nothing is so *natural*, according to all the philosophers, as to resolve something in that same way by which it is composed"), and, finally, (3) simply "solution." For Descartes, too, "analysis" is far more essential than "synthesis"; cf. especially the *Secundae responsiones*,

Ad.-Tann., VII, pp. 155–159, where he says (p. 156): "Ego vero solam Analysim, quae vera et optima via est ad docendum, in Meditationibus meis sum sequutus." (But I have used only analysis, *which is the true and best way of teaching*, in my *Meditations*.) The drawback of synthesis as against analysis consists for Descartes in the fact that "it does not teach the manner in which the thing was *found*" (quia modum quo res fuit inventa non docet — *ibid.*; cf. *Regulae*, Rule IV, Ad.-Tann., X, 375).

236. "Et hic se praebet Geometram Analysta, opus verum efficiundo post alius, similis vero, resolutionem: illic Logistam, potestates quascumque numero exhibitas, sive puras, sive adfectas, resolvendo." (And in the latter case the analyst shows himself to be a geometer by actually carrying out the work in imitation of the like analytic solution, in the former, case, a logistician by resolving whatever powers have been presented numerically, be they simple or conjoined — Chap. VII.)

237. "Itaque artifex Geometra, quanquam Analyticum edoctus, illud dissimulat, et tanquam de opere efficiundo cogitans profert suum syntheticum problema, et explicat: Deinde Logistis auxiliaturus de proportione vel aequalitate in eo adgnita concipit et demonstrat Theorema." (Thus the skillful geometer, though an expert analyst, dissimulates this and presents and explicates his problem as a synthetic one, as if he thought [only] about how to accomplish the work; thereafter, to help the logistician, he constructs and demonstrates the theorem in terms of the proportion or equation recognizable [in the problem]. — Chap. VII.)

238. In this sense the *Effectionum geometricarum canonica recensio* (*Canonical Revision of Solutions Accomplished Geometrically*) and the *Supplementum Geometriae* (both 1593) belong to "exegetic." The last represents a "supplement," insofar as in it, as Vieta says in par. 25 of the eighth chapter of the *Isagoge* (cf. also the beginning of the *Supplementum* itself), "quasi Geometria suppleatur Geometriae defectus" (the defect of geometry is, as it were, supplied by geometry), i.e., by the *neusis* procedure used in this case. Through it this "defect," namely the quadratic or biquadratic equations to which certain planimetric problems normally lead, can be avoided. The corresponding postulate, according to which it is always possible to draw a straight line from a given point which cuts two given straight lines (or a circle and a straight line) in such a way that the segment of the straight line to be drawn which is cut off by the two given lines (or the line and the circle) is of a given length, is an αἴτημα non

δυσμήχανον (a demand not hard to effect), as Vieta, relying on the authority of Nicomedes and especially of Archimedes (cf. Zeuthen, *Die Mathematik im Altertum und im Mittelalter*, p. 41), asserts (*Isagoge*, Chap. VIII, 25, and *Suppl. Geom.*, beginning). To "exegetic" also belongs the treatise *Analytica angularium sectionum in tres partes tributa* on angle sections, which was published posthumously by Anderson in 1615 (ed. van Schooten, pp. 287–304). Cf. also *Isagoge*, Chap. VIII, 24–27.

The *Exegetice in Geometricis* represents a "geometric algebra" in the strict sense. Vieta's precursors in this respect are Bombelli, namely in the two last books of his *Algebra* only recently published by Bortolotti, which contain its "geometric part" (cf. Bortolotti, *Bombelli*, p. 19), and Bonasoni, who expressly entitles his treatise *Algebra Geometrica* (cf. Bortolotti, *Primordi della geometria analitica, L'Algebra geometrica di Paolo Bonasoni nel Mss. 314 della Biblioteca Universitaria di Bologna* [1925], published in the volume of collected works by the same author: *Studi e ricerche sulla storia della matematica in Italia nei secoli* XVI *e* XVII, [1928]). From Bortolotti's excerpts from this treatise, written between 1574 and 1587, we can conclude that Bonasoni (*ibid.*, pp. 14–15) quotes not only Buteo but also Estienne de la Roche (Villafranca). The work of de la Roche (or Stephanus Gallus, as Bonasoni calls him), which makes extensive use of the work *Triparty en la science des nombres*, written in 1484 by Nicolas Chuquet and published only in 1880 by Aristide Marre, must therefore have been well known at that time in Bologna. This invites the conjecture that Bonasoni's compatriot Bombelli, who is explicitly quoted and used by Bonasoni (cf. *ibid.*, pp. 15 and 6), also came, through de la Roche, under the influence of Chuquet. In particular, this would permit a natural explanation for the designation of powers used by Bombelli. (On the relationship of Chuquet to de la Roche, cf. Aristide Marre in the introduction to his Chuquet edition, *Bullettino di bibliografia e di storia delle scienze matematiche e fisiche pubbl. da B. Boncampagni*, XIII [1880], pp. 569–580; also Cantor, II², 347–364 and 371–374). It should also be noted that Bonasoni, as later Vieta, refers to the exclusively "algebraic" solutions of problems by Regiomantanus (cf. Note 223), and proceeds to "supplement" them "geometrically" (Bortolotti, *Primordi del. geom. anal.*, p. 16). Cf. furthermore Marino Ghetaldi's *Variorum problematum collectio*, 1607 (Cantor, II², 809, also Note 275, toward end), and finally Clavius, *Geometrica practica* (1604), lib. VI.

239. Cf. Chap. VIII, 23 "potestatum porro quarumcumque, sive purarum sive (quod nesciverunt veteres, neque novi) adfectarum

tradit Ars resolutionem" (the [analytical] art further yields the solution of all powers whatsoever, whether pure or — and this neither the ancients knew nor the moderns — conjoined; see also Chap. III, 6, wrongly designated 9 in this edition). Such solutions are found in the work *De numerosa potestatum ad Exegesin resolutione* (1600); furthermore in the treatises, published only in 1615 by Anderson, called *De aequationum recognitione et emendatione tractatus duo* (*Two treatises on the discovery and transformation of equations*), ed. van Schooten, pp. 84–158; cf. especially p. 127: "nunc autem circa numerosam Analysin magis esse intentum, nostri est instituti" (but now the attention of our undertaking is to be more directed toward *numerical analysis*), although these treatises are mostly concerned with the second stage of the analytical process, namely the transformation which turns an equation into its canonical or standard form (cf. Ritter, *Revue occidentale philosophique . . .,* X, pp. 390, 396 f.). The treatise *De emendatione* presents the subject matter of the work announced in the *Isagoge* under the title *Ad logisticen speciosam Notae posteriores* and contains a series of formulae ("notae") pertaining to transformations of equations. In particular, it teaches general methods for the solution of equations of the third and fourth degree, partly on the model of Bombelli. The work *Ad logisticen speciosam Notae priores* (first published by Jean de Beaugrand in 1631) represents a collection of elementary general algebraic formulae which correspond to the arithmetical propositions of the second and ninth books of the Euclidean *Elements*.

240. This remark is aimed directly at Diophantus (see P. 170 and also Note 182.)

241. Cf. P. 125. The concepts of "praecepta" (rules) and "exempla," which originally belong to rhetoric, again reveal the influence of Ramus, whose "renewal" of logic consisted precisely in the fact that he tried to develop it from rhetoric. Thus by "logici" we must here understand the followers of the Ramistic reform.

242. Cf. Descartes, *Regulae*, Ad.-Tann., X, 365, 6–9; 372 f.; 405 f.; 439 f. (see Note 279). Cf. also the treatise of Jacobus Acontius, *De Methodo*, 1558, newly edited by H. J. de Vleeschauwer (*Jacob Acontius' Tractaat de Methodo* [Antwerp-Paris, 1932]).

243. This also applies to the use of decimal fractions (cf. Note 292).

244. "Zeteticen autem subtilissime omnium exercuit Diophantus in iis libris qui de re Arithmetica conscripti sunt. Eam vero tanquam per numeros, non etiam per species, quibus tamen usus est, institutam exhibuit, quo sua esset magis admirationi subtilitas et solertia, quando quae Logistae numeroso subtiliora adparent, et abstrusiora, ea

utique specioso familiaria sunt et statim obvia." (Diophantus in those books which he wrote about arithmetic employed zetetics most subtly of all. But he exhibited it as though founded on numbers and not also on species — although he used them — so that his subtlety and skill might be the more admired, since the very things which appear very subtle to a calculator in numbers, and very abstruse, are quite familiar and immediately obvious to a calculator in species.) [Compare the parallel sample problems from the Diophantine *Arithmetic* and Vieta's *Zetetics*, Appendix, Notes 20 and 21.]

245. Descartes, who expresses his distrust even more sharply, is of exactly the same opinion. Pappus and Diophantus, he says (*Regulae*, Ad.-Tann., X, 376 f., Rule IV; cf. Note 215), have kept their knowledge from us out of a "certain injurious cunning" (perniciosa quadam astutia): "Nam sicut multos artifices de suis inventis fecisse compertum est, timuerunt forte, quia facillima erat et simplex [sc. vera Mathesis], ne vulgata vilesceret, malueruntque nobis in eius locum steriles quasdam veritates ex consequentibus acutule demonstratas, tanquam artis suae effectus, ut illos miraremur, exhibere, quam artem ipsam docere, quae plane admirationem sustulisset." (For they perhaps feared, *just as many inventors* have been found to have done in the case of their discoveries, that because the true mathematics was *easy* and *simple* it would become cheapened in becoming *popularized*, and they preferred to exhibit to us in its place *as the results of their art* certain sterile truths, very acutely demonstrated by deduction, so that we might admire those, instead of teaching us *the art itself,* which would have quite subverted our admiration.) Cf. also *ibid.*, p. 373: "Satis enim advertimus veteres Geometras analysi quadam usos fuisse, quam ad omnium problematum resolutionem extentebant, licet eamdem posteris inviderint. Et iam viget Arithmeticae genus quoddam, quod Algebram vocant, ad id praestandum circa numeros, quod veteres circa figuras faciebant." (For we have sufficient evidence that the ancient geometers made use of a *certain analysis*, which they extended to the solution of *all* problems, although they *grudged the same to posterity.* And now there flourishes a certain kind of arithmetic called algebra, *which performs with numbers* that which the ancients did with figures. Cf. furthermore Rule III, beginning.) In the *Secundae Responsiones* (Ad.-Tann., VII, 156) Descartes says that the ancients in their writings made use of "synthesis" but not of "analysis" (!), not because they did not know the latter, "sed, quantum judico, quia ipsam tanti faciebant, ut sibi solis tanquam arcanum quid reservarent" (but, so far as I can judge, because they made so much of it

that they wished to reserve it to themselves alone as a secret). For the rest, Descartes himself acted according to this prescription —he wrote his *Geometry* "obscurely" on purpose; cf. Ad.-Tann., II, 510 ff.; I, 411, also 377 and 478. On popularization, cf. furthermore *Cogitationes privatae*, Ad. Tann., X, 214: "Scientia est velut mulier: quae si pudica apud virum maneat, colitur; si communis fiat, vilescit." (Science is like a woman who, if she remains chastely with a man, is cherished, but if she becomes promiscuous, grows cheap.) In Wallis, who is in this respect of course dependent on Vieta and Descartes, we still read, in reference to the right use of "algebraic" magnitudes: "Hoc autem cum olim vel ignorabant vel non satis attendebant, vel forte dissimulabant, adeoque occultabant Algebristae Veteres . . ." (but since in former times *the ancient algebraists* were ignorant of, or did not pay enough attention to, *or perhaps concealed, or even made a mystery of this* . . . — *Mathesis universalis, Opera* I, 1695, p. 53). Cf. Wallis, *Algebra*, "Praefatio" (*Opera*, II, 1693): "Hanc [sc. Algebram] Graecos olim habuisse non est quod dubitemus, sed studio celatam, nec temere propalandam." (We cannot doubt that the Greeks long ago had this algebra, but it was concealed with care, not indiscreetly broadcast.) And Chap. II (*ibid.*, p. 3.): "Mihi quidem extra omne dubium est, veteribus cognitam fuisse, et usu comprobatam, istiusmodi artem aliquam Investigandi, qualis est ea quam nos *Algebram* dicimus . . . Hanc autem Artem investigandi, Veteres occuluerunt sedulo. . . ." (For me it is certainly beyond doubt that there was known to the ancients, and well-proved in use, a certain skill in just that mode of investigating which we call *algebra*. *The ancients, however, carefully concealed this art of investigating.*) Newton also composed his *Principia* in accordance with this tradition; he told his friend the Rev. Dr. Derham that "to avoid being baited by little smatterers in mathematics, he designedly made his *Principia* abstruse; but yet so as to be understood by able mathematicians who, he imagined, by comprehending his demonstrations would concur with him in his theory" (Portsmouth Collection). Cf. Copernicus in the "Dedicatory Preface" to the *Revolutions of the Heavenly Spheres* (Encyclopedia Britannica [Chicago, 1952], p. 506).

246. Bombelli's generalization of Diophantus' solutions of problems also has as its model the latter's "indeterminate" procedure (cf. Bortolotti, *Bombelli*, pp. 18 f.; 44).

247. His knowledge of Theon is documented by his reference to a remark of Adrastus which occurs in Theon (Hiller, 73, 18 f.; cf. below, Note 253). "Algebraic" magnitudes are equated with

"figurate" numbers also elsewhere in the contemporary mathematical literature. Thus, for instance, we read in the *Fundamentum astronomicum* (1588) of Raimarus Ursus (p. 13^r): "... vel dicto iam modo Geometrice in triangulis aequalium angulorum, vel etiam Arithmetice in numeris Algebraicis seu figuratis" (... be it in the manner actually called geometric in equiangular triangles or even in *numbers either algebraic or figurate*). In Adrianus Metius, *Arithmeticae et geometricae practica* (1611), Book II, Chap. III, pp. 56., the "algebraic" magnitudes are still introduced as "numeri figurati." Cf. also Descartes, *Regulae*, Rule XIV, Ad.-Tann., X, 450 f. (see Note 319).

248. I.e., in continuous proportion: $x:x^2 = x^2:x^3 = x^3:x^4 \ldots$

249. The unknown magnitudes are therefore called "magnitudines scalares" (*scalar* or ladder magnitudes). Their "rungs" or "degrees" (gradus) are:

Latus seu Radix	(Side or Root, i.e., x)
Quadratum	(Square, i.e., x^2)
Cubus	(Cube, i.e., x^3)
Quadrato-quadratum	(Squared-square, i.e., x^4)
Quadrato-cubus	(Squared-cube, i.e., x^5)
Cubo-cubus	(Cubed-cube, i.e., x^6)
Quadrato-quadrato-cubus	(Squared-squared-cube, i.e., x^7)
Quadrato-cubo-cubus	(Squared-cubed-cube, i.e., x^8)
Cubo-cubo-cubus	(Cubed-cubed-cube, i.e., x^9).

This is therefore not the "Anatolian" nomenclature (cf. Note 188) current up to that time and still used by Bombelli (*L'Algebra*, p. 6) and Gosselin, which, in modern terms, arises by multiplication of exponents and goes up to the ninth power, but rather a supplementation of Diophantine nomenclature in which the exponents arise by addition. Vieta adds at the end: "Et ea deinceps serie et methodo denominanda reliqua" (and thence the remaining ones may be designated by this series and method — Chap. II, 3, 4). Gosselin expressly disavows the Diophantine nomenclature; cf. *De Arte magna ...*, pp. 4^v ff., Chap. VI, "De numerorum nominibus" (On the names of numbers).

The genera of the known magnitudes (the "magnitudines comparatae" are ordered correspondingly:

Longitudo latitudove	(Length or breadth)
Planum	(Plane)
Solidum	(Solid)
Plano-planum	(Plane-plane)
Plano-solidum	(Plane-solid).

Solido-solidum
Plano-plano-solidum
Plano-solido-solidum
Solido-solido-solidum

"Et ea deinceps serie et methodo denominanda reliqua." On the series of "scalares" cf. Diophantus 4, 14–16, 5; on the series of "comparatae," Diophantus 2, 18–4, 7, where, however, for Vieta the expression πλευρά (side) serves as the first degree of the "scalares"; see also Note 177. In the fifteenth century there occur, incidentally, two cases in which the Diophantine nomenclature is used; cf. Tropfke, *Geschichte der Elementarmathematik*, II³. 137 f.

250. On the rules of multiplication and division, cf. Diophantus, "Def. IV" (8, 1–10) and "Def. X" (14, 1 f.). Besides these "praecepta," Vieta presents also "*leges Zeteticae*" in the *Isagoge* (Chap. V; see Appendix, P. 339), which refer to elementary operations with equations, namely, in particular, to "*antithesis*" (the transfer of one of the parts of one side of the equation to the other), to "*hypobibasm*" (the reduction of the degree of an equation by the division of all members by the species common to all of them), and to "*parabolism*" (the removal of the coefficient of the "potestas" — cf. Note 256 — or the conversion of the equation into the form of a proportion, of an "analogism"); cf. P. 160, and also Diophantus p. 14, 11 ff.

251. On a possible connection of this notation with the use of letters by Ramus, cf. Cantor, II², 632; 564. Vieta, it should be noted, has a predecessor in his use of letters for unknowns in Johannes Buteo, *Logistica, quae et Arithmetica vulgo dicitur*, 1559 (cf. Tropfke, *Geschichte der Elementarmathematik*, III², 34 and 136), and in Bonasoni (cf. Note 238, and the work of Bortolotti there cited, p. 5).

252. *Isagoge*, Chap. III, 1;
"Si magnitudo magnitudini additur, haec illi homogenea est." (If a magnitude is added to a magnitude, the former is homogeneous with the latter.)
"Si magnitudo magnitudini subducitur, haec illi homogenea est." (If a magnitude is subtracted from a magnitude, the former is homogeneous with the latter.)
"Si magnitudo in magnitudinem ducitur, quae fit huic et illi heterogenea est." (If a magnitude is multiplied with a magnitude, the product is heterogeneous in relation to both the former and the latter.)
"Si magnitudo magnitudini adplicatur, haec illi heterogenea est." (If a magnitude is divided by a magnitude, the latter is heterogeneous in relation to the former.)

"Quibus non attendisse causa fuit multae caliginis et caecutiei veterum Analystarum" (Not to have attended to these things was the cause of much confusion and blindness in the ancient analysts); cf. Note 346.

253. Theon (Hiller) p. 73, 18 f.: τὰ μὲν γὰρ ἀνομογενῆ πῶς ἔχει πρὸς ἄλληλά φησιν Ἄδραστος εἰδέναι ἀδύνατον. (For Adrastus says that it is impossible to know what relations non-homogeneous [magnitudes] have to one another.) Cf. Note 247.

254. Indirectly the "lex homogeneorum" of Vieta is identical with the statement of Adrastus insofar as every *proportion* whose ratios are ratios in the sense of the Euclidean Definition 3 of Book V can be converted into an equation which fulfills Vieta's demand. This may have been just what was in Vieta's mind when he appealed to Adrastus; in other words, he regards the theory of ratios and proportions from the outset in the light of the "theory of equations" understood as a "theory of calculation" (cf. Note 250, the concept of "analogism," and P. 160).

255. Vieta himself says in Chap. VIII, 17: "In numeris homogenea comparationum sunt unitates." (*In the case of numbers* the homogeneous elements of equations are *units*.)

256. The "rung" of this field is determined by the *highest* "rung" of the "scalar" magnitudes (i.e., by the highest exponent or degree of the equation), which is called the "*potestas*" (power). The lower "rungs" are called "gradus parodici ad potestatem" (rungs passed over in the ascent to the power — Chap. II, 5, numbered 8 in this edition). The highest rung of the "scalar" magnitudes corresponds to the genus of that known magnitude which appears as an independent member in the equation and accordingly represents the "unit" of that particular homogeneous field, the "homogeneum comparationis" (homogeneous element of the equation — Chap. VIII, 16); see Appendix, P. 349.

257. The term "symbolum," used for letter signs as well as for connective signs, originated with Vieta himself (cf. Chap. V, 5 [Appendix, P. 340] and IV,). Cf. further in the Dedicatory Letter, (Appendix, P. 317) the definition of the "circuitus" (circle of return) as the "verum et vere physicum symbolum perpetuitatis" (true and truly physical symbol of perpetuity) and, on the other hand, Note 226, end.

258. The expression "formula" in its present-day meaning appears, however, only much later; thus Adrianus Metius, for instance, in *Arithmeticae et geometriae practica* (1611) pp. 64 ff., still understands by "formula operationis" an algorism presented without explicatory text.

259. It was supported by the "Arabic" positional system of ciphers, which had been spreading in the West since the twelfth century and whose "sign" character is much more pronounced than that of the Greek or Roman notation. But it would be a mistake to attempt to understand the origin of the language of symbolic formalism as the final consequence of the introduction of the Arabic sign language. The acceptance of this sign language in the West *itself presupposes a gradual change in the understanding of number*, whose ultimate roots lie too deep for discussion in this study.

260. As far as the usage of mathematicians *before* Vieta is concerned, cf., above all, Chuquet (*Bullettino Boncompagni*, XIII [1880], p. 737), who says of "nombre" that it "est pris Icy *largement*" (is taken here [i.e., in the *Triparty*] in the large sense); by "nombres" we are now to understand not only one, and all the fractions ("tout nombre rout" — "every broken number"), but also those magnitudes marked by a "denomination" (namely an integral exponent), which in Chuquet replace the "cossic" numbers (where, remarkably enough, the ordinary *nombres* receive the *denomination* o, and the series of cossic magnitudes is extended to infinity); see furthermore Gosselin, *De Arte magna . . .*, preface p. ā iiijʳ: ". . . operae precium quoque fuit et numeros quos vocant Cossicos demonstrare (it was worth the effort also to exhibit the numbers which they call cossic): cf. on the other hand, Chap. V, p. 4ʳ: "Numerus in hac arte dicitur omnis quantitas, quam monadibus conflatum intelligimus absque ullo nomine [sc. "latus," "quadratum," etc.]." (In this art every quantity is called a number which, being composed of monads, we understand *apart from any name* [such as "root" or "square"].) "Fractions" and "irrational numbers" also come under the "numeri" thus defined; see Note 305; furthermore Raimarus Ursus as quoted in Note 247.

261 "Causam autem assignare videtur [sc. Diophantus], quod unitates absolutae, unitatis ipsius naturam sapiant. Quemadmodum ergo unitas in quamlibet numerum ducta, producit ipsum eundem numerum, sic et unitates in quamlibet speciem multiplicatae, eandem speciem gignunt." (Diophantus seems to assign as cause that the absolute units have caught something of the nature of unity itself. Therefore, just as when a unit is multiplied with any number whatever, it produces that very number itself, so also when units are multiplied with any species you please, there arises the same species — p. 7 in Bachet's ed. of Diophantus' *Arithmetic*.)

262. The term "coefficient" originated, as is well known, with Vieta himself, who speaks of the "magnitudo coefficiens," i.e., the known

magnitude of a definite "genus" which, when conjoined (adscita) with a "scalar" magnitude one degree lower than the unknown, i.e., "sub parodico ad potestatem gradu" (of an inferior rung passed over in the ascent to the power — cf. Note 256), turns that magnitude into a *species homogenea* (i.e., raises it to the dimension of the "potestas" — *Isagoge*, Chap. III, 6, given as 9 in this edition).

263. Tannery follows Bachet's example; cf. his Diophantus edition, Vol. I, p. 7, note: "nullo signo pro unitate in versione utemur" (we shall use no sign for unity in the translation).

264. Cf. Notes 36 and 81; also Part II, Pp. 130 f.

265. Cf. P. 147; Note 260; also Note 247.

266. In the *Universalium Inspectionum ad Canonem mathematicum liber singularis*, Vieta calls improper fractions "monades non purae" (e.g., pl. 5); later, fractions are called by him simply "numeri fracti" (e.g., *De numerosa potestatum ad Exegesin resolutione* [1600], p. 7=ed. van Schooten, p. 173). "Irrational" numbers he also calls — depending therein on ancient terminology — "numeri asymmetri," in contrast to the "numeri symmetri" (*ibid.*). With irrational numbers in mind he introduces in the *Canon*, pl. 14, under the title "Analogia generalior in numeris irrationalibus" ("The more general proportion of irrational numbers"), the following designations relevant to an approximation procedure: "numerus fere," "id est, dum excedit iustum" (*nearly* the number, that is, when it exceeds the exact number [i.e., a number in need of a negative addition]) and "numerus et amplius," "dum deficit a iusto" (a number and *more*, when it falls short of the exact number [i.e., a number in need of a positive addition]); cf. Aristotle, *Metaphysics Δ* 15, 1021 a 7: καὶ ἔτι (and yet some). But, like Diophantus, Vieta knows no negative numbers, if for no other reason than the fact that they cannot be represented in the geometric "exegetic," and that the parallelism of geometric and arithmetical analysis must always be preserved.

267. Above all, in the work *De numerosa potestatum ad Exegesin resolutione* (cf. Ritter, *Revue occidentale philosophique* ..., X, pp. 383 ff.; Cantor, II², 640 f.). In the eighteenth chapter of *Variorum de rebus mathematicis responsorum, liber VIII*, Vieta, in his computation of π, gives the first example of an infinite (and converging) series of factors (cf. Cantor, II², 594 f.; Ritter, *Revue occidentale philosophique* ..., X, pp. 410 f.).

268. Proclus himself is, to be sure, of the opinion that this μία καὶ ὅλη μαθηματική (one and whole mathematics — p. 44, 2 f.) only ἐφάπτεται τῆς τῶν πρώτων θεωρίας (touches on the study of first

things — p. 19, 24). It represents the σύνδεσμος τῶν μαθημάτων (common bond of objects of learning [i.e., mathematics]), which is spoken of in the *Epinomis* (991 E f.) and which includes not only, as Eratosthenes thought, the general theory of proportion, but precisely all "general" mathematical premisses, that is, all "axioms" of mathematics as well (p. 43, 22 ff.; cf. p. 195, 24 ff.). "Dialectic," as θριγχός (copingstone) of all the sciences (cf. Plato, *Republic* 534 E), is said to hold an even higher rank; the highest rank, finally, belongs to ὁ νοῦς αὐτός (intellect itself — pp. 42–44; cf. p. 9, 19–23).

It should be mentioned here that the ancient tradition itself always preserved the memory of an original connection between the general theory of proportion and arithmetic (or logistic). Thus Eutocius, in his commentary on the *Conics* of Apollonius, remarks on Bk. I, Prop. XI (J.L . Heiberg, *Apollinii Pergaei quae Graece exstant cum commentariis antiquis* [Leipzig, 1891–1893], II, 220): μὴ ταραττέτω δὲ τοὺς ἐντυγχάνοντας τὸ διὰ τῶν ἀριθμητικῶν δεδεῖχθαι τοῦτο· οἵ τε γὰρ παλαιοὶ κέχρηνται ταῖς τοιαύταις ἀποδείξεσι μαθηματικαῖς μᾶλλον οὔσαις ἢ ἀριθμητικαῖς διὰ τὰς ἀναλογίας, καὶ ὅτι τὸ ζητούμενον ἀριθμητικόν ἐστιν. λόγοι γὰρ καὶ πηλικότητες λόγων καὶ πολλαπλασιασμοὶ τοῖς ἀριθμοῖς πρώτως ὑπάρχουσι καὶ δι᾿ αὐτῶν τοῖς μεγέθεσι κατὰ τὸν εἰπόντα· ταῦτα γὰρ τὰ μαθήματα δοκοῦντι εἶμεν ἀδελφά. ("Let it not upset those who happen to notice it that this is demonstrated through numbers; for the ancients used such demonstrations rather as being mathematical [in the sense of involving a *general* theory] than [specifically] as arithmetical, *because of the proportions* and because the thing sought is [actually] arithmetical. *For ratios and sizes of ratios and multiplications primarily exist in numbers* and through these in magnitudes, as he says [namely Archytus; cf. Nicomachus 7, 1 f., Hoche, and Diels, I³, 331, 7 f.]; 'for these mathematical objects seem to be cognate.'" Cf. on this Notes 73 and 12; furthermore Sections 3–5 and Pp. 64 f.).

269. Petrus Ramus uses Proclus-Barocius extensively and with great lack of understanding for the composition of the third book of his *Scholae mathematicae*; cf. *Scholarum mathematicarum libri unus et triginta*, pp. 76 ff.

270. Cf. also Rule XVI (X, 455 f.): "... Advertendum est, ... nos ... hoc in loco non minus abstrahere ab ipsis numeris quam paulo ante a figuris Geometricis, vel quavis alia re." (We must note that in this place we abstract no less from the numbers themselves than a little before we did from geometric figures or anything else.) See Pp. 203 ff. The theory of ἀνάμνησις (recollection) mentioned by Proclus (p. 45, 2 ff.) seems echoed in Descartes' words: "prima quaedam

veritatum semina humanis ingeniis a natura insita . . ." (the first
seeds, as it were, of truth are located in the human mind by nature
— X, p. 376); but his words probably allude rather to the ἔμφυτοι
προλήψεις (innate notions) of the Stoics (cf. P. 198 and Note 316).
An additional source may have been Vanini, *De admirandis naturae
reginae deaeque mortalium arcanis libri IV* (*Four books on the marvelous
mysteries of nature, the queen and goddess of mortals*) [1616], p. 407 (cf. A.
Espinas, "Descartes de seize à vingt-neuf ans," *Séances et travaux de
l'académie des sciences morales et politiques, Compte rendu* [1907], pp.
114 f.), as well as Ficino and Pico della Mirandola (cf. M. Meier,
Descartes und die Renaissance [1914], pp. 26 ff.).

271. Cf. Descartes, *Oeuvres*, Ad.-Tann., XII, 23 and X, 156, note.

272. Cf. Note 275, end.

273. The following works by Cataldi are relevant here: *Trattato dell'-
Algebra proportionale*, etc. (1610); *Algebra discorsiva numerale, et lineale*,
etc. (1618); *Regola della quantità, o cosa di cosa* (1618); *Nuova Algebra
proportionale*, etc. (1619). Cataldi himself is under the influence of
Bombelli.

274. On Descartes' relationship to Stevin see below, Note 306. The
German master reckoner (Rechenmeister) Faulhaber of Ulm
(cf. Note 306, end; also Notes 313 and 319) and Peter Roth of
Nuremberg (cf. Descartes, *Oeuvres*, Ad.-Tann., X, 242) should also
be mentioned in this connection.

275. Cf. especially Descartes, *Oeuvres*, Ad.-Tann., I, 245, 479 f.: II, 82,
193, 524; IV, 228; V, 503 ff. See also Ch. Adam, *Vie et oeuvres de
Descartes* (Vol. XII of the Ad.-Tann. edition, pp. 211 ff.) which,
however, does not do justice to Vieta, and G. Milhaud, *Descartes
savant* (Paris, 1921), pp. 244 ff.

  The dispute concerning the relation of Descartes to Harriot (cf.
Ad.-Tann., II, 456 and 457 ff.) is empty insofar as Harriot himself is
in all essential matters dependent on Vieta, although this may no
longer have been known to Descartes' contemporaries. Walter
Warner, who published Harriot's work *Artis Analyticae Praxis, Ad
aequationes Algebraicas nova, expedita, et generali methodo resolvendas*,
etc. (*Practical handbook of the Analytic Art, for solving Algebraic
Equations by a new, convenient and general method*) in 1631 (see Note
329), does properly emphasize Vieta's importance in founding the
"analytical" art: "Dum vero ille [sc. Vieta] veteris Analytices,
restitutionem, quam sibi proposuit, serio molitus est, non tam eam
restitutam, quam proprijs inuentionibus auctam et exornatam,
tanquam nouam et suam, nobis tradidisse videtur." (Now while

Vieta earnestly toiled for that restoration of ancient analysis which he had set as a task for himself, he seems to have handed on to us not so much that art restored, as increased and adorned with discoveries especially pertaining to it, a new art and his own, as it were. Cf. P. 153 and Note 207.) But at the same time he concedes to Harriot a far larger role in the elaboration of this discipline than he really deserves. Harriot, who takes over Vieta's terminology *in toto*, distinguishes (p. 2, Def. 6–7; also p. 3, Def. 11), quite logically, only two parts of "analytic": "zetetic" (as *analysis* — or *resolutio* — *logica sive discursiva*) and "exegetic" (as *analysis* — or *resolutio* — *operativa*); cf. Pp. 166–168. Of poristic he gives a complicated explication, which is based on Vieta's corresponding definition (see Note 233) and which has, at any rate, the effect of excluding "poristic" from the systematic structure of "analytic" — wherewith the distinction between "theorem" and "problem" is now *finally* dropped (cf. P. 166). "Exegetic" itself is for Harriot, as for Vieta, twofold, except that the former understands the two possible sorts of solution, the "arithmetical" and the "geometrical" (cf. Pp. 167 ff.), on the basis of the two kinds of "logistice," namely the "numerosa" and the "speciosa" (Def. 9). Thus he lays the ground-work for a later misunderstanding of Vieta, according to which his species are supposed to signify "geometric" formations (see P. 171). Harriot is far removed from assigning to "logistice speciosa," and so to "analytic" in general, that fundamental position of a "general" discipline comprehending arithmetic and geometry which it has in Vieta. Furthermore, Harriot seems to be unaware of the Diophantine origin of Vieta's term "species," which he thinks must be derived "ex usu forensi" (from legal usage — p. 1, Def. 1), although we must remember that the source of certain parts of the text of the *Artis Analyticae Praxis* may be the editor, Warner. Wallis, incidentally, agrees with Harriot's opinion, cf. his *Algebra, Opera*, II (1693) p. 70. For Harriot the "Exegetice speciosa" remains therefore confined to linear and quadratic equations (cf. Note 238). He himself treats only "Exegetice numerosa" (p. 3, Def. 10): "Peculiaris est Exegetices huius [sc. numerosae] ars, regulis suis et praeceptis ad praxim instructa, quae in praesenti tractatu, qui totus Exegeticus, est, traduntur." (The art of this [numerical] exegetic is a special one, equipped with its own rules and precepts for practice, which are passed on in the present treatise *which is wholly exegetic*.) Of this "numerical exegetic" (which, strictly speaking, corresponds to Vieta's "rhetic," cf. Note 234) Warner says in his preface that Harriot gave it so new a shape that if Vieta created, as it were, a new analytic, by his "invention of exegetic" (Exegetices inventione; cf. Notes 232 and 233, end), then Harriot, by his "improvement of exegetic" (Exegetice recognitione), produced something like a "new Vieta," namely with respect to the sure and convenient practice of this art. In fact, by simplifying the notation, Harriot had turned "numerical exegetic," even *before* Descartes, into that truly useful "instrument of calculation" (an "ars Mathematicarum omnium instrumentaria" — "an art serving as a tool for all mathematics," as Warner says) it had never really been for Vieta. But the price of this advantage is a neglect of the fundamental issues; the symbolic technique of calculation is already a "matter of course" for Harriot. The simplification of exegetic required, above all, a change in the "logistice speciosa," namely the abandonment of the unwieldy designation of the species in favor of a symbolism employing letters exclusively, a "sola literalis notatio" (notation in letters only), as Warner says. But in introducing this convenient symbolism, Harriot gave up the foundation for Vieta's notation, namely the "lex homogeneorum" (cf. Pp. 173 f. and Pp. 216 ff.). From this point on, the symbolic technique becomes, taken by itself, opaque; Descartes' later "geometric" analysis, in particular, shares this opacity.

Harriot's work had been anticipated in 1630 by the five books *De resolutione et compositione mathematica* of Marino Ghetaldi, which, incidentally, also appeared as an "opus posthumum" although already announced in the *Variorum problematum collectio* of 1607 — Ghetaldi died in 1627; these books resemble Harriot's in depending on Vieta's "analytic," without sharing its presuppositions and fundamental principles. For Ghetaldi, as for Harriot, the "lex homogeneorum" is no longer a controlling factor. It is to be noted that Ghetaldi renders the distinction between "theoretical" and "problematic" analysis as defined by Pappus (cf. Note 218) more correctly than Vieta. But he too makes no actual use of the distinction. For him, as for Descartes, the only essential point is that the algebraic calculation indicates the *way* in which the solution (or the proof) is found in each case: "Etenim Resolutio procedens per species immutabiles, non autem per numeros mutationi, quacunque operatione tractentur, obnoxios, sua vestigia clara relinquit, per quae non est difficilis ad compositionem reditus." (For resolution, proceeding through *unchangeable species* and not through numbers which are *liable to change* by whatever operation they are subjected to, leaves clear traces of itself by which the return to the synthesis is not difficult — Cf. Note 235.)

276. Cf. the remark of his friend Beeckman which is of the same date as the *Regulae* (Ad.-Tann., X, 333): "Dicit idem [i.e., Descartes] se invenisse Algebram generalem . . . ." (Descartes says that he discovered *general Algebra*.)

277. In this connection van Schooten naturally refers to Descartes' *Discours de la méthode* (Ad.-Tann., VI, 551): ". . . Advertebam, illas [sc. particulares scientias Mathematicae], etiamsi circa diversa objecta versarentur, in hoc tamen omnes convenire, quod nihil aliud quam relationes sive proportiones quasdam, quae in iis reperiuntur, examinent." (I noticed that those special mathematical sciences, although they dealt with various objects, all coincided in this, that they considered nothing but certain *relations and proportions* which were found in these — French text, *ibid.*, p. 20.)

278. Cf. Pp. 168 f. See also Descartes' letter to Mersenne of March, 1637 (Ad.-Tann., I, 349): ". . . Ie ne mets pas Traité de la Methode, mais Discours de la methode, ce qui est le mesme que Preface ou Aduis touchant la Methode, pour monstrer que ie n'ay pas dessein de l'enseigner, mais seulement d'en parler. Car comme on peut voir de ce que i'en dis, elle consiste plus en Pratique qu'en Theorie. . . ." (I did not write "Treatise on Method," but "*Discourse on Method*," which is the same as "*Preface or Announcement concerning Method*," to show that it was not my plan to teach it but only to talk of it. For as one can see from what I have said, *it* [i.e., the Method] *consists more of practice than of theory* — cf. Ad.-Tann., I, 370.)

279. This can bε seen with the utmost clarity in the *Regulae*, whose date is about 1628. In the later writings of Descartes this basis of his entire system becomes increasingly veiled. In Rule IV (Ad.-Tann., X, 374) Descartes says in reference to the "alia disciplina" to be substituted for the "Mathematica vulgaris," namely the "Mathesis universalis": "Hanc omni alia nobis humanitus tradita cognitione potiorem, utpote aliarum omnium fontem, esse mihi persuadeo." (*I am persuaded that this is more powerful than all other knowledge passed on to us by human agency, inasmuch as it is the source of all others.*) In Rule XIV (*ibid.*, 439 f.) we read: ". . . In omni ratiocinatione per comparationem tantum veritatem praecise [cognoscimus]. Ver. gr., hic: omne A est B, omne B est C, ergo omne A est C; comparantur inter se quaesitum et datum, nempe A et C, secundum hoc quod utrumque sit B, etc. Sed quia, ut saepe jam monuimus, syllogismorum formae nihil juvant ad rerum veritatem percipiendam, proderit lectori, si illis plane rejectis, concipiat omnem omnino cognitionem, quae non habetur per simplicem et purum unius rei solitariae

intuitum haberi per comparationem duorum aut plurium inter se. Et quidem tota fere rationis humanae industria in hac operatione praeparanda consistit." (In all reasoning we know the truth precisely *only by comparison*. For example: all *A* is *B*, all *B* is *C*, therefore all *A* is *C*; here what is *sought* and what is *given* are compared with each other, namely *A* and *C* are compared in respect to each being *B*, etc. But because, as we have often warned, the forms of the syllogism are of no aid in perceiving the truth of things, it will be of profit to the reader to reject them outright and to conceive *of all knowledge* which is not gotten by simple and pure intuition of a single separate thing [but cf. Note 315] entirely *as being gotten by comparison of two or more things with each other. Indeed almost the whole task of human reason consists in preparing for this operation.*) For if the truth is not seen "immediately," the "assistance of an *art*" (artis adjumento) is needed to get to the truth. And this art consists essentially in transforming the "relations or proportions" (habitudines sive proportiones; cf. Rule XVIII, 462, 11: relatio sive habitudo; Rule VI, 385, 1 f.: proportiones sive habitudines) according to which a "nature common" (natura communis) in each case to "what is sought" (quaesitum) and "what is given" (datum) distributes itself over what is sought and what is given, so "that an equality between what is sought and something else which is known appears clearly" (ut aequalitas inter quaesitum, et aliquid quod sit cognitum, clare videatur — 440). Cf. 447: "... Velimus duntaxat proportiones quantumcumque involutas eo reducere, ut illud, quod est ignotum, aequale cuidam cognito reperiatur." (We wish simply to reduce proportions, however involved, to that point where that which is unknown is found to be equal to something known.) *This holds for all possible investigations and questions* (quaestiones omnes). Cf. Rule VI, 384 f.; also the beginning of Rule XVII; furthermore the designation of this art as "Algebra generalis," transmitted by Beeckman (X, 333), through which Descartes "ad perfectam Geometriae scientiam pervenit, imo qua ad omnem cognitionem humanam pervenire potest" (attained a perfected science of geometry, by which, what is more, it is possible to attain all human knowledge — X, 331–332); cf. the title planned for the *Discours de la méthode*, as Descartes gives it in a letter to Mersenne in March 1636 (I, 339): "Le projet d'une Science universelle qui puisse éleuer nostre nature à son plus haut degré de perfection. Plus la Dioptrique, les Meteores, et la Geometrie; où les plus curieuses Matieres que l'Autheur ait pû choisir, pour rendre preuue de la Science universelle qu'il propose, sont expliquées en telle sorte, que ceux mesmes qui n'ont point estudié les peuuent entendre." (The project of a universal science which is able to raise our nature to its highest degree of perfection. In addition the *Dioptrics*, the *Meteors* and the *Geometry*, where the most intriguing matters which the author was able to choose for the purpose of giving samples of the universal science which he proposes, are explicated in such a way that even those who have never studied can understand them.) Cf. also Note 308. In the *Regulae* Descartes is not yet completely clear concerning the type of symbolic representation to be used (see Section 12 B, especially Note 309 and Pp. 203 ff.).

With the passages quoted from the *Regulae* Nicolaus of Cusa, *De docta ignorantia* (*On learned ignorance*), Chap. 1, might be compared: "Omnes autem investigantes in comparatione praesuppositi certi proportionabiliter incertum iudicant; comparativa igitur est omnis inquisitio, medio proportionis utens." (Indeed all who conduct investigations judge what is uncertain by comparing it with pre-suppositions which are certain in the manner of a proportion. Therefore every inquiry is comparative and uses proportion [i.e., analogy] as a means.)

280. Cf. Ad. Trendelenburg, "Über Leibnizens Entwurf einer allgemeinen Characteristik," *Abhandlungen der Akademie der Wissenschaften zu Berlin*, 1856= *Historische Beiträge zur Philosophie* (1867), III, pp. 1–47.

281. A comprehensive review of his life and his works is given by H. Bosmans in the article "Stevin" of the Belgian *Biographie Nationale*, Vol. XXIII (1924), pp. 884–938.

282. Like Bombelli, by whose *Algebra* he was strongly influenced, he was especially expert in hydraulic techniques, and held an office appropriate to this interest under Maurice of Orange. Besides this he was quartermaster-general of the Netherlands army and comptroller of finances (cf. Bosmans, p. 888).

283. An exception is formed by the greater part of the *Arithmetique* written in French (1585) and, in particular, also by the *Appendice Algebraique* (1594), which contains a general rule for the solution of numerical equations of any degree (ed. Girard, I, pp. 88 f.). The *Appendice Algebraique* was later translated into Flemish by Stevin himself (cf. Bosmans, pp. 900 ff.). The writings of Stevin contained in the French edition of his complete works published in 1634 by Girard, can, textually speaking, be divided into three groups: (1) those written in or translated into French by Stevin himself, to which belongs, above all, the *Arithmetique*; (2) those translated by

Jean Tuning under the supervision of Stevin, namely a part of the
Memoires Mathematiques, intended for Maurice of Orange. (These
appeared for the first time in a Flemish, a French, and a Latin version
in 1608, but printing had already begun in 1605. Of the Latin
translation the larger part is by Willebrord Snell, a small part by
Hugo Grotius); (3) the works translated by the editor, Girard,
himself, especially *l'Art Ponderaire* (*The Art of Weighing*, i.e.,
Statics), furthermore the *Geographie* and the *Astronomie*. Girard's
translation is fairly reliable; his own additions are indicated as such;
Stevin's thought is, in any case, rendered exactly, although Girard
permits himself some abridgments (cf. Bosmans, pp. 889; 914 ff.;
924 ff.) We quote the *Arithmetique* according to the first edition of
1585 (*L'Arithmetique de Simon Stevin de Bruges*, Leyden), other works
according to Girard's edition of 1634. [An English translation: *The
Principal Works of Simon Stevin*, Vol. I, General Introduction,
Mechanics; Vol. II A, B, Mathematics; Vol. III, Astronomy and
Navigation — edited by E. Crone, E. J. Dijksterhuis, R. J. Forbes,
M. G. J. Minnaert, A. Pannekoek; Amsterdam, 1955–1961.]

284. The *Geographie* forms the second part of the *Cosmographie*, whose
first (trigonometric) part contains the *Doctrine des Triangles* and
whose third part contains the *Astronomie*. The third book of the
Astronomie is devoted to a presentation of the Copernican system.

285. On this point, as on many others, Stevin refers to communications
from Joseph Scaliger.

286. Cf. the third book of the *Geographie*, beginning, II, p. 137, col. 1.
Stevin says that he owes the matter of this book to a work by
Petrus Nonius, namely *De crepusculis liber unus* (1542) which, in
turn, represents an amplification of the corresponding work by
Alhazen (i.e., Ibn al-Haitham, *De crepusculis et nubium ascensionibus
liber unus* [*One book on dawn and dusk and the Risings of Clouds*] in
Opticae thesaurus Alhazeni Arabis libri septem, nunc primi editi . . . a
Federico Risnero [Basel, 1572], pp. 283–288).

287. "... nonobstant il est incertain qui il a esté, de quelle nation, et en
quel temps il a vescu, combien qu'il soit tenu fort ancien" (in spite of
the fact that it is unsettled who he was, of what nation, at what time
he lived, though he is held to be very ancient — p. 109, col. 2).
With reference to the art of making gold involved in alchemy,
Stevin says: "tel abus se devroit rapporter aux abuseurs de ceste
inespuisable science, mais non pas à elle" (such an abuse should be
laid at the door of the abusers of *that inexhaustible science* [i.e.,
alchemy], not at its door — *ibid.*) In Stevin's view this science —

actually he means chemistry — was destined to take the place of the traditional school physics.

288. None less than Hugo Grotius, who knew Stevin personally, shared his views on the "wise age." Stevin requested of him a kind of memorandum on the *testimonia* in favor of the existence of such an age. It is printed in the Girard edition, II, p. 110. Cf. also Bacon, *The New Organon*, I, Aphorism 122.

289. "C'est icy un poinct qui ne me fait gueres bien presumer de pouvoir quelque jour encor parvenir à ce siecle sage." (This is a point which scarcely makes me hopeful that some day we may yet arrive at this wise age — p. 112, col. 1.)

290. "Donc le Grec estant tel, que par iceluy on apprend les Mathematiques, doit estre tenu pour un bon langage." (Since Greek is a language such that one learns mathematics by means of it, it must be held to be a good language — p. 113, col. 2.)

291. For this reason Stevin taught mathematics in his own language in Leyden, where he held the title "Professor in de Duytsche Mathematik" — Professor of Dutch Mathematics (cf. Note 306).

292. His *Disme* (*The Art of Tenths*), in which he expounds calculation with decimal fractions and demands furthermore that the decimal system be used for all measures, rests ultimately on the notion of the universal applicability of the "progression decuple," the decimal progression, which forms the basis of the Arabic positional system and which Stevin emphasizes again and again (cf. II, p. 108, col. 2; *Arithmetique*, p. 139). Incidentally, Vieta too (following Regiomontanus who, in turn, depends on Peurbach) uses decimal fractions in his *Canon Mathematicus* (cf. Ritter, *Revue occidentale philosophique* . . . , X, pp. 251; 361; see also Tropfke, *Geschichte der Elementarmathematik*, I³, 172–177). But it was due to Stevin that calculation with decimal fractions came into general use.

293. At the end of his *Arithmetique*, pp. 202 f., Stevin summarizes those assertions made in his work which cannot be reconciled with the usual views under the title of "Theses mathematiques". Thesis 1 is that mentioned in the text. The "Theses" were to be more precisely demonstrated in a special work — cf. *Arithmetique*, pp. 203 and 5ᵛ.

294. Which therefore cannot have anything to do with the concept of *hyle* in Domninus (cf. Part I, Pp. 32 f.).

295. *Arithmetique*, p. 3ʳ: "O heure infortunée en laquelle fuit premierement produicte ceste definition du principe du nombre! O cause de difficulté et d'obscurité de ce qui en la Nature est facile et clair!" (O

unfortunate hour in which that definition of the beginning of number was first brought forth! O cause of difficulty and obscurity of that which is easy and clear in Nature!)

296. Stevin must here be thinking not only of the old definition of the monad as a στιγμὴ ἄθετος (point without position, cf., e.g., Aristotle, *Metaphysics* M 8, 1084 b 25–27), but also of the representation of the figurate numbers, ∴, ∷, ⫶, etc. But, characteristically, he says in respect to this Greek mode of presentation: "et estoient iceux poincts mis en usage entre leurs chiffres" (*and these points were used* [by the Greeks] *among their ciphers* — p. 108, col. 2). Now as far as the sign "." for o is concerned, it is, indeed, to be found among the Arabs, but demonstrably only from the ninth century A.D. on; also among the Indians in the so-called "Book of Calculation" of Bakhṣhāli, which is likewise to be dated to a very late time (after the seventh century A.D., at any rate, and perhaps as late as the twelfth). The sign o which occurs in the "Table of Chords" in Ptolemy is probably only an abbreviation of οὐδέν (nothing); cf. Tropfke, *Geschichte der Elementarmathematik*, I³, 29; 40 f.; 22 f.; 18; 25. For a general discussion of the question of the origin of zero, cf. Neugebauer, *Vorlesungen über die Geschichte der antiken mathematischen Wissenschaften*, Vol. I, *Vorgriechische Mathematik* (Berlin, 1934), p. 42; also p. 5.

297. In the *Arithmetique* (Def. III, p. 5ᵛ; cf. also p. 3ᵛ) Stevin, deferring to traditional terminology, still calls zero "commencement de nombre," but says explicitly in his *Geographie*: "et puis que o est appellé au siecle sage *poinct* nous luy donnerons aussi ce nom *poinct de nombre*, en difference du *poinct Geometrien*, et delaisserons ce premier nom *commencement*, que nous avons eu en usage jusques à present" (and since o was called *point* in the wise age, we too will give it the name *number point*, as distinguished from *geometric point*, and we will leave off using that previous name "*beginning*," which we have been using up till now — p. 108, col. 2).

Whereas, for instance, Peurbach, at the beginning of his *Algorithmus Magistri Georgij Peurbachij in integris* (*The Algorism* [i.e., reckoning scheme] *for whole numbers of Master George Peurbach*), still says explicitly: "Unitas autem non est numerus: sed principium numeri. Unde ipsa habet se in Arithmetica ad numerum sicut punctum in Geometria ad magnitudinem." (The unit, however, is not a number, but the beginning of number. Wherefore it has that relation to number in arithmetic which a point has to magnitude in geometry — cf. Cantor, II², 180 f.) Joh. Buteo, *Logistica, quae et Arithmetica vulgo dicitur* (1559), p. 8, already says: "Et quamvis monas non sit numerus, in omni tamen Logistica ratione vim et effectum parem numeris obtinet." (And although the monad is not a number, yet in all calculational contexts it gains a force and effect equal to numbers.) Cf. also Note 217.

298. This is also the reason for the well-known shift in meaning of the word "cifra" or "chiffre," which was borrowed from the Arabic; signifying at first only zero, it gradually comes to be the common designation for all ten "ciphers" (cf. Tropfke, *Geschichte der Elementarmathematik*, I³, 9 ff.; 14 ff.).

299. The symbolic reinterpretation of geometric formations is herewith accomplished; however, it becomes clearly visible only in Descartes (cf. this Section, B).

300. *Arithmetique* p. 3ᵛ f.: "Comme la ligne *AB* ne se peut augmenter par addition du poinct *C*, ainsi ne se peut le nombre *D* 6, augmenter par l'addition de *E* 0, car aioustant 0 a 6 ils ne sont ensemble que 6.

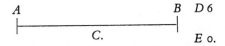

Mais si l'on concede que *AB* soit prolongée iusques au poinct *C*, ainsi que *AC* soit une continue ligne, alors *AB* s'augmente par l'aide du poinct *C*; Et semblablement si l'on concede que *D* 6, soit prolongé iusques en *E* 0, ainsi que DE 60 soit un continue numbre faisant soixante, alors *D* 6 s'augmente par l'aide du nul 0.

(As the line *AB* cannot be augmented by the addition of a point *C*, so the number *D* 6 cannot be augmented by the addition of *E* 0, for if 0 forms a sum with 6 they are together only six.

But if one admits that *AB* is prolonged to the point *C*, so that *AC* is a continuous line, then *AB* is augmented by means of the point *C*. And similarly, if one admits that *D* 6 is prolonged through the number E 0, so that *DE* 60 is a continuous number making 60, then *D* 6 is augmented by means of nought 0.)

It can be seen that the *ciphers as such* have taken the place of the traditional lengths of measurement. This argument is a telling one only if by 6 and 60 are understood the decimal fractions 0.6 and 0.60, which Stevin writes $\overset{①}{6}$ and $\overset{①}{6}\,\overset{②}{0}$.

301. Def. VII: "Nombre entier est unité, ou composée multitude d'uni-
     tez." (*A whole number* is a unit or a multitude composed of units.)
     Def. X: "Nombre rompu, est partie ou parties de nombre entier."
     (A *fractional number* is a part or parts of a whole number.) Cf.
     Euclid VII, Defs. 3 and 4.

302. Cf. the opinion of Peletier, *De occulta parte numerorum, quam
     Algebram vocant, Libri duo* (1560), Bk. II, Chap. II, entitled "Numeri
     irrationales sintne numeri, an non et cuiusmodi sint" ("Whether
     irrational numbers are numbers or not and of what sort they are").
     What they really represent, he says, "id tanquam in perpetuis
     tenebris delitescit" (this must lie hidden, as it were, in perpetual
     darkness). They are, in any case, "something," and it is certain that
     their *use is necessary* (necessarium usum), especially "in laying off
     continuous magnitudes" (praesertim in Continuorum dimensioni-
     bus). They are, just like genuine "absolute" numbers, subject to
     "rules" (praeceptiones). Their relation to the "absolute" numbers
     may be compared with that of the animals to man: "Habent igitur
     numeri Irrationales cum Absolutis obscuram quandam mutemque
     communicationem, non secus quam cum hominibus, Bruta: quae
     praeter id quod sentiunt, suo etiam modo ratiocinantur." (Irrational
     numbers, then, have a certain obscure and mute communication
     with absolute numbers, not differently from that which brutes, who
     besides having sense impressions, do, in their own way, even reason,
     have with men.) All in all, they are inexplicable (inexplicabiles) and
     have only a kind of shadow existence. They must not be counted
     among the numbers, but their being must rather be understood as
     altogether contained in their "designation" (appellatio).

303. Proof (cf. Pp. 191–192): "La partie est de la mesme matiere que son
     entier; Racine de 8 est partie de son quarré 8: Doncques $\sqrt{8}$ est de
     la mesme matiere que 8: Mais la matiere de 8 est nombre; Doncques
     la matiere de $\sqrt{8}$ est nombre: Et par consequent $\sqrt{8}$ est nombre."
     (The part is of the same material as the whole; the root of 8 is part
     of its square 8; therefore $\sqrt{8}$ is of the same material as 8. But the
     material of 8 is number; therefore the material of $\sqrt{8}$ is number:
     and consequently $\sqrt{8}$ is number — p. 31.)
     Albert Girard, Stevin's pupil, goes even further. In his *Invention
     nouvelle en l'algèbre* (1629), he says (pp. 13 f.): "Notez qu'on appelle
     un nombre tant les radicaux simples, comme est $\sqrt{2}$, ou $\sqrt{5071}$, que
     les multinomes, commes les binomes $2+\sqrt{5}$, item $7-\sqrt{48}$, item
     $\sqrt{26}-5$, comme les trinomes $4+\sqrt{2}-\sqrt{17}$, et autres multinomes,
     car ce qui lié par les signes soit+ soit − ne font qu'un nombre."
     (Note that we call *a number* all the simple radicals, like $\sqrt{2}$ or $\sqrt{5071}$,

the polynomials, like the binomials $2 + \sqrt{5}$, or $7 - \sqrt{48}$, or $\sqrt{26} - 5$, the trinomials $4 + \sqrt{2} - \sqrt{17}$, and other polynomials; for *whatever is joined by the signs $+$ or $-$ cannot but make a number*.) However, Def. XXVI of Stevin's *Arithmetique* had already stated: "Multinomie algebraique est un nombre consistant de plusieurs diverses quantitez." (An algebraic polynomial is a *number* consisting of several different quantities.)

304. This mode of designation, which represents a modification of that of Bombelli, and which can perhaps be traced back to Chuquet (cf. Note 238, Note 260), is used extensively by Stevin in his very free translation of the four first books of Diophantus (*Arithmetique*, pp. 431–642).

305. As early as 1484, Chuquet says in the beginning of his *Triparty* (*Bullettino Boncompagni*, XIII [1880], p. 593): "Nombrer si est le nombre en lentendement conceu par figures communes artifitielement representer ou de paroles perceptiblement exprimer." (To count means to represent artificially by common signs or to express perceptibly by words the number conceived in the understanding.) In the work by R. Gemma Frisius (whom Stevin, incidentally, mentions explicitly in another connection, cf. Cantor, II², 614) called *Arithmeticae practicae methodus facilis*, which first appeared in 1540, later had numerous republications (by Peletier, among others), and which was translated into French by Forcadel in 1582, we read in the beginning of the "Pars prima": "Numerare, est cuiusvis propositi numeri valorem exprimere, atque etiam quemcunque datum numerum suis characteribus adsignare." (To number is to express the value of any number proposed and so to refer any number given to its proper sign.) Gosselin (*De Arte magna . . .*, p. 2^r f.) in 1577 formulates the same in the following way: "Numerare, est quamcunque numeri qualitatem cum aliquo characteris vel figurae genere representare." (To number is to represent any quality of a number by some kind of sign or figure.) Among the Greeks this was done by points and dashes (puncta vel lineolae), but it may also, as among the Hebrews and the Romans, be done by means of letters. Only "the Arabs thought of and invented more convenient signs and an easier way of counting" (. . . excogitarunt et invenerunt Arabes faciliores characteres, facilioremque numerandi viam . . .). Gosselin gives the following classification of "numbers" (p. 4^r): "integer numerus et absolutus" (whole and absolute number, e.g., 8, 9, 5), "particula numeri" (part of a number, e.g., $\frac{1}{2}$, $\frac{1}{3}$), "latus numeri" (side of a number, e.g., latus $8 = 2$, latus $9 = 3$, i.e., rational roots) and "surdus numerus" (surd number, e.g., latus 7, latus 5,

latus $\frac{1}{2}$). Clavius also, in the *Epitome Arithmeticae practicae* (1584), says (p. 6): "Numeratio est cuiusvis numeri propositi per proprios characteres, ac figuras descriptio, atque expressio." (Denumeration is the description and expression of any number proposed by means of proper signs or figures.)

306. The very few remarks about Stevin which are to be found in the writings of Descartes himself concern his *Statics* (cf. Note 283): Ad.-Tann., II, 247, also 252; IV, 696. Furthermore Stevin's expertness in the subject of harmonics is impugned (I, 331); Descartes, however, adds explicitly that Stevin "ne lassoit pas d'estre habile en autre chose" (did not lack skill in other matters). On the other hand, it is possible, even likely, that Descartes saw Stevin in the last years of his life, and he may even have been instructed by him; for it was during those very years, from 1618 to 1619, that Descartes, then still "un homme qui ne sçait que très peu de chose" (a man who knows only very little — I, 24, 15), served under Maurice of Orange who had his officers instructed in the building of fortifications, drawing, etc., by expert teachers, among whom Stevin must certainly be numbered (cf. Note 282). Thus Descartes, in a letter to Beeckman on January 24, 1619, (X, 152, 2 f.), explicitly says that at the moment he is occupying himself "with drawing, military architecture and, above all, with the Flemish language" (in Pictura, Architectura militari, et praecipue sermone Belgico); here we must recall the important role Stevin assigned to Flemish in the advancement of the sciences (see Pp. 189 f.; cf. G. Cohen, *Ecrivains français en Hollande dans la première moitié du XVIIᵉ siècle* [Paris, 1920] pp. 372 f. and 381). Stevin probably taught in Flemish, just as he had at one time held the title of "Professor in de Duytsche Mathematik" at Leyden (*ibid.*, p. 381). But even if the conjecture concerning a personal meeting between Descartes and Stevin were incorrect, there could nevertheless be no doubt that Descartes must have come across traces of Stevin everywhere in Holland: Stevin was acquainted with Scaliger and Grotius, who were then the most famous scholars in Holland (cf. Notes 285 and 288); Constantin Huygens as well as Golius, with whom Descartes corresponded, were in communication with Girard, the most important pupil and the editor of Stevin, who published an algebraic work of his own in 1629 (*Invention nouvelle en l'algèbre*, cf. Cantor, II², 787 ff.; the title, incidentally, continues: "tant pour la solution des equations, que pour recognoistre le nombre des solutions qu'elles reçoivent, avec plusieurs choses qui sont necessaires à la perfection de ceste divine science" — "for the solution of equations as well as the discovery

of the number of solutions which they permit, together with many things which are necessary for the perfection of that *divine science*"; cf. P. 181). Girard, like Descartes, was also interested in problems of mechanics, optics, and music theory (cf. Note 283, and Cohen, *Écrivains français* . . . , p. 341). Snell recommends Stevin to Beeckman, Descartes' friend, probably even before 1611 (cf. Ad.-Tann., X, 29 and Note 283); in 1618 this same Beeckman addresses to Descartes a question based on Stevin's *Statics* (cf. Ad.-Tann., X, 228; see P. Duhem, *Les origines de la statique* [Paris, 1905], I, pp. 280–282, and Milhaud, *Descartes savant* [1921] pp. 35 f.). We may therefore assume that Descartes "knew" Stevin's *Arithmetique* in the sense in which Descartes was wont to "know" and use books. *Here he might find, above all, a complete and effective assimilation of the "numerical" and the "geometric" realm* (see Pp. 194 ff.). Thus his understanding of "powers" (cf. Rule XVI, Ad.-Tann., X, 456 f.) coincides to a large extent with that of Stevin, who is, to be sure, in turn dependent on Bombelli (see Note 304); cf., however, also Clavius, *Algebra* (1608) chap. II, pp. 9–10, whose presentation especially as far as the use of the concept of "exponentes" and, "exponere" is concerned, depends in this point on that of Stifel (cf. Tropfke, *Geschichte der Elementarmathematik*, II³, 151). We might mention in addition that the master reckoner Faulhaber, whom Descartes met in Ulm in 1619–1620, cites as his sources in his *Neuer arithmetischer Wegweiser* (1617) besides some more or less insignificant German and Swiss masters, only Gemma Frisius and Stevin.

307. Cf. P. 183 and Note 245; furthermore Rule VII, beginning. Cf. on this whole section, above all, L. Liard, *Descartes* (1882) chap. I, "La mathématique universelle," especially pp. 44 ff. In the following section the *Regulae* are quoted according to the Adam-Tannery edition, Vol. X.

308. The first indication of this conception is the letter of March 26, 1619, to Beeckman, where a "scientia penitus nova" (a completely new science) is mentioned, "qua generaliter solvi possint quaestiones omnes, quae in quolibet genere quantitatis tam continuae quam discretae, possunt proponi" (by which all problems which can be proposed in terms of any kind of quantity, whether continuous or discrete, can be generally solved); Beeckman makes the marginal note: "Ars generalis ad omnes quaestiones solvenda quaesita" (sought: a universal art for the solution of all problems — X, 156 f.). Cf. also the following remark of Beeckman (X, 52): "Dicit [Picto, i.e., Descartes] . . . se nunquam hominem [or: neminem] reperisse, praeter me, qui hoc modo, quo ego gaudeo, studendi utatur,

accurateque cum Mathematica Physicam jungat. Neque etiam ego, praeter illum, nemini locutus sum hujusmodi studij." (Descartes says . . . that he has never found a man [or: anyone], except me, who employs that mode of study in which I take pleasure, and who *neatly brings together physics with mathematics.* Nor have I on my part spoken of this kind of study to anyone but him.) It does look as if the "scientia mirabilis," the "wonderful science," whose foundations Descartes declares he discovered on November 10, 1619 (X, 179 and 216), is precisely the *mathesis universalis now conceived as applicable to physics.* (On the "inventum mirabile," the "wonderful discovery" of November 11, 1620, referred to in X, 179, on the other hand, cf. Milhaud, *Descartes savant* [1921], Chap. IV, pp. 89 ff.). *This* was the issue which compelled Descartes to develop his metaphysics, a metaphysics in which, to be sure, the actual points of departure of his "system" came to be increasingly consigned to oblivion.

309. In the *Regulae* representation by straight lines, as also by points, is only a special case; cf. Rule XIV, end; Rule XV, and Rule XVIII; furthermore, Pp. 203 ff.; cf. also Beeckman's note: "Algebrae Des Cartes specimen quoddam" (an example of Descartes' algebra — Ad.-Tann., X, 333–335); Milhaud, *Descartes savant*, p. 70, note 71. On the *Regulae* in general, see *ibid.*, Chap. III, 1.

310. In general Descartes' "method" grows out of a desire to justify the *place* which he assigns to algebra. The point of view of "methodical" cognition is therefore secondary for the *original* identification of the "general" mathematical object with extension having figure. But since everything depends on the *justification* of this identification, the "method" gradually gains a more and more central significance, *while its rules are borrowed from the "mathesis universalis" itself*; thus the road of "inventio," which the "mathesis universalis" understood as "general algebra" follows, is discovered to be the way of cognition generally most appropriate to the human understanding (cf. Note 279). In this sense the "Regulae ad directionem ingenii" ("Rules for the direction of the mind") are indeed identical both with the "rules" of the "mathesis universalis" and with those of the "method" as such (cf. Milhaud, *Descartes savant*, p. 69; cf. also *Cogitationes privatae*, X, 217: "Dicta sapientum ad paucissimas quasdam regulas generales possunt reduci." (The pronouncements of the wise can be reduced to a very few general rules.)

311. Cf. J. von Arnim, *Stoicorum veterum fragmenta* (Leipzig, 1921), I, frs. 59 and 66; II, frag. 56. On this, see A. F. Bonhöffer, *Epictet und die Stoa, Untersuchungen zur stoischen Philosophie* (Stuttgart, 1890), pp.

161, 163 f., 178 f., also pp. 184 ff.; furthermore O. Rieth, "Grund-begriffe der stoischen Ethik," *Problemata*, fasc. 9, 1933, p. 114 and M. Meier, *Descartes und die Renaissance* (1914), p. 63. See also Notes 316 and 326.

312. Cf. *Principia philosophiae*, II, 8, also I, 55.

313. Descartes is here thinking of contemporary works like those of the "Rosicrucians," for instance: Joh. Faulhaber, *Ansa inauditae et mirabilis novae artis*, etc. (*Clue to an incredible and wonderful new art*), 1618, German version, 1613; *Numerus figuratus, sive Arithmetica analytica arte mirabili inaudita nova constans*, etc. (*Figurate number, or analytic arithmetic, derived by a wonderful, incredible, new art*), 1614; *Mysterium Arithmeticum*, 1615, anonymous; *Miracula Arithmetica*, 1622, in German (another "continuation" of the *Arithmetischer Wegweiser* [*Arithmetic guide*] which first appeared in 1615), etc.; the Latin text of the writings of Faulhaber was probably due to Joh. Remmelin, cf. A. G. Kästner, *Geschichte der Mathematik* (Göttingen, 1799), III, pp. 29–34, 111–152; furthermore Ad.-Tann., X, 252–255. But he may also be thinking of the somewhat older "Arithmologies" of Joach. Camerarius the Younger, and Johannes Lauterbach (cf. Christianus Primkius, α. ω. *Arithmologia Sacro-Profana* [Liegnitz, 1659]; see furthermore Part I, P. 66), as well as of the work *Numerorum Mysteria* (first edition 1584 under the title *Mysticae numerorum significationis liber*) by Petrus Bungus (Pietro Bongo), which circulated in many editions, the last in 1618.

314. Cf. the *Commentarii collegii conimbricensis* (1592) used by Descartes at La Flèche, namely *Physics* I, 1, 4 (quoted after É. Gilson, *Index scolastico-cartésien* [Paris, 1912], p. 167): ". . . Res Mathematicae vel cogitatione abjunguntur a materia sensibili tantum, vel etiam ab intelligibili. Si priori modo, ad Geometram pertinent; si posteriori ad Arithmeticum." (Mathematical things are separated by thought either from their sensible material only, or *also from their intelligible material*. If they come about in the first way, they belong to geom-etry, if in the latter, *to arithmetic*.) Thus, while the ancient and medieval tradition assigned a "materia intelligibilis" (a ὕλη νοητή or φανταστή) to the object of arithmetic as well, namely the "pure" units as such (cf. Note 45 and P. 104), here a special level of "ab-straction," one on which even the *res numeratae*, i.e., the monads counted, the units, may be "disregarded," is ascribed to arithmetic; this is clearly in the spirit of the new concept of "number" and in line with the self-interpretation of the contemporary "reckoners" (especially the "algebraists") mentioned by Descartes. Descartes and, as far as we can see, only Descartes, struggles to fix the exact

meaning of such an "abstraction" — although this holds only for his thinking in the *Regulae*.

315. The *actio intellectus* which consists in "grasping" such a *res simplex* is called by Descartes an *intuitus*, literally, an insight: "mentis purae et attentae non dubius conceptus" (the unwavering grasp of a clear and attentive mind — Rule III, 368, 18). On this occasion he expressly points out that the *intuitus* can, and even must, extend also to "discursive" matter, i.e., also to "relations" and "proportions" (*ibid.*, 369).

316. When the relation between *imaginatio* and *intellectus purus* in Descartes is considered in terms of the tradition, his *imaginatio* must be assigned the role of Aristotle's νοῦς παθητικός and φθαρτός (passive and destructible intellect) in contrast to the νοῦς ποιητικός (active intellect), the latter alone being described by Aristotle as ἀΐδιος (everlasting), namely as χωριστὸς καὶ ἀπαθὴς καὶ ἀμιγής (separable and impassible and unmixed — *On the Soul* Γ 5); yet, on the other hand, Descartes' *imaginatio* derives also, if only indirectly, from Aristotle's *phantasia*, since it is essentially a version of the Stoic concept of *phantasia* which depends on that of Aristotle (*On the Soul* Γ 3, and *On Memory*), cf. J. Freudenthal, *Über den Begriff des Wortes ΦΑΝΤΑΣΙΑ bei Aristoteles* (Göttingen, 1863), especially pp. 29 f. On the influence of the Stoa on Descartes, cf. Brochard, "Descartes stoicien," *Revue philosophique*, IX (1880) and, above all, Dilthey, *Gesammelte Schriften* (Berlin-Leipzig, 1914), II, 294–296; on the general significance of the Stoa for the sixteenth and seventeenth century see *ibid.* pp. 93, 153 ff., 174 ff., 181, 255 ff., 261 ff., 279 ff., 285 ff., 441 ff., 486 ff. As far as the *Regulae* are concerned, they are, of course, directly in the tradition of the "Studium bonae mentis" (Ad.-Tann., X, 191–204) — their general theme is precisely the "bona mens" (well-judging mind), also called by Descartes "sapientia universalis" (universal wisdom, cf. Rule I, 360, 19 f.), a concept which Descartes apparently borrowed from Justus Lipsius, the great mediator of Stoic ideas who, in turn, probably owed it to Seneca. On this question see É. Gilson, *Études sur la role de la pensée médiévale dans la formation du système cartésien* (Paris, 1930) pp. 265 f.; cf. also F. A. Trendelenburg, *Historische Beiträge zur Philosophie* (Berlin, 1867) III, p. 396. The concept of the "bona mens" is, incidentally, closely related to, if not identical, with the "recta ratio" (right reason); for instance, Epictetus, *Dissertationes* IV, 8, 12, says: τί τέλος [sc. τοῦ φιλοσόφου]; . . . τὸ ὀρθὸν ἔχειν λόγον. (What is the end of the philosopher? . . . to possess *right reason*.) Cf. II, 8, 2,

where the οὐσία θεοῦ (being of god) is characterized as *nous*, *episteme*, and *logos orthos* (right reason); on the other hand, the concept of the "recta ratio" serves as a characteristic sign of "*ars*"; thus we read in Thomas Aquinas, *Summa Theologica* II, quaest. 57, art. 3, that: "Ars nihil aliud est quam ratio recta aliquorum operum faciendorum." (Art is nothing but sound judgment about any works to be carried out.)

From the point of view of tradition, Descartes' central difficulty is to be characterized as follows: He understands the *intellectus purus*, in effect, in the sense of the Stoic ἡγημονικόν (leading or guiding faculty) — and thus "somatically," but *at the same time* he interprets it in the Peripatetic-Thomistic sense as νοῦς ποιητικός (active intellect) — and thus as an "extraworldly" faculty. On the first, see especially Rule XII, 415 f., where the *vis cognoscens* (cognitive power), according as it "applies itself" (se applicat) to different faculties, carries out different "functions" and is correspondingly called by different names, namely "intellectus purus," "imaginatio," "memoria," "sensus"; see also Sextus, *Against the Mathematicians* VII, 307: . . . ἡ αὐτὴ δύναμις κατ' ἄλλο μέν ἐστι νοῦς, κατ' ἄλλο δὲ αἴσθησις (the same faculty is in one respect intellect, in another sensation), quoted from Bonhöffer, *Epictet und die Stoa* (1890), p. 99 (a general reference to this work should be made in this context); see furthermore Bonhöffer, "Zur stoischen Psychologie," *Philologus*, LIV (1895), p. 416; for the conception of the "aiding" — or better, "serving" — capacity of the different "parts of the soul" as contrasted with the *intellectus purus* as the "leading" and "ruling" faculty (Rule XII, beginning) compare also Stobaeus, *Eclogue* I, 41, 25 (Meineke, 252, 1–3): . . . τὸ ἡγεμονικὸν ὡς ἂν ἄρχοντος χώραν ἔχειν ὑπετίθεντο [sc. Stoici], τὰ δὲ ἄλλα μέρη ἐν ὑπηρέτου τάξει ἀπεδίδοσαν. (. . . The Stoics suppose the leading faculty to have the place, as it were, of a ruler, the other parts they assign *to the order of servant*. . . .) Cf. also Plato, *Philebus* 27 A. On the intellect as ruling faculty, cf., for example, Gilson, *Index scolastico-cartésien*, p. 95, no. 160. This sort of intermingling of Aristotelian (as well as Platonic) with Stoic doctrine occurs already in some of the Stoics themselves, e.g., in Marcus Aurelius (cf. Bonhöffer, *Epictet und die Stoa*, pp. 41 and 32) and also in Galen, whom Descartes certainly used, whether directly, or indirectly through Telesio. On this whole matter see Dilthey, *Gesammelte Schriften*, II, pp. 290 ff.; thus, in particular, assertions such as that the origin of the nerves lies in the brain or that the transfer of "impressions" from the *sensus externus* to the *sensus communis* or *phantasia* — and therefore also the reverse sequence! — occurs "in a moment" (Rule XII, 413 f.; see also *Dioptrics*, chap. IV, 5, Ad.-Tann., VI, 598; French text, *ibid.*, 111; cf. Note 326) go back to Galen, e.g., *On the Teachings of Hippocrates and Plato*, passim, especially pp. 208 f., 644 f., 656 (ed. I. Mueller [1874]); when he was writing the *Regulae*, Descartes apparently did not yet know Vesalius, since he devoted himself to the study of anatomy only from 1629 on (see Ad.-Tann., I, 102, 18 and II, 525); cf. furthermore Galileo, *Dialogue on the Two Great Systems of the World*, Second Day, beginning (Edizione nazionale, VII, 133 f.). In this connection it is of great significance that in the *Regulae* Descartes calls the *vis cognoscens* — which is at the same time the *bona mens* — "*in-genium*": "proprie autem ingenium appellatur" (but it is properly called "ingenium" — Rule XII, 416, 8); cf. also the title of the work "Regulae ad directionem *ingenii.*" For its name can mean that it is either itself the *generator* of "ideas" in the "imagination" or that it is, conversely, *impregnated* by the "ideas" already present in the imagination, namely those brought about by the senses: "cum modo ideas in phantasia novas format, modo jam factis incumbit" (when it either *forms* new ideas in the imagination or *hatches* those already formed). Thus the interaction of *intellectus* and *imaginatio* is here really interpreted on the model of Aristotelian *physis* (nature), the crucial point being that the *intellectus* is assigned the role not only of the *eidos* (form) but also of the *hyle* (material): "interdum patitur, interdum agit" (sometimes it is affected, sometimes it acts — 415, 23 f.). In other words, the "extraworldliness" of the *intellectus* does *not determine its mode of being* — the *intellectus*, the "res cogitans" (knowing thing), "*is*" in no other sense than the "body," the "res extensa" (extended thing). What characterizes both in the same way is their "unrelatedness." Because of this it is, on the one hand, impossible to understand how they can "come together," while, on the other, they *are* for Descartes now and always already *together*, namely on the level of "corporeality."

317. Cf. Baillet, *La vie de Monsieur Descartes* (1691) II, 486–487 (Ad.-Tann., X, 202 f.): "... M. Descartes appelloit les études d'imagination, *méditation*; et celles d'entendement, *contemplation*. C'est là qu'il rapportoit toutes les sciences, mais principalement celles qu'il appelloit *cardinales* ou *originales*, comme la vraye Philosophie, qui dépend de l'entendement, et la vraye Mathématique, qui dépend de l'imagination." (M. Descartes called exercises of the imagination "meditation," and those of the understanding "contemplation." It is to this distinction that he referred all the sciences, but principally

those which he called "cardinal" or "original," such as true philosophy, which depends on the understanding, and *true mathematics, which depends on the imagination*.)

318. Cf. Rule XIV, 442, 17 f.: "Per extensionem intelligimus, illud omne quod habet longitudinem, latitudinem et profunditatem. . . ." (By extension we understand all that which has length, breadth, and depth.)

319. Here it is important that the "figures" appear as "numbers" only through the "*mediating unit*" (mediante unitate) and that the *unit* itself is understood as "unit of measurement" (mensura). The "measure" is here, to be sure, applied only to symbolic formations; but this does not change the fact that Descartes basically retains the traditional — Peripatetic — understanding of the *hen* (cf. Part I, Section 8). He *fails* to pursue his own, different, original notion, according to which the unit is a "res simplex communis" and therefore a *res simplex intellectualis* (see P. 201); cf. also Aristotle, *Physics* Δ 14, 223 a 21–29.

Descartes, incidentally, from the very beginning distinguishes "two kinds of figures" (duo genera figurarum), namely those which represent a "multitudo," e.g.,

$$\therefore\!\cdot$$

i.e., a "numerus triangularis" (→ triangular number, cf. Part I, Pp. 28 f., 33, 55), or

PATER

FILIUS   FILIA

i.e., a genealogical "arbor quae alicujus prosapiam explicat" (tree which displays someone's family relations), and on the other hand, those which represent a "magnitudo," e.g.,

△, □, etc.

(Rule XIV, 450 f.). In both cases he has in mind completely traditional *nonsymbolic* formations. The symbolic "figure" which forms the object of the *mathesis universalis* is what is "common to" such figures, that is, "figurality" itself, or, in other words, the "second intention" of figure (cf. Pp. 47 f.), understood, however, as a "first intention." Thus in the passage of the *Regulae* here cited there appears once more the connection between "algebraic" magnitudes and "figurate" numbers, that is, Pythagorean *eide* of numbers; cf. Part I, Pp. 55 f. and 68 f.; furthermore Part II, P. 175. It is possible that in this respect Descartes is dependent on Adrianus Metius (cf., Note 247), whose lectures he in fact attended in 1629 in Franeker (cf. Cohen, *Écrivains français . . .* , pp. 436 f.). On the other hand, he may have been guided by a reminiscence of the writings of Faulhaber (see Note 306, end, and Note 313) who, beginning with the appearance of his *Arithmetischer cubiccossischer Lustgarten* (*Arithmetic Pleasure Garden of the Cubic Unknown*) in 1604, interlinked the study of "polygonal" and "pyramidal numbers" with the "cossic art"; cf. Descartes' treatise *De solidorum elementis* (*On the Elements of Solids* — Ad.-Tann., X, 265–276); also *ibid.*, 252 f., on which see Milhaud, *Descartes Savant*, pp. 84–87; finally also Ad.-Tann., I, 277 f.)

320. In the *Regulae* Descartes uses the lower case letters *a*, *b*, *c* . . . , for known magnitudes, the capitals *A*, *B*, *C* . . . , for the unknowns (Rule XVI, 455, 10 ff.).

321. On this see Gilson, *Index*, pp. 138 f., no. 226; cf. also P. Boutroux, "L'imagination et les mathématiques selon Descartes" (1900), *Université de Paris, Bibliothèque de la Faculté des lettres*, X, pp. 16 f.

322. Cf. Oughtred, *Arithmeticae in numeris et speciebus institutio quae tum logisticae, tum analyticae, atque adeo totius Mathematicae, quasi clavis est*, 1631 (see Note 329), dedicatory letter, p. A^{3r}: ". . . Ut ipsas res clarius intuerer, propositiones et demonstrationes verborum integumentis exutas, brevibus tantum symbolis ac notis oculis etiam ipsis uno obtutu perspiciendas designavi. . . ." (So that I should see the things themselves more clearly, I stripped the covering of words from the propositions and demonstrations [sc. of the great ancient geometers], and I signified them by *brief symbols* and *signs* only, so that they could be discerned at one glance even of the eyes alone.) Oughtred is dependent above all on Vieta, but probably also on Stevin.

323. Besides the traditional ancient Pythagorean eidetic representation (see Note 319) that of the "latitudines formarum" (breadths of forms) by Nicholas Orême (middle of the fourteenth century) indubitably forms an essential source for the figural symbolism of Descartes. See the beginning of the anonymous work *De latitudinibus formarum secundum doctrinam magistri Nicolai Horem*, which is a short treatment of a part of Orême's extensive work, *De uniformitate et difformitate intensionum* (*On the uniformity and nonuniformity of intensities*), on which see P. Duhem, *Études sur Léonard de Vinci*

(Paris, 1913), III, pp. 376 ff. and 399 f.; furthermore, Wieleitner, "Der 'Tractatus de latitudinibus formarum' des Oresme," *Bibliotheca Mathematica*, 3d series, Vol. XIII (1912–1913), pp. 115–145; same author, "Über den Funktionsbegriff und die graphische Darstellung bei Oresme," *Bibl. Math.*, 3d series, Vol. XIV (1913–1914), pp. 193–243; more recently, E. Borchert, "Die Lehre von der Bewegung bei Nicolaus Oresme," *Beiträge zur Geschichte der Philosophie und Theologie des Mittelalters*, Vol. XXXI, fasc. 3 (1934), p. 19 f. and 92–100; [also A. Maier, *An der Grenze von Scholastik und Naturwissenschaft* (Rome, 1952), Pt. III.]. The author of the present work was able to use the editions of 1486 (Padua) and of 1515 (Vienna), the latter of which belonged to Tycho Brahe: "Formarum quia latitudines multipliciter variantur multiplices varietates difficilime discernuntur: nisi ad figuras geometricas quodammodo referuntur. Ideo premissis quibusdam diuisionibus latitudinum cum diffinitionibus suis: species infinitas earumdem ad figurarum species infinitas applicabo ex quibus propositum clarius apparebit." (Because the breadths of forms vary in manifold fashion, their manifold varieties are distinguished with very great difficulty, unless they are somehow referred to *geometrical figures*. Accordingly, when I have first laid down certain classes of breadth and have given their definitions, *I shall connect the infinite species of breadths with the infinite species of these same figures*, whereby what has been set forth will be more clearly evident. — Translation after C. G. Wallis, *An Abstract of Nicholas Orême's Treatise on the Breadth of Forms* [St. John's Bookstore, Annapolis: 1941].) Compare this with Descartes' Rule XII (413, 11 ff.): "Quid igitur sequetur incommodi, si . . . concipiamus diversitatem, quae est inter album, coeruleum, rubrum, etc., veluti illam, quae est inter has aut similes figuras, etc.? Idemque de omnibus dici potest, cum figurarum infinitam multitudinem omnibus rerum sensibilium differentijs exprimendis sufficere sit certum." (What disadvantage would therefore result if we conceived the diversity which exists between white, blue, red, etc., just like that which exists between these or similar *figures* [Descartes here gives a drawing], etc.? The same can be said of *all* cases, since *it is certain that the infinite multitude of figures is sufficient for expressing all the differences of sensible things*.) Orême, incidentally, besides motions, changes, temperature, etc., also explicitly mentions colors (see later). Cf. furthermore the beginning of the treatise *De uniformitate et difformitate intensionum*, cited here according to the text given by Borchert, *Beiträge zur Geschichte der Philosophie und Theologie*, XXXI, fasc. 3, pp. 92–93, notes 175 and 177: "Omnis res mensurabilis exceptis numeris ymaginatur ad modum quantitatis continue. Ideo oportet pro eius mensuratione ymaginari puncta, lineas et super- ficies aut istorum proprietates. In quibus ut uult philosophus mensura et proportio per prius reperitur. In alijs autem cogno- scuntur in similitudine dum per intellectum referuntur ad ista. . . ." (Every measurable thing except numbers is pictured in the mode of a continuous quantity. Therefore *it is necessary, in order to measure it, to imagine points, lines and surfaces or their properties*, in which, as the philosopher [Aristotle] contends, measure and proportion are primarily to be found. In other objects, however, they are under- stood by analogy, since they are referred to the former by the intellect.) See also *ibid.*, pp. 94 f., note 180: ". . . Uniformitas eius [sc. qualitatis] atque difformitas cicius, facilius et clarius perpen- duntur quando in figura sensibili aliquod simile describitur quod ab ymaginatione velociter et perfecte capitur et quando in exemplo visibili declaratur . . . multum enim iuuat ad cognitionem rerum ymaginatio figurarum." (The uniformity of one and the non- uniformity of another quality are considered more quickly, easily and clearly when something similar to it, which can be caught quickly and completely by the imagination, is represented in a sensible figure and when it is expressed in a visible example, *for pictorialization in figures is a great aid to the understanding of things*.)

That Descartes in some sense "knew" of the "quantitative" representation of "qualities" as it had been developed by Orême and had since been used time and again (cf. Duhem, *Études*, III, 399– 405; 481 ff.; especially also 502–504) is proved by his treatment of the question concerning the free fall of bodies in a vacuum, posed to him at the end of 1618 by Beeckman (Ad.-Tann., X, 75–78; cf. 219–221). He conceives of the "intensity" (speed) of the motion in each case as a straight line which is perpendicular to another straight line which in turn corresponds to the total distance (not time!) traversed by the falling body. *Thus the motion as such is represented by a "figure,"* namely here by an (isosceles) right-angled triangle. It is easy to see that Descartes is not quite clear about the implications of the question (cf. Duhem, *Études*, III, 566–574; Wieleitner, "Das Gesetz vom freien Falle in der Scholastik, bei Descartes und Galilei," *Zeitschrift für mathematischen und naturwissenschaftlichen Unterricht*, XLV (1914), pp. 216–223; Milhaud, *Descartes savant*, pp. 26–34); while Beeckman is searching for the relation of the *distances* traversed in free fall to the *times* necessary to traverse them, Descartes' argument results in *speeds* being related to *distances* traversed. But if in Descartes' diagram the distance line is taken as a time line, which is what Beeckman actually does in his interpretation of Descartes' answer without noticing that he is thus correcting him (cf. Ad.-Tann., X, 58–61), then the right solution of the problem results. Descartes continues to miss this fact (see especially Ad.-Tann., I, 71–75 and 394 f.; Milhaud, *Descartes savant,* pp. 30–32; Wieleitner, *Zeitschrift . . .* , XLV, p. 222 [A. Koyré "La loi de la chute des corps, Descartes et Galilée," *Études Galiléennes,* (Paris, 1939), II, pp. 102–119]); so in 1604 Galileo too (*Ed. naz.*, X, 115 f. and VIII, 373 f.; cf. also VIII, 203 and 208 ff.; furthermore VII, 248 ff.) confuses the time with the distance traversed in a very similar fashion (cf. Duhem, *Études,* III, 562–566; Wieleitner *Zeitschrift . . .* , XLV, 223–228). To understand this situation one must bear in mind that neither Descartes nor Galileo are *immediately* concerned so much with grasping a "motion" with the aid of a line-("coordinate") system, where the proper choice of "variables" would of course be crucial, as with the *possibility of representing motion by figures generally.* Cf. Ad.-Tann., X, 220 5–9: "Ut autem hujus scientiae fundamenta jaciam, motus ubique aequalis linea repraesentabitur, vel superficie rectangula, vel parallelogrammo, vel parallelepipedo; quod augetur ab una causa, triangulo; a duabus, pyramide . . . , a tribus alijs figuris." (In order that I may lay the foundations of this science, uniform motion will always be represented by a line, or a rectangular plane, or a parallelogram or a parallelopiped; what is increased by one cause, by a triangle, by two, by a pyramid . . . by three, by other figures.) Cf. on this Wieleitner, *Zeitschr. f. math. . . .* , XLV, pp. 221–222; furthermore Duhem, *Études,* III, 386–388 and also 517–519. And Descartes, it should be noted, refers in precisely this context to his "*Algebra geometrica*" (X, 78, 23) to the presentation of which, by then accompanied with a claim to greater universality, the later *Regulae* are devoted.

For Orême too the "geometrization" of the concept of the "intensio formae" = excessus gradualis = latitudo gradualis (intensity of form = increase of degree = breadth of degree — *De uniformitate et difformitate . . .* , Part II, Chap. II, Supposition 3; cf. Supp. 9) serves exclusively to *represent* "qualities" or "intensities" *in figures* (cf. Part II, Chap. III) — a fact which contradicts the usual interpretation of Orême and should be emphasized, as Wieleitner has done in the articles in the *Bibliotheca Mathematica* cited earlier. These qualities or intensities may be either "formae permanentes," such as colors (colores), temperatures, i.e., warmth (calores), etc., or "formae successivae," such as changes (alterationes) and local motions (motus locales), cf. Part II, Chap. II, Supp. 13 and the concluding remarks of Chap. III. Supposition 9 (Part II, Chap. II) states: "extensio forme ymaginanda est per lineam rectam: intensio vero per figuram planam super rectam consurgentem" (the extension of the form is to be pictured by means of a straight line — *but the intensity by means of a plane figure* erected upon the straight line); cf. Supp. 12: to the "intensio totalis forme date" (whole intensity of the given form [i.e., the "integral"]) corresponds the "superficies super rectam lineam collocata" (plane surface located on the straight line); but Orême also knew about representation by means of solid figures (Wieleitner, *Bibl. Math.*, 3d series, Vol. XIV, pp. 204 and 214-216; Duhem, *Études*, III, 386-388); note also the traditional titles of his work *De figuratione potentiarum et mensurarum difformitatum* (*On the rendition by* figures *of the nonuniform powers and measures*) and *De configuracionibus qualitatum* (*On the configurations of qualities*). This "figura repraesentativa" changes according to the *kind of change of intensity* involved. In the case of a change of intensity which is everywhere the same with itself, namely a "uniformis latitudinis variatio" (uniform variation of breadth), a "latitudo uniformiter difformis" (uniformly nonuniform breadth) results, which is represented by a right-angled triangle. This holds true in particular of uniformly nonuniform *motion* (cf. Duhem, *Études*, III, 388 ff.; Wieleitner, *Bibl. Math.*, 3d series, Vol. XIV, pp. 222 ff.). That this motion is actualized in the special case of the uniformly accelerated motion of bodies in free fall was a fact generally known to the nominalistic school since Albert of Saxony (cf. Duhem, *Études*, III, 309-314 [M. Clagett, *The Science of Mechanics in the Middle Ages* (Madison, 1959), pp. 565-569 and passim; C. Wilson, *William Heytesbury, Medieval Logic and the Rise of Mathematical Physics* (Madison, 1960), Chap. 4]). But this fact was not connected with Orême's graphic representation until much later (see Duhem, *Études*, III, 556 ff.). At any rate Galileo, Beeckman, and Descartes had a settled tradition available even on this point. Furthermore, Orême conceives of something as a "quantum," i.e., as something "quantitative" or as a "magnitudo," when it is subject to "more and less" and has "the capacity of exceeding and falling short" (cf. Part 1, Note 109), exactly as does Descartes (Rule XIV, 440): "quod recipit majus et minus" (what is susceptible of more and less — see Pp. 208 f.); compare with this Orême, Part II, Chap. II, Supp. 2: "Omne quod excessu graduali excedit aliud vel exceditur ab alio est imaginandum per modum quantitatis." (Everything which exceeds another by a difference of degree or is exceeded by another is to be pictured in the mode of quantity.) This means whatever can be subjected to ratios and therefore also to proportions (Orême, II, II, Supp. 1): "Omnia que secundum aliquam proporcionem [= rationem] se habent adinuicem ratione participant quantitatis." (All things which are related to one another in any *ratio* share the nature of quantity.) Here a *quantitas* is understood to exist either "truly" or "figuratively" (vere vel ymaginative). Descartes' concept of "dimension" in the *Regulae* (447–449; cf. on this also Beeckman's notes, X, 334; furthermore II, 542) likewise appears to be directly connected with Orême's doctrine (II, II, Supps. 5, 6, 7, and 8). Orême himself, especially in his predilection for the concept of *ratio* (= proportio) belongs to the tradition for which the work of Bradwardine, the "doctor profundus," is definitive [see Thomas of Bradwardine, *Tractatus de Proportionibus*, ed. and trans. H. Lamar Crosby, Jr., (Madison, 1961)]. Thus, for instance, the distinction between "formae permanentes" and "formae successivae" goes back to Bradwardine (cf. Cantor, II², 113 ff.; 118 ff.; furthermore Duhem, III, 294 ff.; selections from the pertinent texts in Max Curtze, "Über die Handschrift R. 4⁰. 2, *Problematum Euclidis explicatio*, der Königlichen Gymnasialbibliothek zu Thorn," *Zeitschrift für Mathematik und Physik*, XIII, Supplement (1868), Section 11: "Geometria Bradwardini" and 9: "Algorismus proportionum magistri Nicolay Orem," pp. 81–84; 65–79 and 101–104; cf. also Note 227).

What is peculiar to Descartes' representation of motion as contrasted with that of the tradition is the fact that he sees a direct connection between his figural representation and figurate numbers. For instance, Beeckman's premise "quod semel movetur, semper movetur, in vacuo" (in a vacuum what is once in motion is always in motion), or "mota semel numquam quiescunt, nisi impediantur" (things once moved never come to rest unless checked — cf. Ad.-Tann., X, 60, 78, and 219) leads him to the assertion that motion (more exactly, the motive force) always represents a "numerus triangularis" and is precisely for this reason to be rendered by a "figura triangularis" (X, 76, 2–4). Here Descartes seems to come dangerously close to the opinion of Leonardo da Vinci (Duhem, III, 512 f. and Tannery, *Mém. scient.*, VI, 482) and of Baliani (Cantor, II², 698 f.), according to which in uniformly accelerated motion distances traversed in successive moments of time are to each other as the members of the series of natural numbers, 1, 2, 3, 4. . . . Yet Descartes succeeds in doing something which is, as against Orême's procedure, entirely new, namely to pass, with the aid of the concept of "indivisibles" (cf. Duhem, II, 1909, pp. 7 ff.) directly to the series $\frac{1}{2}$, $\frac{3}{2}$, $\frac{5}{2}$, $\frac{7}{2}$ ..., which he must have before he can use the triangular "figure" at all.

Finally, as far as the use of coordinates in the later *Geometry* of Descartes is concerned, it has no connection with the "qualitative geometry" of Orême; rather, it depends directly on the corresponding procedure in Apollonius, which, as is well known, is also the case for Fermat. Operation with "coordinates," however, has a much deeper foundation in Descartes than in Fermat, namely in the idea of a general algebra (and thus also, indirectly, in the doctrine of Orême) discussed before. This is why Descartes gives a much better account of the meaning of the "generalization" to which he and Fermat subject Apollonius' procedure; cf. Fermat's remarks on this "generalization" in his *Isagoge ad locos planos et solidos*, *Oeuvres*, ed. Tannery-Henry, I, pp. 91, 93 and 99.

324. Strictly: as the intended object of an *intentio secunda*, cf. Eustachius a Sancto Paulo (Gilson, *Index*, p. 107): "... quid sit secunda intentio. Respondetur secundam intentionem, si vim nominis spectes, esse ipsam mentis operationem, qua secundario [i.e., actu signato] tendit in rem jam antea cognitam, quatenus cognita est." (... what is a second intention. The answer is that a second intention, if you look at the meaning of the name, is *that very operation of the mind* by which it secondarily [i.e., in a reflective act] intends a thing already conceived before, insofar as it has been conceived.) There follows the definition according to "usus" (usage) previously quoted in the text, in which the intended object itself is understood by the term.

325. Cf. Rule XII, 414: "... Concipiendum est, ... hanc phantasiam esse veram partem corporis, et tantae magnitudinis, ut diversae ejus portiones plures figuras ad invicem distinctas induere possint. ..." (This imaginative organ must be conceived as a *true part of the body* and as of such a *size* as to permit its different *portions* to assume several *figures* distinct from one another.) Furthermore, Rule XIV, 442: "... nihil omnino facilius ab imaginatione nostra [percipitur]" (absolutely nothing is more easily perceived by our imagination) than extension.

326. Within this complex of thought it is possible to make a precise distinction between the traditional part and what is specifically Descartes' own and therefore "new." (Dilthey, *Gesammelte Schriften*, II, p. 295 seems to have completely overlooked this.) The doctrine of "impressions" ($\tau\upsilon\pi\acute{\omega}\sigma\epsilon\iota\varsigma$ — *typosis*), especially in relation to the *imaginatio*, is, to begin with, again unquestionably Stoic: $\phi\alpha\nu\tau\alpha\sigma\acute{\iota}\alpha$ $o\mathring{\upsilon}\nu$ $\mathring{\epsilon}\sigma\tau\acute{\iota}$ $\kappa\alpha\tau'$ $\alpha\mathring{\upsilon}\tau o\grave{\upsilon}\varsigma$ [sc. $\tau o\grave{\upsilon}\varsigma$ $\Sigma\tau o\ddot{\iota}\kappa o\acute{\upsilon}\varsigma$] $\tau\acute{\upsilon}\pi\omega\sigma\iota\varsigma$ $\mathring{\epsilon}\nu$ $\psi\upsilon\chi\widehat{\eta}$. (Imagi-

nation is, then, according to the Stoics, an *impression made in the soul.*
— cf. Sextus, *Against the Mathematicians* VII, 228; also VII, 372 and
VIII, 400=Arnim, *Stoic. vet. fragmenta*, I, fr. 484, furthermore
Sextus, *ibid.* VII, 236=Arnim, I, fr. 58.) Prior to this it goes back to
Plato and Aristotle (*On Memory*, 1, 450 a 30–32): ἡ γὰρ γιγνομένη
κίνησις ἐνσημαίνεται οἷον τύπον τινὰ τοῦ αἰσθήματος, καθάπερ οἱ
σφραγιζόμενοι τοῖς δακτυλίοις. (For as this movement occurs, it
makes a mark *like an impression* of the object of sense, just as when
people make seal impressions with seal rings — cf. *On the Soul* B 12,
424 a 17 ff.) On this movement see *ibid.*, Γ 3, 429 a 1 f.: . . . ἡ
φαντασία ἂν εἴη κίνησις ὑπὸ τῆς αἰσθήσεως τῆς κατ᾽ ἐνέργειαν
γιγνομένη. (Imagination must be a *movement* occurring under the
actual exercise of sense perception.) Cf. also Plato, *Theaetetus* 191 C
ff., *Philebus* 33 D ff., 39 A ff. However, in ancient philosophy *typosis*
is *always* understood in such a way that the "impression" in the soul
is not taken literally but is simply regarded in each case as some sort
of counterpart to the "looks" of the thing *precisely as it presents itself
to our* ("*external*") *senses.* Even in Cleanthes, who takes *typosis* quite
literally, namely as κατὰ εἰσοχήν τε καὶ ἐξοχήν, ὥσπερ καὶ ⟨τὴν⟩ διὰ
τῶν δακτυλίων γιγνομένην τοῦ κηροῦ τύπωσιν (embossed and im-
printed, just like the impression made in wax by seal-rings —
Arnim, I, fr. 484), the impression of the *eidos* of the particular thing
concerned is intended. The other Stoics are at one with Chrysippos
and Zeno (and surely also with Aristotle) in understanding the term
typosis entirely as a mere metaphor, a position which Descartes
explicitly opposes (cf. also *Dioptrics*, Chap. IV, 6–7: Ad.-Tann., VI,
599 f.; French text, *ibid.*, pp. 112 ff.; furthermore *Principia philosophiae* IV, 197–198). But, above all, the Cartesian conception of
this process is completely original in reducing *everything* perceptible
by the (external) senses, that is, besides the "things" themselves also
their colors, warmth, coldness, hardness, roughness, sweetness, etc.,
to "figures," which are supposed to represent the *true* "nature" of
the "things" or "forces" or "properties" in question, namely
precisely that nature which is inaccessible to the external senses:
"Quid . . . sequetur incommodi, si . . . abstrahamus ab omni alio,
quam quod habeat [sc. color] figurae naturam?" (What incon-
venience would result if we were to abstract from everything other
than the figured nature which color [which is taken here as an
example] has? — Rule XII, Ad.-Tann., X, 413.) The variety of these
"things," "forces," or "properties" is then shown to be rooted in
the variety of the infinitely many possible "figures," of which
Descartes presents some in a drawing (*ibid.*). It is precisely this *true*

nature of the parts of the world which the intellect "sees" when it "turns toward" the "impressions" "in" or "on" the part of the brain which *is* the "phantasia vel imaginatio." This conception of the processes of perception, which Descartes here introduces only as a *suppositio*, underwent a number of changes in the further development of his thought, especially through the stronger emphasis on the "movement" of the "figures" which is finally converted into "movement" within the pineal gland (cf. incidentally, Aristotle, *On the Soul* Γ 3, 428 b 10 ff.), and through the doctrine of the "spiritus animales" (vital spirits), which itself goes back to ancient sources, especially to Galen. But these original assumptions, especially those underlying the understanding of *extensio*, remain not only the basis of Descartes' later writings, but are, in fact — admittedly or unadmittedly — presupposed by all modern physiology and physiological psychology even to this day (cf., among others, the pregnant formulations by Euler, *Lettres à une princesse d'Allemagne sur divers sujets de physique et de philosophie*, II [Leipzig, 1770], Letters 94 and 96, pp. 63 ff.; 72 ff.). The fact that for Descartes the difference between "phantasia" and "sensus communis" becomes obscured later on is not unimportant, cf. *Meditations* VI, Ad.-Tann., VII, 86, 16 ff., and also *The World, Treatise on Man*, Ad.-Tann., XI, 174 ff.; cf. Aristotle, *On Memory* 1, 450 a 10 f.: ... τὸ φάντασμα τῆς κοινῆς αἰσθήσεως πάθος ἐστιν (the appearing image is an affection of the common sense); furthermore the Aristotelian doctrine according to which the κοινὴ αἴσθησις (common sense) is essentially related to whatever "*koina*" there are: *kinesis* (motion), *stasis* (rest), *schema* (figure), *megethos* (magnitude), *arithmos* (number), *hen* (one) — *On the Soul* Γ 1, 425 a 14 ff. and B 6, 418 a 16–19; cf. Note 209. Here Descartes, in agreement with the contemporary psychology (cf. Gilson, *Index*, p. 267), obviously understands both the "phantasia" and the "sensus communis" to be two faculties of the *sensus internus* (internal sense) which are distinguished from the "sensus externi" (external senses), but only by a "diversitas officiorum" (diversity of offices performed), "non natura et specie" (not in nature or kind).

327. Cf. *Cogitationes privatae*, X, 217, 12 ". . . imaginatio utitur figuris ad corpora concipienda . . ." (the imagination uses figures for conceiving bodies); the continuation of this passage shows the whole significance which Descartes, at least in his early period, accorded to the "vis imaginationis" (faculty of imagination); cf. also Note 317.

328. Cf. also Descartes' *Principia philosophiae*, II, 64. Descartes' further development leads him to find the guarantee for the certainty of

"clear and distinct," that is, above all, *mathematical* cognition — and consequently also the guarantee of the possibility of a "true physics" — via a detour into metaphysics ultimately in God. Hence the importance of the *imaginatio* in the Cartesian system declines more and more, with the result that its position, especially with respect to the conception of *extensio*, becomes so much the more questionable. The character of *extensio* itself, however, remains essentially untouched by this development.

329. Both works appeared in 1631, but were written earlier: Harriot died in 1621, Oughtred in 1660, but Wallis declares explicitly that Harriot wrote his book *later* than Oughtred or at about the same time (cf. Wallis, *Mathesis Universalis*, *Opera* [Oxford, 1695], p. 55; also *Algebra*, *Opera* [Oxford, 1693], Preface, pp. a 3$^{r \ et \ v}$ and 136 ff.). The exact title of the *Clavis mathematicae* is: *Arithmeticae in numeris et speciebus institutio quae tum logisticae, tum analyticae, atque adeo totius mathematicae, quasi clavis est (Fundamentals of the Arithmetic of Numbers and Species which is, as it were, a key both to logistic and to analytic, and even to the whole of mathematics).* On Oughtred's work see Cantor, II², 720 f. and Wallis, *Algebra, Opera*, II, pp. 71–73; furthermore, Note 322.

330. I quote from the complete edition undertaken by Wallis himself (1693–1699). The *Mathesis universalis* is contained in Vol. I (1695). Its exact title is: *Mathesis universalis sive, Arithmeticum opus integrum, tum Philologice, tum Mathematice traditum, Arithmeticam tum Numerosam, tum Speciosam sive Symbolicam complectens, sive Calculum Geometricum; tum etiam Rationum proportionumve traditionem; Logarithmorum item Doctrinam; aliaque, quae Capitum Syllabus indicabit (Universal Mathematics, or the whole arithmetical enterprise as handed down both by the Philosophical and the Mathematical Tradition, embracing both Numerical and Specious or Symbolic Arithmetic as well as the Geometric method of calculation; furthermore the traditional theory of Ratio and Proportion, also the Doctrine of Logarithms; and other matters which will be indicated by the Table of Chapter Headings).*

331. Wallis has in mind Descartes' *Geometria*, which was published, together with commentaries and the "remarks" of Florimond de Beaune, by van Schooten in 1649; furthermore van Schooten's *Principia Matheseos universalis seu introductio ad Geometriae Methodum Renati Des Cartes (Principles of Universal Mathematics or An Introduction to the Method of Geometry of René Descartes)*, edited by Bartholinus (first edition in 1651).

332. He says in this connection: "Nihil absurdi esse [respondeo] majorum inventis addere; praesertim in Mathematicis. Nec hoc illorum

laudibus quicquam detrahit: nam et illi prioribus addiderunt." (I answer that there is nothing absurd in adding to the discoveries of the ancients, especially in mathematics. Nor does it detract in any way from the praise due to them, for they also added to the work of their predecessors.) Cf. Note 233, end.

333. This distinction goes back to one current, especially in the nominalistic school, since Albert of Saxony, that of the "maximum in quod sic" (maximum at which it [remains] such as it is) and the "minimum in quod non" (minimum at which it is not [yet itself]); cf. Duhem, *Études sur Léonard de Vinci*, II, 26 ff. The occasion for this distinction is given by the commentaries on Aristotle's *On the Heavens* A 11; [cf. Wilson, *William Heytesbury*, Chap. 3].

334. Cf. P. 157, Vieta's remark on the "geometric" character of ancient "algebra."

335. Cf. the introduction of the law of homogeneity by Vieta, *Isagoge* Chap. III, 1 (see Pp. 172–174, and Appendix, P. 324) and Descartes, *Regulae*, Ad.-Tann., X, 456 f.

336. Cf. Pp. 192 f., the opinion of Stevin, which Wallis is clearly following here.

337. Cf. the first Cartesian sketch of "geometric algebra" in the *Regulae*.

338. First Wallis shows (pp. 52–53) that within the positional system, whether we employ a decimal, a quaternary, or any other system as a basis, when we arrange "the decuples, quadruples and triples [i.e., the bases 10, 4, 3] in proportion, they have just the same value":

$$\text{(In proportione}\begin{cases}\text{decupla} & 2,7 \\ \quad[\text{i.e., } 2.10^1+7=27] \\ \text{quadrupla} & 1,2,3 \\ \quad[\text{i.e., } 1.4^2+2.4^1+3=27] \\ \text{tripla} & 1,0,0,0 \\ \quad[\text{i.e., } 1.3^3+0.3^2+0.3^1+0=27]\end{cases}\begin{matrix}\text{tantundem} \\ \text{valent),}\end{matrix}$$

and that the positional base units, which follow each other in the sequence, form a geometric progression, that is, a progression having "gradus ascendentes et descendentes" (ascending and descending degrees; cf. Vieta, *Isagoge*, Chaps. I and III; also Descartes, *Regulae*, Ad.-Tann., X, 457, 2–3, and above, Note 292). Then he says that it should be emphasized that: "Universam Artem *Algebrae* sive *Analytices* hoc uno quasi fundamento niti: Atque si haec, quae de Gradibus (in quacunque ratione Ascendentibus et Descendentibus) diximus, satis intelligantur; magnam exinde affulgere lucem ad *Potestates*(quas vocant)*Algebricas*intelligendas, et rite tractandas. Nam revera, quod nobis nunc est *gradus* (sive Ascendens, sive Descendens) *primus, secundus, tertius,* etc. illud est Algebristis *Latus, Quadratus, Cubus,* etc., vel *potestas prima, secunda, tertia,* etc. . . . ." (The universal art of "algebra" or "analytic" rests on this as on a foundation; and if these things which we have said about degrees, ascending or descending in whatever order, are well understood, a great light flashes forth for the understanding and for the correct handling of so-called "algebraic powers." For what we now know as a "first," "second," "third" degree, etc., whether ascending or descending, that is the very thing the algebraists call a "side," "square," "cube," etc., or a "first," "second," "third" "power," etc.) On the expression "potestas," cf. Note 256. In the generalized sense in which Wallis uses it, it stems from Oughtred (cf. Tropfke, *Geschichte der Elementarmathematik,* II³, 161).

339. Cf. Note 255. Wallis, incidentally, gives as examples: 2 hundreds = 20 tens, and 2 thousands = 20 hundreds (cf. Pp. 131 and 143).

340. The words *bracketed* by Wallis in the text above and omitted by me are: "vel saltem ad unitatem vere rationem habeant" (or at least are really in a ratio with the unit). They invalidate the meaning of the main sentence and show directly the ambiguity which the expression "numeri proprie dicti" (numbers properly so called) has in Wallis, see Pp. 220 ff.

341. Inexplicitly, this is the case already in Harriot and Oughtred on the one hand and in Stevin and Descartes on the other.

342. Chap. XXIII treats Euclid II in the same way.

343. Chap. XXV, p. 134: ". . . comparatorum, alterum *Antecedens* dici solet, alterum *Consequens* . . . Puta si A ad B comparatur; A dicitur Anticedens, B Consequens . . ." (of compared numbers, one is usually called the "antecedent," the other the "consequent"; think of A as compared to B; A is called the antecedent, B the consequent). This terminology goes back to Leonardo of Pisa. (Cf. Tropfke, *Geschichte der Elementarmathematik,* III², 18).

344. On the term "exponens" cf. Note 306, toward end.

345. "Dimension" is here understood exactly as in Descartes (*Regulae,* X, 447–449), except that Descartes does not stress the dimensionlessness of the *numeri* themselves as emphatically as Wallis does. The reason for this is that for Descartes the "figures" with which the *mathesis universalis* operates are exactly as symbolic as the "numbers" (cf. Pp. 203–206). In Wallis, on the other hand, as in Vieta, the traditional conception of geometric structures is preserved, as it is, incidentally, also later. The peculiar place which geometry holds

within symbolic mathematics was of essential significance for the development of mathematical *physics*.

346. Corresponding to the Greek term παραβολή (parabola, that which falls alongside); cf., e.g., Zeuthen, *Die Mathematik im Altertum und Mittelalter*, p. 38. Such *applicationes* "nonnisi καταχρηστικῶς Divisiones vocantur" (are not called divisions except by a misuse of language). Note that Vieta, who refuses to recognize the law of homogeneity as valid precisely in the case of division, also uses the term "adplicare" (to apply) for "dividing" (cf. P. 171 and Note 252, furthermore *Isagoge*, Chap. IV, Precept IV, Appendix, P. 335 and P. 334, Note 39).

347. The third case: division of a magnitude (of determinate dimensions) by a (dimensionless) *numerus* "non tam divisio est ... quam multi-plicatio (nempe quantitas A non tam dividitur per 2, quam multi-plicatur per $\frac{1}{2}$) quippe non quaeritur, quoties *numerus* 2 contineatur in *magnitudine* A (quod absurdum esset) sed datae quantitati A, alia in data ratione quaeritur; quod *Multiplicationis* est, potius quam *Divisionis* opus; quippe quae in Multiplicatione Ratio datur, in Divisione quaeritur" (is not so much division as multiplication, for clearly the quantity A is not so much divided by 2 as multiplied by $\frac{1}{2}$; since what is sought is, of course, not how many times the "number" 2 is contained in the "magnitude" A — which is absurd — but rather for a given quantity A, another in a given ratio is sought; and this is the business of "multiplication" rather than "division," for the ratio given in multiplication is the very one sought in division—pp. 135–136).

348. The immediate continuation of this passage is: "totumque Euclidis Elementum quintum Arithmeticum esse, utut speciatim de Magnitudinibus efferantur propositiones, quae interim non minus recte de Quantitatibus simpliciter quibusvis efferi possent, quo sensu apud Euclidem μεγέθη intelligenda sunt" (and the whole fifth book of Euclid is arithmetic, however specifically, as if concerned with [geometric] magnitudes, propositions may be presented, proposi-tions which could meanwhile be carried out just as correctly for any quantities desired in general, and this is the sense in which the word "magnitudes" must be understood in Euclid).

Appendix

INTRODUCTION TO THE ANALYTICAL ART

by François Viète (Vieta)

To the Illustrious Princess Mélusine,
Catherine of Parthenay,
Most Pious Mother of the Lords of Rohan,
I, François Viète of Fontenay,
Pledge Honor and Obedience.[1]

O Princess Mélusine,[2] *most pious mother of the lords of Rohan,*
the Bretons extol the noble family and ancient ancestry of the
house of Rohan, which I do not think could be matched on the
whole earth by any other more ancient and illustrious on
account of more legitimate possessions or more authentic
monuments. They will acknowledge your children as the original
stock and as the descendants of the royal blood of Conan, as
those who by God's will escaped the yoke of the invader
Nominhoë; and they will be confident that this noble race will
last as long as they, while going about the quarries, woods, and
ponds of your domain of Salles, see engraved in marbles, oaks,
and scales of fish the insignia of the golden rhomboids which it

[1] This translation is based primarily on the text of the *Isagoge* as republished with annotations in the Francisci Vietae *Opera Mathematica*, ed. F. van Schooten (Leyden, 1646) pp. 1–12, and as much as possible of its style has been preserved. The original edition was also consulted; its full title is: F. Vietae, *In Artem Analyticem* [sic!] *Isagoge*, Seorsim excussa ab *opere restitutae Mathematicae Analyseos, seu*, Algebra Nova (*Introduction to the Analytical Art*, excerpted as a separate piece from the *opus* of the restored Mathematical Analysis, or *The New Algebra* [Tours, 1591]; cf. Pp. 151 and 153).

The passages in small italics are the editor's annotations printed in the 1646 edition. The passages in square brackets, as well as the footnotes, have been added by the translator. This translation was made in 1955 at St. John's College in Annapolis — J. Winfree Smith.

[2] Catherine of Parthenay (1554–1631) was an ardent Huguenot. After her first husband was killed in the Massacre of St. Bartholomew, she married René of Rohan in Brittany and had by him five children, the eldest of whom, Henri of Rohan, became the famous leader of the

wears.[3] *For by their own religious lore (cabalâ), the Bretons
will testify that as it was granted of His sole favor by God most
great and most good to the prayers of St. Mériadec, a former
prince of the family, so also now it is granted to me, who*

Huguenots. Vieta had supervised her education and remained her friend
and adviser all his life.

Catherine herself was descended from the family of Lusignan, whose
ancestral seat was the château of Lusignan, fifteen miles from Poitiers.
The legendary ancestress of the family was a fairy named Mélusine. The
name was originally Mère des Lusignans, then became Mère Lusigne,
afterwards Merlusine, and finally Mélusine. Mélusine had the remarkable
ability to turn the lower part of her body into a serpent every Saturday.
When she married Raymond, it was on the condition that he would never
see her on Saturday. He broke the agreement, whereupon she turned
completely into a serpent, escaped by the window, and disappeared, only
to reappear on the occasion of the death of the lords of Lusignan, when
she would utter strange cries of grief. Mélusine was a beneficent fairy
and, according to the legend, built the château of Lusignan and many
others for her husband.

A Hugh of Lusignan went on the crusade of 1100–1101. Another
member of the family, Hugh the Brown, went as a pilgrim to the
Holy Land in 1164. In the last quarter of the twelfth century his son
Guy became king of Jerusalem and ruler of Cyprus, where his brother's
descendants reigned as kings until 1475. In the middle of the fourteenth
century, some of the Lusignans of Cyprus went off and made themselves
rulers of Armenia, where they held sway from 1342 to 1375. A branch of
the family continued in Poitou during the thirteenth century and ruled La
Marche until 1303. Hugh of La Marche, whose betrothed wife, Isabel of
Angoulême, was seized by King John of England and made his queen,
was a nephew of Guy of Lusignan. After John's death Hugh married her
and had by her a number of sons who were, therefore, half-brothers of
Henry III of England.

The family of René of Rohan owned extensive domains in Brittany,
including those of Porhoët and Léon. They were descended from the
ancient kings of Brittany the first of whom was Conan Mériadec
(409). Judicaël, Eudon, and Erech, whom Vieta mentions, were all kings
of Brittany. St. Mériadec, a descendant of Conan, lived in the seventh
century and was bishop of Vannes.

[3] A reference to the coat of arms of the Rohan family.

marvel at few things, to marvel time and again at the strange
warblings of birds and other remarkable things around the
sanctuary, which long ago was his, constructed in the midst of
woods and pleasant groves. I, of Fontenay in Poitou, a regular
inhabitant of the banks of the Vendée, cherish the name (nomen)
and the majesty (numen) of Mélusine and her descendants of the
castle constructed long ago by the divine Mélusine, of whom by
Raymond you are the blessed progeny. And I also add a prophecy
(omen). However, I do not for this purpose oppose to the
Judicaëls, the Eudons, and the Erechs of the house of Rohan your
Guys, Godfreds, Hughs, and Bruns; nor to their Breton kings,
princes in Léon, counts in Porhoët, do I oppose your kings of
Cyprus, your princes of Antioch and Armenia, your counts of
Angoulême and La Marche nor to their Isabel, daughter of the
Scot, nor to Isabel of Navarre, do I oppose your Isabel, mother
of English kings and of your ancestors of Lusignan. But rather I
piously recall and judge that it happened auspiciously and as if by
decree of destiny that the goddess Mélusine in gratitude for the
help received from René of Rohan, since he had strenuously
defended her castle of Lusignan when it was besieged at the
instigation of the Guises, forthwith bestowed on him you, her
own and Raymond's offspring and heir, and the rule of the
family of Rohan. Raymond himself, to be sure, was descended
from the family of Rohan, and now the offspring of Raymond
and Mélusine were returned to that source from which they first
began; thus it will hardly perish, for this circle is a true and truly
physical symbol of perpetuity. But even less will your virtues
perish in this cyclical restitution of the beginning. And just as our
ancestors, in their own idiom, which was then being adopted,
called your ancestress "Fairy Mélusine" because of her venerable
appearance and her rare and remarkable gifts of mind, so posterity

will invoke you as heavenly goddess (δῖαν θεάων) and will address
you as queen (πότνιαν), as trustworthy ruler (κεδνήν), and with
a more worthy epithet, if any occurs.[4] And may the fruits of our
nightly labor be pleasing to her, so that she may credit them
where they are owed, to you and to your most dear sister
Françoise of Rohan, duchess of Nîmes and of Loudinois. For the
benefits which you and she bestowed on me in most unhappy
times are infinite. How can I adequately commemorate that you
delivered me from brigand's chains and from the jaws of death
and that, in a word, you helped me with your solicitude and
generosity as often as my needs and misfortunes prompted you?
I owe my life, or if there is anything dearer to me than life,
entirely to you; and now, O divine Mélusine, I owe to you
especially the whole study of Mathematics, to which I have been
spurred on both by your love for it and by the very great skill you
have in that art, nay more, the comprehensive knowledge in all
sciences (Encyclopaedia) which can never be sufficiently admired
in one of your sex who is of so royal and noble a race. O
princess most to be revered, those things which are new are wont
in the beginning to be set forth rudely and formlessly and must
then be polished and perfected in succeeding centuries. Behold,
the art which I present is new, but in truth so old, so spoiled and
defiled by the barbarians, that I considered it necessary, in order
to introduce an entirely new form into it, to think out and publish
a new vocabulary, having gotten rid of all its pseudo-technical
terms (pseudo-categorematis) lest it should retain its filth and
continue to stink in the old way, but since till now ears have
been little accustomed to it, it will be hardly avoidable that

---

[4] These are all Homeric epithets, applied in Homer to gods and heroes.
Cf. *Iliad* XVIII, 388; XIX, 6; *Odyssey* V, 215; XIII, 291; XX, 11; XIV,
170.

*many will be offended and frightened away at the very threshold.
And yet underneath the Algebra or Almucabala which they
lauded and called " the great art," all Mathematicians recognized
that incomparable gold lay hidden, though they used to find very
little. There were those who vowed hecatombs and made
sacrifices to the Muses and Apollo if any one would solve some
one problem or other of the order of such problems as we solve
freely by the score, since our art is the surest finder of all things
mathematical.*[5] Now that the thing has come to pass, will they
be bound by their vows? *However, it would be right for me
not now to commend my own wares, but in all moderation yours
and those which have been acquired and renewed through your
beneficence, to bear witness to my desire that whatever glory is
due on accout of the felicity of your rule should not be snatched
away. For it is not the same in mathematics as in other studies,
that everyone's opinion is free and free his judgment. Here things
are done by rule and effort, and neither the persuasions of
rhetoricians nor the pleadings of lawyers are of use. The metal
which I bring forth yields the kind of gold which they wanted
for so long a time. Either that gold is alchemical and faked or it
is genuine and true. If it is alchemical, it will surely vanish into
smoke, or certainly by the royal touchstone. If on the contrary it
is genuine, as it surely is (for I am not one who fights against
nature [φυσιομάχος]), I yet do not accuse of deceit those who,
with every expectation of seeing their work rewarded, enticed
others into digging that gold out of mines hitherto inaccessible and
barred by the watchful custody of flame-spouting dragons and*

---

[5] According to legend, Pythagoras sacrificed an ox upon the discovery
of the famous Pythagorean theorem. Vieta introduces Theorem III of
Chapter IX of his *Ad Problema Adriani Romani Responsum* with the words
"Moved by the beauty of this discovery, O divine Mélusine, I have
sacrificed to you a hundred sheep in place of one Pythagorean ox."

*other poisonous and deadly serpents, but I fairly ask and expect
that they should at least not refuse the support of their authority
(which I esteem) against the ignorance or impudence of men who
calumniate and detract from another's praise. Therefore, my
princess, hold your own work dear and bless it with your
blessedness, having referred everything to the supreme ruler of
rulers whom you most religiously reverence in soul and in truth
(ἐν ψυχῇ καὶ ἀληθείᾳ), with the praise and glory of all praises.
From the marshes of the Isles de Mont of your most dear sister,
in the second year of our most Christian and august King
Henry IV, most zealous and most just punisher of the enemies
of the state and the murderers of Christ (χριστοκτόνων).*

Chapter I
On the definition and division of analysis and those things which are of use to zetetics

In mathematics there is a certain way of seeking the truth,
a way which Plato is said first to have discovered,[6] and
which was called "analysis" by Theon and was defined by
him as "taking the thing sought as granted and proceeding
by means of what follows to a truth that is uncontested"; so,
on the other hand, "synthesis" is "taking the thing that is
granted and proceeding by means of what follows to the
conclusion and comprehension of the thing sought."[7] And
although the ancients set forth a twofold analysis,[8] the
zetetic (ζητητική) and the poristic (ποριστική), to which
Theon's definition particularly refers, it is nevertheless
fitting that there be established also a third kind, which may
be called rhetic or exegetic (ῥητική ἢ ἐξηγητική), so that there

---

[6] See Note 218.
[7] See Note 217.
[8] See P. 155.

is a zetetic art by which is found the equation or proportion between the magnitude that is being sought and those that are given, a poristic art by which from the equation or proportion the truth of the theorem set up is investigated, and an exegetic art by which from the equation set up or the proportion there is produced the magnitude itself which is being sought. And thus, the whole threefold analytical art, claiming for itself this office, may be defined as the science of right finding in mathematics. Now what truly pertains to the zetetic art is established by the art of logic through syllogisms and enthymemes, the foundations of which are those very stipulations (symbola)[9] by which equations and proportions are arrived at, which stipulations must be derived from common notions as well as from theorems that are demonstrated by the power of analysis itself. In the zetetic art, however, the form of proceeding is peculiar to the art itself, inasmuch as the zetetic art does not employ its logic on numbers — which was the tediousness of the ancient analysts — but uses its logic through a logistic which in a new way has to do with species.[10] This logistic is much

[9] See Note 226.

[10] See P. 165. Although Diophantus seems the most likely source for Vieta's use of the word "species," the Reverend John Wallis in his *A Treatise of Algebra* (London, 1685), p. 66, advances another theory: "The name of Specious Arithmetick is given to it (I presume) with respect to a sense wherein the civilians use the word *Species*; for whereas it is usual with our Common Lawyers to put *Cases* in the name of John-an-Oaks and John-a-Stiles or John-a-Down, and the like (by which names they mean any person indefinitely who may be so concern'd) and of later times (for brevity sake) of J.O. and J.S. or J.D. (or yet more shortly) of A, B, C, etc. In like manner, the Civilians make use of the Names of Titus, Sempronius, Caius, and Mevius or the like, to represent indefinitely, any person in such circumstances. And cases so propounded they call *Species*. Now with respect hereunto, Vieta (accustomed to the language of the Civil Law) did give, I suppose, the Name of *Species* to the letters A, B, C, etc., made use of by him to represent indefinitely any Number or Quantity, so circumstanced as the occasion required. And accordingly, the accommodation of Arithmetical Operations to Numbers or other

more successful and powerful than the numerical one for comparing magnitudes[11] with one another in equations, once the law of homogeneity has been established; and hence there has been set up for that purpose a series or ladder, hallowed by custom, of magnitudes ascending or descending by their own nature from genus to genus, by which ladder the degrees and genera of magnitudes in equations may be designated and distinguished.

Chapter II
On the stipulations (symbola) governing equations and proportions

The Analytical Art assumes as manifest the better known stipulations governing equations and proportions which are to be found in the *Elements*, such as are:[12]

1. The whole is equal to its parts.

2. Things which are equal to the same thing are equal to each other.

3. If equals are added to equals, the sums are equal.

4. If equals are subtracted from equals, the remainders are equal.

5. If equals are multiplied by equals, the products are equal.

Quantities thus designed by *Symbols* or *Species*, was called *Arithmetica Speciosa* or *Specious Arithmetick*; the word *Species* signifying what we otherwise call *Notes, Marks, Symbols,* or *Characters,* made use of for the compendious expressing or designation of Numbers or other Quantities."

Wallis' theory derives credence from the fact that Vieta was a jurist. It may be, of course, that the word "species" as used by Vieta is meant to contain something of the meaning of the Diophantine *eide* and also something of the juridical meaning. (Cf. Note 275, P. 281.)

[11] See Pp. 156 and 157.

[12] See Note 226.

6. If equals are divided by equals, the results are equal.

7. If any magnitudes are proportional directly, they are proportional inversely and alternately [i.e., if $a:b::c:d$, then $b:a::d:c$ and $a:c::b:d$].

8. If like proportionals are added to like proportionals, the sums are proportional [i.e., if $a:b::c:d$, then $a+c:b+d::a:b$].

9. If like proportionals are subtracted from like proportionals, the remainders are proportional [i.e., if $a:b::c:d$, then $a-c:b-d::a:b$].

10. If proportionals are multiplied by proportionals, the products are proportional [i.e., if $a:b::c:d$ and $e:f::g:h$, then $ae:bf::cg:dh$].

For when proportionals are multiplied by proportionals, the same ratios are being compounded. Now it was commonly received by the ancient geometers that ratios which are compounded of the same ratios are the same with each other, as is seen everywhere in Apollonius, Pappus, and the other geometers. But the compounding of ratios is effected by the multiplication of the antecedents and the consequents, respectively, as is clear from those things that Euclid shows in the twenty-third proposition of the sixth book and the fifth proposition of the eighth book of the Elements.

11. If proportionals are divided by proportionals, the results are proportional [i.e., if $a:b::c:d$ and $e:f::g:h$, then $a/e:b/f::c/g:d/h$].

For when proportionals are divided by proportionals from the same ratios other same ratios are separated, and just as by multiplication ratios are compounded together, so by division one ratio is separated from another; for division undoes what multiplication, as shown, does.

12. The equation or ratio is not changed by a common multiplier or divisor [i.e., $ma:mb::a:b$ and $a/m:b/m::a:b$].

13. Products under the several segments are equal to the product under the whole [i.e., $ab+ac=a(b+c)$].

14. Products obtained by a succession of magnitudes, or quotients obtained by a succession of divisors, are equal, no

matter in what order the multiplication or division is done [i.e., $a.b = b.a$ and $(a/b)/c = (a/c)/b$].

But the paramount stipulation governing equations and proportions and the one that is all-important in analysis is:

15. If there be three or four magnitudes and the result of the multiplication of the extreme terms is equal to the result of the multiplication of the mean by itself or to the product of the means, then those magnitudes are proportional [i.e., if $ab = cd$, then $a:c::d:b$; or if $ab = c^2$, then $a:c::c:b$]. And conversely.

16. If there be three or four magnitudes, and as the first is to the second, so that second, or else some third, is to another, the product of the extreme terms will be equal to the product of the means [i.e., if $a:b::c:d$, then $ad = bc$; and if $a:b::b:c$, then $ac = b^2$].

And so, a proportion can be called the composition (constitutio) of an equation, an equation the resolution (resolutio) of a proportion.

Chapter III
Concerning the law of homogeneity and the degrees and genera of the magnitudes that are compared[13] (comparatarum)

The supreme and everlasting law of equations or proportions, which is called the law of homogeneity because it is conceived with respect to homogeneous magnitudes, is this:

1. Only homogeneous magnitudes are to be compared (comparari) with one another.

---

[13] Comparison (comparatio) means, on the one hand, adding and subtracting magnitudes to form algebraic expressions and, on the other, equating magnitudes or expressions with one another. Cf. Descartes, *Rules for the Direction of the Mind*, eds. Haldane and Ross (Dover, 1955), p. 55, Rule XIV.

For, as Adrastus[14] said, it is impossible to know how heterogeneous magnitudes may be conjoined.

And so, if a magnitude is added to a magnitude, it is homogeneous with it.

If a magnitude is multiplied by a magnitude, the product is heterogeneous in relation to both.

If a magnitude is divided by a magnitude, it is heterogeneous in relation to it.

Not to have considered these things was the cause of the darkness and blindness of the ancient analysts.

2. Magnitudes which by their own nature ascend and descend proportionally from genus to genus may be called "ladder-rungs."[15]

3. The first of the ladder magnitudes is "side" (latus) or "root" (radix).

The second is "square" (quadratum).

The third is "cube" (cubus).

The fourth is "squared-square" (quadrato-quadratum).

The fifth is "squared-cube" (quadrato-cubus).

The sixth is "cubed-cube" (cubo-cubus).

The seventh is "squared-squared-cube" (quadrato-quadrato-cubus).

The eighth is "squared-cubed-cube" (quadrato-cubo-cubus).

The ninth is "cubed-cubed-cube" (cubo-cubo-cubus).

And those remaining may be denominated from these by this series and method.

4. The genera of the compared (comparatarum) magnitudes, so that they may be equated in an orderly way to the ladder magnitudes, are:[16]

---

[14] See P. 173 and Notes 253, 254.

[15] See Note 248.

[16] The equated magnitudes would be not simply known magnitudes which we nowadays would designate by such letters as a, b, or c and which by the law of homogeneity would have to be understood as "lengths" or "planes" or "solids" according as they are equated with x

The first, "length" (longitudo) or "breadth" (latitudo),
The second, "plane" (planum),
The third, "solid" (solidum),
The fourth, "plane-plane" (plano-planum),
The fifth, "plane-solid" (plano-solidum),
The sixth, "solid-solid" (solido-solidum),
The seventh, "plane-plane-solid" (plano-plano-solidum),
The eighth, "plane-solid-solid" (plano-solido-solidum),
The ninth, "solid-solid-solid" (solido-solido-solidum).

And the remaining ones may be denominated from these by this series and method.

5. Of ladder magnitudes, the higher degree in relation to the "side" (latus), as the lowest and that to which the compared magnitude corresponds, is called the "power" (potestas). The other, lower, ladder magnitudes are called degrees "on the way (parodici) to the power" [translated simply by "lower"].

6. The power is pure when it is free from "conjoined" magnitudes. If the power is joined with a magnitude which is the product of a lower rung and a coefficient, it is a "conjoined" power [x^5 is a pure power; $x^5 + ax^4$ is a "conjoined" power].

Pure powers are: "square," "cube," "squared-square," "squared-cube," "cubed-cube," etc.

Conjoined powers are:

At the second rung: a "square" together with a "plane" which is the product of a "side" and a "length" or "breadth" [$x^2 + ax$];

At the third rung:

(i) a "cube" together with a "solid" which is the product of a "square" and a "length" or "breadth" [$x^3 + ax^2$],

(ii) a "cube" together with a "solid" which is the product of a "side"

---

or x^2 or x^3, but also, as appears in the sequel, products of known and unknown magnitudes. Thus, ax^2 would be a product of a "length" and a "square." It would itself be a "solid" and might be equated with x^3. See Note 249 and P. 172.

and a "*plane*" [$x^3 + bx$, *where* b *is understood as a* "*plane*" *magnitude*],

(iii) *a* "*cube*" *together with a* "*solid*" *which may be either the product of a* "*square*" *and a* "*length*" *or* "*breadth*" *or the product of a* "*side*" *and a* "*plane*" [$x^3 + c$, *where* c *is understood as a* "*solid*" *magnitude;* c *can equal* mx^2 *or* dx, *where* d *is a* "*plane*" *magnitude*];

At the fourth rung:

(i) *a* "*squared-square*" *together with a* "*plane-plane*" *which is the product of a* "*cube*" *and a* "*length*" *or* "*breadth*" [$x^4 + ax^3$],

(ii) *a* "*squared-square*" *together with a* "*plane-plane*" *which is the product of a* "*square*" *and a* "*plane*" [$x^4 + bx^2$, *where* b *is a* "*plane*"],

(iii) *a* "*squared-square*" *together with a* "*plane-plane*" *which is the product of a* "*side*" *and a* "*solid*" [$x^4 + cx$, *where* c *is a solid*"],

(iv) *a* "*squared-square*" *together with a* "*plane-plane*" *which is either the product of a* "*cube*" *and a* "*length*" *or* "*breadth*" *or the product of a* "*square*" *and a* "*plane*" [$x^4 + c$, *where* c *is a* "*plane-plane*"; c *can be equal to* mx^3, *or to* dx^2, *where* d *is a* "*plane*"],

(v) *a* "*squared-square*" *together with a* "*plane-plane*" *which is either the product of a* "*cube*" *and a* "*length*" *or* "*breadth*" *or the product of a* "*side*" *and a* "*solid*" [$x^4 + c$, *where* c *is a* "*plane-plane*"; c *can equal* mx^3 *or* dx *where* d *is a* "*solid*"],

(vi) *a* "*squared-square*" *together with a* "*plane-plane*" *which is either the product of a* "*square*" *and a* "*plane*" *or of a* "*side*" *and a* "*solid*" [$x^4 + c$, *where* c *is a* "*plane-plane*"; c *can equal* mx^2 *where* m *is a* "*plane*" *or* dx, *where* d *is a* "*solid*"],

(vii) *a* "*squared-square*" *together with a* "*plane-plane*" *which is either the product of a* "*cube*" *and a* "*length*" *or* "*breadth*," *or of a* "*square*" *and a* "*plane*," *or of a* "*side*" *and a* "*solid*" [$x^4 + c$, *where* c *is a* "*plane-plane*"; c *can equal* mx^3 *or* dx^2, *where* d *is a* "*plane*," *or* ex, *where* e *is a* "*solid*"].

In the same order the conjoined powers at the remaining rungs of the ladder may be found. But if we want to know how many genera of conjoined powers are at each rung, let there be taken a number less by unity than that term which is produced by geometric progression from unity in the double ratio [$1:2::2:4::4:8$, *etc.*] *and which has the same ordinal position as the power under consideration. Thus, if one wants to know how many conjoined powers are at the rung of the* "*squared-square*," *i.e., at the fourth rung, the fourth term of the geometric progression, namely 8, must be taken, from which, when unit has been taken away, 7 remains. And so, there are at the fourth rung as many conjoined powers as we have just enumerated. By this procedure it will be found that at the rung of the "squared-cube," i.e., the fifth rung, there are fifteen genera of conjoined powers.*

7. Coefficient[17] magnitudes which multiply ladder magnitudes that are lower in relation to a certain power and thus produce a homogeneous magnitude to be added to that power shall be called "subrungs."

The "subrungs" are "lengths" or "breadths," "plane," "solid," "plane-plane," etc. Thus, if there be a "squared-square" to which there is joined a "plane-plane" which is the product of a "side" and a "solid," the "solid" magnitude will be the "sub-rung"; and in relation to the "squared-square" the "side" will be a lower ladder magnitude. [In the expression $x^4 + cx$ it is apparent that x^4 is a "squared-square"; cx is a "plane-plane," being the product of the "side" x and the "solid" c; c, then, is the subrung, and x is in relation to x^4 a lower ladder magnitude.] Or if there be a "squared-square" together with a "plane-plane," which is either the product of a "square" and a "plane" or the product of a "side" and a "solid," the "plane" and the "solid" will be "subrung" magnitudes; and in relation to the "squared-square" the "square" and the "side" will be lower ladder magnitudes. [Thus we may have $x^4 + cx$ as above or $x^4 + cx^2$, where cx^2 is a "plane-plane" and c is understood as a "plane." Then the "plane" c is the "subrung," and the "square" x^2 is a lower ladder magnitude in relation to x^4.]

Chapter IV
On the precepts of the reckoning by species

The numeral reckoning (logistice numerosa) operates with numbers; the reckoning by species (logistice speciosa) operates with species or forms of things,[18] as, for example, with the letters of the alphabet.

[17] See Note 262.
[18] See Pp. 166 and 171.

Diophantus has handled the numerical reckoning in the thirteen books of the Arithmetic, *of which only the first six are extant, but which are now available in Greek and Latin and elucidated by the very learned commentaries[19] of a most illustrious man, Claude Bachet [de Méziriac]. But Vieta has produced the reckoning by species in the five books of the* Zetetics, *which he has arranged chiefly from selected problems of Diophantus, some of which he solves by a method peculiar to himself. Wherefore, if you wish to discern profitably the distinction between the two kinds of reckoning, you must consult Diophantus and Vieta together, and the zetetics of the latter must by viewed along with the arithmetical problems of the former; it is in order that I may lighten for you the labor of this task that I shall briefly note the zetetics which have been taken from the problems of Diophantus (see P. 330).*

There are four canonical precepts for reckoning by species (logistices speciosae).

Precept I
To add a magnitude to a magnitude

Let there be two magnitudes A and B. It is required to add the one to the other.

But, since heterogeneous magnitudes cannot be conjoined, those which are proposed to be added to one another are two homogeneous magnitudes. That one of them is greater or less than the other does not imply that they are of different genera. Therefore, they may be fittingly added[20] by means of the sign for coupling or addition; and, put together, they will be A "plus" B, if they are simple "lengths" or "breadths."

But if they stand higher on the aforesaid ladder or if they share a genus with those that stand higher, they will be designated by the appropriate denominations, as, for instance, we may say, "A square 'plus' B plane" or "A cube 'plus' B solid," and similarly in other cases.

The analysts, however, are accustomed to indicate the performance of addition by the symbol $+$.[21],

---

[19] See Pp. 176–178.

[20] See Pp. 176 ff.

[21] See Note 153.

Diophantus		Vieta	
Book of the Arithmetic	Problem[22]	Book of the Zetetics	Problem[23]
I	1	I	1
	4		2
	2		3
	7		4
	9		5
	5		7
	6		8
II	8, 9	IV	1
	10		2, 3
	11		6
	12		7
	13		8
	14		9
V	8		11
III	7, 8	V	1
	9		3
	10		4
	11		5
	12		7
	13		8
V	9		9
IV	34		13

[22] This problem (Tannery, p. 16) is reproduced here as an example. For Diophantus' signs see Pp. 141–147.

To divide a given number into two numbers with a given difference.

So, let the given number be $\bar{\rho}$ [one hundred], and let the difference be $\overset{\circ}{M}\bar{\mu}$ (forty units).

To find the numbers.

Let the less be taken as $\varsigma\bar{\alpha}$ [one unknown]. Then the greater will be $\varsigma\bar{\alpha}\overset{\circ}{M}\bar{\mu}$ [one unknown and forty units]. Then both together become $\varsigma\bar{\beta}\overset{\circ}{M}\bar{\mu}$ [two unknowns and forty units]. But they have been given as $\overset{\circ}{M}\bar{\rho}$ [one hundred units.]

$M\bar{\rho}$ [one hundred units], then, are equal to $\varsigma\bar{\beta}\overset{\circ}{M}\bar{\mu}$ [two unknowns and forty units].

Precept II
To subtract a magnitude from a magnitude

Let there be two magnitudes A and B, and let the former be greater than the latter. It is required to subtract the less from the greater.

Since, then, a magnitude is to be subtracted from a magnitude, but heterogeneous magnitudes cannot be conjoined, those which are proposed are two homogeneous magnitudes. That one of them is greater and the other less does not imply that they are of different genera. Therefore, subtraction may be fittingly effected by means of the sign of the disjoining or removal[24] of the less

[24] "Removal" here translates a juridical term "multa" which means a fine, and is preserved in the English word "mulct."

And, taking like things from like: I take $\dot{M}\bar{\mu}$ [forty units] from the $\bar{\rho}$ [one hundred] and likewise $\bar{\mu}$ [forty] from the β [two] numbers and $\bar{\mu}$ [forty] units. The $\varsigma\beta$ [two unknowns] are left equal to $\dot{M}\bar{\xi}$ [sixty units]. Then, each ς [unknown] becomes $\dot{M}\bar{\lambda}$ [thirty units].

As to the actual numbers required: the less will be $\dot{M}\bar{\lambda}$ [thirty units] and the greater $\dot{M}\bar{o}$ [seventy units], and the proof is clear.

[23] This problem (*Opera Mathematica*, p. 42) is reproduced here as an example.

Given the difference of two "sides" and their sum, to find the "sides."

Let the difference B of the two "sides" be given, and also let their sum D be given.

It is required to find the "sides."

Let the less "side" be A; then the greater will be $A+B$. Therefore, the sum of the "sides" will be $A2+B$. But the same sum is given as D. Wherefore, $A2+B$ is equal to D. And, by antithesis, $A2$ will be equal to $D-B$, and if they are all halved, A will be equal to $D\frac{1}{2}-B\frac{1}{2}$.

Or, let the greater "side" be E. Then the less will be $E-B$. Therefore, the sum of the "sides" will be $E2-B$. But the same sum is given as D. Therefore, $E2-B$ will be equal to D, and by antithesis, $E2$ will be equal to $D+B$, and if they are all halved, E will be equal to $D\frac{1}{2}+B\frac{1}{2}$.

Therefore, with the difference of two "sides" given and their sum, the "sides" are found.

For, indeed, half the sum of the "sides" minus half their difference is equal to the less "side," and half their sum plus half their difference is equal to the greater.

Which very thing the zetesis shows.

Let B be 40 and D 100. Then A becomes 30 and E becomes 70.

from the greater; and disjoined, they will be A "minus" B, if they are simple "lengths" or "breadths."

But if they stand higher on the aforesaid ladder or if they share a genus with those that stand higher, they will be designated by the appropriate denominations, as, for example, we may say "A square 'minus' B plane" or "A cube 'minus' B solid," and similarly in the other cases.

Nor will it be done differently if the magnitude which is subtracted is itself conjoined with some magnitude, since the whole and the parts are not to be judged by separate laws; thus, if "B 'plus' D" is to be subtracted from A, the remainder will be "A 'minus' B, 'minus' D," the magnitudes B and D having been subtracted one by one.

But if D is already subtracted from B and "B 'minus' D" is to be subtracted from A, the result will be "A 'minus' B 'plus' D," because in the subtraction of the whole magnitude B that which is subtracted exceeds by the magnitude D what was to have been subtracted. Therefore, it must be made up by the addition of that magnitude D.

The analysts, however, are accustomed to indicate the performance of the removal by means of the symbol —. And this is "defect" ($\lambda\epsilon\hat{\iota}\psi\iota\varsigma$) in Diophantus, as the performance of addition is "presence" ($\tilde{\upsilon}\pi\alpha\rho\xi\iota\varsigma$).

But when it is not said which magnitude is greater or less, and yet the subtraction must be made, the sign of the difference is: $=$, i.e., when the less is undetermined; as, if "A square" and "B plane" are the proposed magnitudes, the difference will be: "A square$=B$ plane," or "B plane$=A$ square."[25]

Precept III
To multiply a magnitude by a magnitude

Let there be two magnitudes A and B. It is required to multiply the one by the other.

Since, then, a magnitude is to be multiplied by a magnitude,

---

[25] The introduction of negative quantities makes it unnecessary for us to distinguish the two minus signs. The second of these signs, which is now used to signify equality, was so used as early as 1557 by Robert Recorde in his *The Whetstone of Witte*. Vieta has no symbol for equality.

they will by their multiplication produce a magnitude heterogeneous in relation to each of them; and therefore, their product
will rightly be designated by the word "*in*" or "*sub*," as, for
example, "*A* in B," by which it will be signified that the one has
been multiplied by the other; or as others say, that a magnitude is
produced "under" *A* and *B*, and that simply, if *A* and *B* are simple
"lengths" or "breadths."[26]

But if they stand higher on the ladder or if they share in genus
with magnitudes that stand higher, it is agreed to add the names
themselves of the ladder magnitudes or of those that share in their
genus, as, for example, "*A* square in *B*" or "*A* square in *B* plane"
or "*A* square in *B* solid," and similarly in the other cases.

If, however, the magnitudes to be multiplied, or one of them,
be of two or more names, nothing different happens in the
operation.[27] Since the whole is equal to its parts, therefore also
the products under the segments of some magnitude are equal to
the product under the whole. And when the positive name[28]
(nomen adfirmatum) of a magnitude is multiplied by a name also
positive of another magnitude, the product will be positive, and
when it is multiplied by a negative name (nomen negatum), the
product will be negative.

From which precept it also follows that by the multiplication
of negative names by each other a positive product is produced,
as when "$A = B$" is multiplied by "$D = G$" [giving $DA - DB - GA + GB$], since the product of the positive *A* and the negative *G*
is negative, which means that too much is removed or taken
away, inasmuch as *A* is, inaccurately, brought forward (producta)
as a magnitude to be multiplied [as a whole, i.e., in the factor
$(-AG)$], and since, similarly, the product of the negative *B* and
the positive *D* is negative, which again means that too much is
removed, inasmuch as *D* is, inaccurately, brought forward as a
magnitude to be multiplied [i.e., in $(-DB)$]. Therefore, by way
of compensation, when the negative *B* is multiplied by the negative *G*, the product is positive.

[26] See P. 171.

[27] That is, $a(u+v) + a(x+y) = au + av + ax + ay$; or $a(x+y+z) = ax + ay + az$ (cf. Euclid II, 1).

[28] The "names," i.e., the signs themselves, are multiplied together *as if*
they were particular numbers.

The denominations of products made by magnitudes ascending proportionally from genus to genus are related to one another in precisely the following way:

A "side" multiplied by itself produces a "square."

A "side" multiplied by a "square" produces" a "cube."

A "side" multiplied by a "cube" produces a "squared-square."

A "side" multiplied by a "squared-square" produces a "squared-cube."

And interchangeably, i.e., a "square" multiplied by a "side" produces a "cube," a "cube" multiplied by a "side" produces a "squared-square," etc.

Again,

A "square" multiplied by itself produces a "squared-square."

A "square" multiplied by a "cube" produces a "squared-cube."

A "square" multiplied by a "squared-square" produces a "cubed-cube."

And interchangeably.

Again,

A "cube" multiplied by itself produces a "cubed-cube."

A "cube" multiplied by a "squared-square" produces a "squared-squared-cube."

A "cube" multiplied by a "squared-cube" produces a "squared-cubed-cube."

A "cube multiplied by a "cubed-cube" produces a "cubed-cubed-cube."

And interchangeably, and so on in that order.

In like manner, among the homogeneous magnitudes,

A "breadth" multiplied by a "length" produces a "plane."

A "breadth" multiplied by a "plane" produces a "solid."

A "breadth" multiplied by a "solid" produces a "plane-plane."

A "breadth" multiplied by a "plane-plane" produces a "plane-solid."

A "breadth" multiplied by a "plane-solid" produces a "solid-solid."

And interchangeably.

A "plane" multiplied by a "plane" produces a "plane-plane."

A "plane" multiplied by a "solid" produces a "plane-solid."

A "plane" multiplied by a "plane-plane" produces a "solid-solid."

And interchangeably.

A "solid" multiplied by a "solid" produces a "solid-solid."

A "solid" multiplied by a "plane-plane" produces a "plane-plane-solid."

A "solid" multiplied by a "plane-solid" produces a "plane-solid-solid."

A "solid" multiplied by a "solid-solid" produces a "solid-solid-solid."

And interchangeably, and so on in that order.

Precept IV

To divide a magnitude by a magnitude

Let there be two magnitudes *A* and *B*. It is required to divide the one by the other.

Since, then, a magnitude is to be divided by a magnitude, namely higher ones by lower ones, i.e., magnitudes of one kind by magnitudes of another kind, the proposed magnitudes are different in kind. Let *A*, if you will, be a "length" and *B* "plane." And then let a horizontal line appropriately stand between the higher magnitude *B* which is being divided and the lower *A* by which the division is made.

But the magnitudes themselves, i.e. the resultant quotients, will be denominated in accordance with their own rungs at which they are fixed or to which they have been reduced in the ladder of proportional or homogeneous magnitudes, as, for example:

(*B* plane)/*A*, by which symbol the "length" which results from the division of "*B* plane" by "*A* length" may be signified.

And if *B* is given as a "cube" and *A* as a "plane," the result will be (*B* cube)/*A*, by which symbol the "length" which results from the division of "*B* cube" by "*A* plane" may be signified.

And if *B* is assumed to be a "cube" and *A* a "length," the result will be (*B* cube)/*A*, by which symbol the "plane" which arises from the division of "*B* cube" by *A* may be signified, and so on in that order, *in infinitum*.

Nor will anything different be observed among binomial or polynomial magnitudes.

The denominations of the magnitudes that arise from dividing by magnitudes that ascend proportionally by degrees from genus to genus are related to one another in precisely the following way:

> A "square" divided by a "side" gives a "side."
> A "cube" divided by a "side" gives a "square."
> A "squared-square" divided by a "side" gives a "cube."
> A "squared-cube" divided by a "side" gives a "squared-square."
> A "cubed-cube" divided by a "side" gives a "squared-cube."
> And interchangeably, i.e., a "cube" divided by a "square" gives a "side," a "squared-square" divided by a "cube" gives a "side," etc.

> Again,
> A "squared-square" divided by a "square" gives a "square."
> A "squared-cube" divided by a "square" gives a "cube."
> A "cubed-cube" divided by a "square" gives a "squared-square."
> And interchangeably.

> Again,
> A "cubed-cube" divided by a "cube" gives a "squared-square" [sic!].

A "squared-cubed-cube" divided by a "cube" gives a "squared-cube."

A "cubed-cubed-cube" divided by a "cube" gives a "cubed-cube."

And interchangeably, and so on in that order.

In like manner, among the homogeneous magnitudes,

A "plane" divided by a "breadth" gives a "length."

A "solid" divided by a "breadth" gives a "plane."

A "plane-plane" divided by a "breadth" gives a "solid."

A "plane-solid" divided by a "breadth" gives a "plane-plane."

A "solid-solid" divided by a "breadth" gives a "plane-solid."

And interchangeably.

A "plane-plane" divided by a "plane" gives a "plane."

A "plane-solid" divided by a "plane" gives a "solid."

A "solid-solid" divided by a "plane" gives a "plane-plane."

And interchangeably.

A "solid-solid" divided by a "solid" gives a "solid."

A "plane-plane-solid" divided by a "solid" gives a "plane-plane."

A "plane-solid-solid" divided by a "solid" gives a "plane-solid."

A "solid-solid-solid" divided by a "solid" gives a "solid-solid."

And interchangeably, and so on in that order.

Moreover, if the magnitude that is being divided be the sum, difference, product, or quotient of other magnitudes, nothing prevents the aforesaid precepts from applying to the division, it being noted that, when the magnitude that is being divided, whatever may be its rung, is the product of some magnitude and a magnitude that is the same as the divisor, nothing either in genus or value is added to or taken away from the factor that is not the same as the divisor and that also arises from the division, since what multiplication does division undoes: for example, $(B \text{ in } A)/B$ is A, and $(B \text{ in } A \text{ plane})/B$ is "A plane."

And thus, in the case of additions, let it be required to add Z to A plane/B, The sum will be

$$\frac{(A \text{ plane}) + (Z \text{ in } B)}{B}$$

[i.e., $a^2/b + z = (a^2 + zb)/b$].

Or let it be required to add (Z square)/G to (A plane)/B. The sum will be

$$\frac{(G \text{ in } A \text{ plane}) + (B \text{ in } Z \text{ square})}{B \text{ in } G}.$$

In the case of subtractions, let it be required to subtract Z from (A plane)/B. The remainder will be

$$\frac{(A \text{ plane}) - (Z \text{ in } B)}{B}.$$

Or, let it be required to subtract (Z square)/G from (A plane)/B The remainder will be

$$\frac{(A \text{ plane in } G) - (Z \text{ square in } B)}{B \text{ in } G}.$$

In the case of multiplications, let it be required to multiply (A plane)/B by B. The result will be A plane.

Or let it be required to multiply (A plane)/B by Z. The result will be (A plane in Z)/B.

Or, finally, let it be required to multiply (A plane)/B by (Z square)/G. The result will be [(A plane)/(B in G)] in Z square.

In the case of division, let it be required to divide (A cube)/B by D. Each magnitude having been multiplied by B, the result will be (A cube)/(B in D) [i.e., $(a^3/b)/d = (a^3b/b)/bd = a^3/bd$].

Or let it be required to divide B in G by (A plane)/D. Each magnitude having been multiplied by D, the result will be (B in G in D)/A plane.

Or, finally, let it be required to divide (B cube)/Z by (A cube)/(D plane). The result will be (B cube in D plane)/(Z in A cube).

Chapter V
Concerning the laws of zetetics

The way to do zetetics is, in general, encompassed in the following laws:

1. If it is a "length" that is being sought, but the equation or proportion is hidden under the wrappings[29] of what is given in the problem, let the unknown which is being sought be a "side."

2. If it is a "plane" that is being sought, but the equation or proportion is hidden under the wrappings of what is given in the problem, let the unknown which is being sought be a "square."

3. If it is a "solid" that is being sought, but the equation or proportion is hidden under the wrappings of what is given in the problem, let the unknown which is being sought be a "cube."

Accordingly, that magnitude which is being sought will by its own nature ascend or descend through the several rungs of the magnitudes that are compared or equated with it.

4. Let the magnitudes that are given, as well as those that are being sought, be assimilated and compared (in accordance with the condition dictated by the problem) by adding, subtracting, multiplying, and dividing, the constant law of homogeneity being everywhere observed.

Accordingly, it is clear that finally something will be found which is equal to the magnitude that is being sought or to the power to which it ascends and that that will consist either entirely of given magnitudes or partly of given magnitudes

---

[29] In solving a problem by algebra the equation may not emerge immediately from the given conditions. It may take some reflection before one sees the equation that satisfies the conditions. This is what Vieta means when he speaks of the equation as "hidden under the wrappings of what is given in the problem" (cf. Descartes, *Geometry*, [Dover, 1954], pp. 6–8).

and partly of the unknown which is being sought or of magnitudes lower than it on the ladder.[30]

5. In order that this work may be assisted by some art, let the given magnitudes be distinguished from the undetermined unknowns by a constant, everlasting and very clear symbol, as, for intance, by designating the unknown magnitude by means of the letter A or some other vowel E, I, O, U, or Y, and the given magnitudes by means of the letters B, G, and D or the other consonants.[31]

6. Products composed entirely of given magnitudes may be added to one another, or subtracted from one another, according to the sign of their conjunction, and may merge into one product, which shall be the homogeneous element of the equation, i.e., the element under a given measure; and it shall constitute one side of the equation.[32]

7. In like manner, products composed of given magnitudes and of the same lower ladder magnitude may be added one

[30] In the equation $x^2 = ab$, x^2 is the magnitude which is being sought; ab is equal to it, and is a product entirely of given magnitudes. In the equation $x^3 = ax^2$, x^3 is the unknown which is being sought; ax^2 is equal to it and is a product partly of the given magnitude a and partly of a magnitude lower than x^3 on the ladder, namely x^2.

[31] This, of course, differs from the convention of present-day algebra, according to which the letters at the end of the alphabet $(x, y, z, \ldots)$ are used to represent unknowns and the letters at the beginning $(a, b, c, \ldots)$ represent knowns. Thomas Harriot in his *Artis analyticae praxis* (1631) followed Vieta in using vowels for unknowns and consonants for known quantities, except that he substituted small letters for Vieta's capitals. Descartes in his *Geometry* (1637) introduced the system we use; cf. Note 275.

[32] In Vieta's symbols, "B in C" and "D in F" would be products composed entirely of given magnitudes, which products may be added to or subtracted from one another by means of the plus sign or the minus sign. When so added or subtracted, they become "B in $C + D$ in F" or "B in $C - D$ in F." If we were then to form the equation "A square is equal to B in $C + D$ in F," "B in $C + D$ in F" would be homogeneous with "A square." Since "B in C" and "D in F" belong to the rank of "planes," the unit measure is given as a "plane" unit. In modern notation this would be $x^2 = ab + cd$.

to another, or subtracted one from another, according to the sign of their conjunction, and may merge into one product which shall be the element homogeneous in conjunction, or the element under the rung of the lower ladder magnitude.[33]

8. Elements which are homogeneous under the rungs of lower ladder magnitudes shall accompany the power with which they are conjoined, and, along with that power, shall constitute one side of the equation. And thus, the element that is homogeneous under a given measure will be equated to a power designated in its own genus or order; simply, if that power is free from all conjunction with other magnitudes, but if magnitudes homogeneous in conjunction accompany it, which magnitudes are indicated both by the symbol of the conjunction and by the rung of the lower ladder magnitudes, then the magnitude homogeneous under a given measure will be equated not only to it, but to it along with the magnitudes that are products of rungs and coefficient magnitudes.[34]

[33] For example, "A square in B" and "A square in C" would be products composed of the given magnitudes B and C and of the same lower ladder magnitude "A square." They may be conjoined by means of the plus sign or the minus sign, and then we get either "A square in B+A square in C" or "A square in B−A square in C." Each is a product under the lower ladder magnitude "A square," which is lower in relation to the rung of the product. "A square in B+A square in C" or "A square in B−A square in C" is called the element homogeneous in conjunction because it is regarded as something to be "conjoined" with a pure power "A cube" with which it is homogeneous.

In modern notation this would be ax^2+bx^2 or ax^2-bx^2, either of which binomials would be the element homogeneous in conjunction because it would be considered as "conjoined" with x^3 to make $x^3+ax^2+bx^2$, $x^3+ax^2-bx^2$, etc.

[34] "A square in G" and "A square in H" are elements homogeneous under the rung "A square." They accompany the power "A cube," are conjoined with it by addition or subtraction, and with it constitute one side of the equation. "B plane in C+D solid" is an element homogeneous under a given measure. It might be equated to "A cube" simply, in which case we would have the equation: "A cube is equal to B plane in C+D solid"; or "A cube" might be accompanied by "A square in

9. And, therefore, if the element that is homogeneous under a given measure happens to be mingled with the element that is homogeneous in conjunction, there shall be antithesis.[35]

There is antithesis when positively or negatively conjoined magnitudes cross from one side of the equation to the other under the opposite signs of conjunction, by which operation the equation is not changed. But that must now be demonstrated.

Proposition I
An equation is not changed by antithesis

Let it be given that "*A* square 'minus' *D* plane" is equal to "*G* square 'minus' *B* in *A*."

I say that "*A* square 'plus' *B* in *A*" is equal to "*G* square 'plus' *D* plane" and that by this transposition under opposite signs of conjunction the equation is not changed. For since "*A* square 'minus' *D* plane" is equal to "*G* square 'minus' *B* in *A*," let there be added to both sides "*D* plane 'plus' *B* in *A*." Therefore, from the common notion, "*A* square 'minus' *D* plane 'plus' *D* plane 'plus' *B* in *A*" is equal to "*G* square 'minus' *B* in *A* 'plus' *D* plane 'plus' *B* in *A*." Now the negative conjunction on the same side of an equation cancels the positive. On the one side, the conjunction of "*D* plane" vanishes; on the other, the conjunction of

---

G − *A* square in *H*," a magnitude homogeneous in conjunction, in which case we might have this equation: "*A* cube + *A* square in *G* − *A* square in *H* is equal to *B* plane in *C* + *D* solid."

In modern notation the last equation would be $x^3 + ax^2 - bx^2 = cd + e$, where *c* is understood as a "plane," *d* as a "length," and *e* as a "solid."

[35] "Antithesis" means the transposition of terms from one side of the equation to the other, with accompanying change of sign. Thus we might have the equation: "*A* cube + *A* square in *G* − *A* square in *H* − *B* plane in *C* + *F* plane in *K* is equal to *D* solid." By antithesis we could infer the equation: "*A* cube + *A* square in *G* − *A* square in *H* is equal to *B* plane in *C* − *F* plane in *K* + *D* solid."

In modern notation, from $x^3 + ax^2 - bx^2 - cd + ef = g$ we get by antithesis $x^3 + ax^2 - bx^2 = cd - ef + g$. We understand *c* and *e* as "planes" and *g* as a "solid."

"*B* in *A*," and there will remain: "*A* square 'plus' *B* in *A*" is equal to "*G* square 'plus' *D* plane."[36]

10. And if it happens that all the magnitudes have as a factor a certain rung, and therefore that the homogeneous element determined by the over-all measure does not immediately appear, there shall be a hypobibasm.[37]

Hypobibasm is the like lowering of the power and of the lower ladder magnitudes, the order of the ladder being observed, until the homogeneous element determined by the lower rung coincides with the over-all homogeneity according to which the magnitudes that remain are equated, by which operation the equation is not changed. But that must now be demonstrated.

The operation of hypobibasm differs from parabolism only in this, that in the case of hypobibasm each side of the equation is divided by an unknown quantity, but in the case of parabolism each side is divided by a known quantity, as is clear from the examples presented by the author.

Proposition II
An equation is not changed by hypobibasm

Let it be given that "*A* cube 'plus' *B* in *A* square" is equal to "*Z* plane in *A*."

I say that, by hypobibasm, "*A* square 'plus' *B* in *A*" is equal to "*Z* plane."

For that means to have divided all the "solids" by a common divisor, by which it is certain that the equation is not changed.[38]

11. And if it happens that the higher rung to which the unknown magnitude ascends does not subsist by itself but is

[36] In modern notation, we are given that $x^2 - d = y^2 - bx$. We want to show that $x^2 + bx = y^2 + d$. We add to both sides $d + bx$ and get $x^2 - d + d + bx = y^2 - bx + d + bx$ or $x^2 + bx = y^2 + d$. Here d, of course, is understood as "*d* plane."

[37] "Hypobibasm" means dividing both sides of the equation by the unknown. It comes from the verb ὑποβιβάζω, "to lower." Division "lowers" a magnitude from a higher rung to a lower rung.

[38] In modern notation: If $x^3 + bx^2 = cx$, $x^2 + bx = c$. Here c is thought of as a "plane" so that cx is a solid; then it may be said that all the "solids" are divided by the common divisor x.

multiplied by some given magnitude, parabolism[39] may be effected.

There is parabolism whenever the homogeneous magnitudes of which an equation is composed are divided by a given magnitude which in the equation appears as multiplied by the higher rung of the unknown magnitude, so that that rung assumes the name of the power, and in that power the final equation remains. But this must now be demonstrated.

Proposition III
An equation is not changed by parabolism

Let it be given that "*B* in *A* square 'plus' *D* plane in *A*" is equal to "*Z* solid."

I say that by parabolism "*A* square 'plus' [(*D* plane)/*B*] in *A*" is equal to "*Z* solid"/*B*. For that means to have divided all the "solids" by the common divisor *B*, by which it is certain that the equation is not changed.[40]

12. And then the equation shall be thought to be expressed clearly and shall be called "well ordered": it must be capable of being referred to a proportion, the following condition (cautio) especially being satisfied: the product of the extremes must correspond to the power together with the conjoined homogeneous elements; the product of the means must correspond to the homogeneous element under the given measure.[41]

13. Whence also an ordered proportion may be defined as a series of three or four magnitudes, so expressed in terms either simple or conjoined that all are given except that

[39] "Parabolism" means dividing both sides of the equation by a known quantity. It comes from the verb παραβάλλω, "to apply," i.e., to divide, as when an area of *a* units is applied to a length of *b* units, the breadth of the figure will give the the quotient *a/b*. See Note 346.

[40] In modern notation: If $bx^2 + cx = d$, $x^2 + cx/b = d/b$ where c is thought of as a "plane" and d as a "solid."

[41] Thus, if we have the equation "*A* square +*B* in *A* is equal to *C* in *D* + *C* in *E*," then it follows that *A* is to *C* as "*D*+*E*" is to "*A*+*B*."

which is being sought, or else the power and the lower ladder magnitudes.[42]

14. Finally, when the equation has been thus ordered, or the proportion thus ordered, let it be considered that zetetics has performed its function.[43]

Diophantus in those books which concern arithmetic employed zetetics most subtly of all. But he presented it as if established by means of numbers and not also by species (which, nevertheless, he used), in order that his subtlety and skill might be the more admired; inasmuch as those things that seem more subtle and more hidden to him who uses the reckoning by numbers (logistice numerosa) are quite common and immediately obvious to him who uses the reckoning by species (logistice speciosa).[44]

Chapter VI
Concerning the investigation of theorems by means of the poristic art

When the zetesis has been completed, the analyst turns from hypothesis to thesis and presents theorems of his own finding, theorems that obey the regulations of the art and are subject to the laws 'κατὰ παντός, καθ' αὐτό, καθόλου πρῶτον,'[45] which theorems, although they have from the

[42] That is, an ordered proportion would be one of the type given in the preceding note or one of the type $x^2 + ax : b :: c : d + x$, which yields the equation $x^3 + dx^2 + adx + ax^2 = bc$. This law and the preceding indicate that an ordered proportion is one that yields an equation in one unknown.

[43] Cf. Ch. I and P. 170.

[44] Cf. Pp. 165, 170, and Notes 244, 245.

[45] Cf. Note 235. These rules, as applied to the propositions of "poristic" (such, for example, as "*A* cube is equal to *B* plane in *C*") would seem to mean: (1) that the predicate must be "true of every instance" to which the subject is understood to refer, (2) that it must be predicated "essentially" of the subject, which would be the same as the law of homogeneity, and (3) that the predicate must be completely convertible with the subject, as is the case when the predicate is "commensurately universal"

zetesis their demonstration and firmness, are subjected to the law of synthesis, which is considered a more logical way of demonstrating; and whenever there is occasion, they are proved through it, yet by the great miracle of the art of finding. And for this reason, the steps of the analysis are retraced, which retracing is itself also analytical; and yet not in virtue of the reckoning by species (logistice speciosa), which has already performed its assigned duty. But if something unfamiliar has been hit upon and is proposed for proof, or if something has been presented by chance the truth of which must be weighed and investigated, then the way of poristic has first to be tried, from which it is easy to return to the synthesis; examples of this have been offered by Theon in the *Elements*, by Apollonius of Perga in the *Conics* and also by Archimedes himself in various books.[46]

Chapter VII
Concerning the function of the rhetic art

When the equation of the magnitude which is being sought has been set in order, the rhetic or exegetic ($\dot{\rho}\eta\tau\iota\kappa\dot{\eta}$ $\ddot{\eta}$ $\dot{\epsilon}\xi\eta\gamma\eta\tau\iota\kappa\dot{\eta}$) art, which is to be considered as the remaining part of the analytical art and as one which pertains principally to the application of the art (since the two others are concerned more with general patterns than with precepts, as one must by right concede to the logicians), performs its function both in regard to numbers if the problem concerns a magnitude that is to be expressed by number, and in regard to lengths, surfaces, and solids if it is necessary to show the magnitude itself. And, in the latter case, the analyst appears as a geometer by actually carrying out the work in imitation of the like analytical solution; in the former case, he appears

---

with the subject. Regarding the last rule, we may remark that if one is to form the "synthesis" from the "analysis" by reversing or retracing the steps, each statement must be completely convertible.

[46] See P. 166 and Notes 233, 275.

as a logistician by resolving whatever powers have been presented numerically, whether simple powers or conjoined. Whether it be in arithmetic or geometry, he produces some specimens of his own [analytic] art according to the conditions of the equation that has been found or of the proportion that has been derived in an orderly way from it.

In fact, not every geometrical solution is a neat one, for particular problems have their own elegances. But that solution is preferred to others which does not derive the synthetic operation from the equation, but derives the equation from the synthesis, while the synthesis proves itself. Thus the skillful geometer, though a learned analyst, conceals this fact and presents and explicates his problem as a synthetic one, as if thinking merely about the demonstration that is to be accomplished; then, by way of helping the logisticians, he constructs and proves a theorem having to do with a proposition or an equation perceived in that synthetic problem.[47]

Chapter VIII
The symbolism in equations and and the epilogue to the art

1. In analysis the name equation is understood simply as referring to an equality properly set in order by means of the zetesis.

2. And so, an equation is the coupling (comparatio) of an unknown magnitude with a known.

3. The unknown magnitude is a root or power.

4. Again, a power is either simple or conjoined.

5. Conjunction exists either through subtraction or addition.

6. When an element homogeneous in conjunction is subtracted from a power, the subtraction is direct.[48]

[47] See Pp. 166–168 and Note 234.

[48] For example, "*A* cube − *A* square in *G* − *A* square in *H*," or in modern notation, $x^3 - (gx^2 + hx^2)$.

7. When, on the contrary, the power is subtracted from the element homogeneous in conjunction, the subtraction is inverse.[49]

8. The measuring subrung is the measure itself of the rung of the element homogeneous in conjunction.[50]

9. But it is necessary to designate the rank of the power, the rank of the lower rungs, and also the quality or sign of the conjunction. Also the coefficient subrung magnitudes must be given.

10. The first lower ladder magnitude is the root which is being sought. The last is that which is lower than the power by one rung of the ladder. This is customarily understood by the name "epanaphora."[51]

Thus "square" is the epanaphora of "cube," "cube" of "squared-square," "squared-square" of "squared-cube," and so on in the same series in infinitum.

11. A lower ladder magnitude is the reciprocal of a lower ladder magnitude when a power is produced through the multiplication of one by the other. Thus, the coefficient magnitude is the reciprocal of that rung which it sustains.

As, for example, if there should be a "side" which is a lower ladder magnitude in relation to the "cube," the reciprocal rung will be the "square." But a "plane" multiplied by a "side" will be a reciprocal magnitude in relation to the "side," since the "solid" is produced from the "side" multiplied by the "plane," the "solid" being itself a magnitude of the same rung as the "cube."

[49] For example, "A square in $G + A$ square in $H - A$ cube," or in modern notation $(gx^2 + hx^2) - x^3$.

[50] This would seem to mean that in the case of "A cube $+ A$ square in $G + A$ square in H," where "A square in $G + A$ square in H" is the element homogeneous in conjunction, it is "A square" which, while being the subrung of "A cube," measures the whole element "A square in $G + A$ square in H."

[51] In modern notation, x^2 is the "epanaphora" of x^3, x^3 of x^4, x^4 of x^5, etc. The word "epanaphora" is from ἐπαναφέρω, "to carry up to, refer to."

12. After the root the lower ladder magnitudes progressing by "length" are the same ones that are designated on the ladder.

13. After the root the lower ladder magnitudes progressing by "plane" are:

"Square"	"*Plane.*"
"Squared-square"	"*Square* of the plane."
"Cubed-cube"	"*Cube* of the plane.'

And so on successively in that order.

14. After the root the lower ladder magnitudes progressing by "solid" are:

"Cube"	"*Solid.*"
"Cubed-cube"	"Square of the *solid.*"
"Cubed-cubed-cube"	"Cube of the *solid.*"

15. "Square," "Squared-square," "Squared-cubed cube," and those magnitudes which are produced from these continuously in this order are simple middle powers; the rest are manifold.

Thus, the simple middle powers can also be defined in such a way that they will be those the exponents of which progress in the geometrical subduplicate ratio. So powers of the second degree, of the fourth, of the eighth, of the sixteenth, will be simple middle powers. The remaining powers, standing in the intermediate degrees, are manifold.[52]

16. A known magnitude with which the others are equated is the homogeneous element of the equation.

As, for example, if "A cube + A in B square is equal to B in Z plane," "B in Z plane" will be the homogeneous element of the equation.

[52] If, of the series of ladder magnitudes we consider x, x^2, x^3, x^4, x^5, x^6, x^7, x^8, x^9, x^{10}, x^{11}, x^{12}, x^{13}, x^{14}, x^{15}, x^{16} . . . and then the exponents of the powers x^2, x^4, x^8, x^{16} . . ., we see that the exponents are in geometrical progression: $2:4::4:8::8:16$. They may be said to progress in the subduplicate ratio in that $2:4$ is the subduplicate of $2:8$, $4:8$ is the subduplicate of $4:16$, etc.

"A *cube*" will be the power to which the unknown magnitude which
is being sought ascends by its own nature
"A in B *square*" will be the element homogeneous in conjunction.
"A" is the lower ladder magnitude.
"B *square*" is a subrung magnitude or "parabola."[53]

17. In the case of numbers, the homogeneous elements of
equations are units.[54]

18. When a "root" that is being sought is, while remaining on its own base, equated to a given homogeneous
magnitude, the equation is simple absolutely.[55]

19. When the power of a "root" that is being sought,
being free from all conjunction, is equated to a given
homogeneous magnitude, the equation is simple ladder-
wise.[56]

20. If the power of a "root" that is being sought is joined
with magnitudes at the designated rung accompanied by
their given coefficients and if it is equated to a given magnitude, the equation is polynomial in proportion to the
multitude and variety of the conjunction.[57]

21. A power can be involved in as many conjunctions as
there are ladder magnitudes lower in relation to that power.

Thus, a "square" can be conjoined with a magnitude at
the rung of the "side"; a "cube" with magnitudes at the
rungs of the "side" and the "square"; a "squared-square"

---

[53] "Parabola" here means the result of application or the quotient
resulting from division by the unknown.

[54] In the equation "A cube is equal to B plane in C," the homogeneous
element of the equation is "B plane in C" and its unit is a "solid" unit.
We could not have "A cube is equal to B plane," for the unit of "B plane"
is a "plane" unit. It is different with equations involving numbers rather
than species. In the equation "A is equal to 9." "A" becomes a number,
and the units of all numbers are the same in kind as long as the numbers
are pure numbers.

[55] "A is equal to B," or $x = a$ in modern notation.

[56] "A cube is equal to B solid," or $x^3 = a$ in modern notation, a being
understood as a "solid."

[57] "A cube + B in A square − C plane in A is equal to D solid" or, in
modern notation, $x^3 + ax^2 - bx = c$, where b is a "plane" and c is a "solid."

with magnitudes at the rungs of the "side," the "square," and the "cube"; a "squared-cube" with magnitudes at the rungs of the "side," the "square," the "cube," and the "squared-square"; and so on in that series *in infinitum*.

22. Proportions are distinguished from one another and receive their nomenclature from the kinds of equations into which they are resolved.

23. With a view to exegetic in arithmetic the trained analyst is taught:

> *To add a number to a number.*
> *To subtract a number from a number.*
> *To multiply a number by a number.*
> *To divide a number by a number.*

The analytical art, furthermore, yields the resolution of all possible powers whether they be pure or (a thing of which both ancients and moderns have been ignorant) conjoined with other magnitudes.[58]

24. With a view to exegetic in geometry the analytical art selects and enumerates more regular procedures by which equations of "sides" and "squares" may be completely interpreted.[59]

25. With a view to "cube" and "squared-square," in order that the deficiency of geometry may be supplied as if by geometry, the analytical art postulates that

[58] This is the program of Vieta's work *De Numerosa Potestatum Purarum, atque Adfectarum ad Exegesin Resolutione Tractatus,* (*Opera Mathematica,* p. 163, cf. Note 210). The "numerical resolution of powers" means the solution of equations that have numerical solutions, such as the equation $x^2 = 2916$ (Problem I, p. 165, of the first section of the *De Numerosa,* which section has to do with pure powers), or $x^2 + 7x = 60{,}750$ (Problem I, p. 174, of the second section, which has to do with conjoined powers).

[59] This is the program of Vieta's *Effectionum Geometricarum Canonica Recensio* (*Opera Mathematica,* pp. 229 ff.), at the beginning of which he says, "The geometrical procedure by which all equations *which do not exceed the measure of squares* may be rightly interpreted I enumerate as follows...."

A straight line can be drawn from any point across any two lines in such a way that the intercept between these two lines will be equal to a given distance, any possible intercept having been predefined.

This being granted (it is, indeed, a postulate not difficult to fulfil), analysis skilfully solves the more famous problems which have hitherto been called irrational: the mesographicum, the problem of the division of an angle into three equal parts, the finding of the side of the heptagon, and as many others as fall into the formulas of equations in which "cubes" are equated to "solids," "squared-squares" to "plane-planes," whether simply or with some conjunction.[60]

26. Since all magnitudes are lines, surfaces, or solids, what great use could be made in human affairs of proportions involving triplicate or even quadruplicate ratios, if not perhaps in divisions of angles, so that we might obtain the angles from the sides of the figures or the sides from the angles?

27. Therefore, analysis, whether with a view to arithmetic or to geometry, discloses the mystery, known hitherto by no one, of the division of angles, and it teaches how:

When the ratio of the angles is given, to find the ratio of the sides.

To make an angle to be in the same ratio to an angle that a number is to a number.

28. It does not equate a straight line to a curve, because an

[60] This is the program of Vieta's *Supplementum Geometriae* (*Opera Mathematica*, pp. 240 ff.), which begins with a restatement of the postulate about the intercept and which contains Vieta's solutions of the three problems here mentioned. Propositions V–VII of the *Supplementum Geometriae* contain the solution of the problem of the mesographicum; this was the problem of finding two mean proportionals to two given straight lines, and its solution immediately yields the solution of the problem of doubling the cube. Proposition IX contains the solution of the problem of the trisection of an angle. Proposition XXIV contains the solution of the problem of finding the side of the regular heptagon which is to be inscribed in a given circle.

angle is something in between a straight line and a plane figure. Thus, the law of homogeneity seems to oppose it.

29. Finally, the analytical art, having at last been put into the threefold form of zetetic, poristic, and exegetic, appropriates to itself by right the proud problem of problems, which is:

TO LEAVE NO PROBLEM UNSOLVED.[61]

[61] The capital letters are Vieta's; see P. 185.

Index of Names*

*Italics refer to Note numbers. Numbers in parentheses following Note numbers refer to pages on which names occur.

Index of Topics*

Abstraction, 48, 103*f.*, 120, 128*ff.*, 198*ff.*, 207, 214, 220, 224

Algebra, 4, 62, 122, 126*f.*, 133, 146*ff.*, 153*f.*, 157*f.*, 167, 181, 183*f.*, 188, 197*f.*, 206, 210, 214*ff.*, 319, *201, 211, 212, 225, 245* (273), *279* (284), *338*

Analysis and synthesis, 90, 154*ff.*, 160, 163*ff.*, 179*f.*, 320*ff.*, 347*ff.*, *131, 217, 218, 220, 235-237, 239* (271), *245* (274)

Arithmetic and logistic, 6*f.*, 10*ff.*, 17*ff.*, 31*f.*, 37*ff.*, 49, 54*ff.*, 70, 129, 133, 135*ff.*, 223

Arithmos and number, vii, 7*ff.*, 32*ff.*, 46*ff.*, 61*ff.*, 70, 89*ff.*, 101*ff.*, 124, 136*f.*, 147, 174*ff.*, 185, 198*ff.*, 216*ff.*, 223*f.*, *260*

Art (*techne*), 18*f.*, 73*f.*, 122, 125, 166, 169*f.*, 181, 183*ff.*, *132*

Being, 64*ff.*, 82*ff.*, 93, 95*ff.*, 101, 175, 179, 207, *119*

*Italics refer to Note numbers. Numbers in parentheses following Note numbers refer to pages on which topics occur.

Christians, 129

Conceptualization (intentionality), vii, 48*f.*, 63, 118*ff.*, 192, 207*ff.*, *324*

Cosmology, 8, 38, 64*ff.*, 69*f.*, 152

Eidos and *species*, 7, 16, 27, 32*f.*, 55*f.*, 68*f.*, 80*f.*, 89*ff.*, 106, 143*ff.*, 165*ff.*, 185, 206, 216, 321 (*10*), 328*ff.*, *323* (301)

Eleatics, *102*

Equation, 133, 160, 168, 322, 347, *181, 182, 221, 225, 254*

Fractions, 7, 39*ff.*, 132, 136*ff.*, 195, 219*f.*, *172, 260, 266*

Geometry, 33*f.*, 72, 162*f.*, 167*f.*, 172, 189, 203*f.*, 206, 211, 214*ff.*, *229, 238, 247, 306* (293), *323* (301)

Homogeneity, 23, 54, 78, 172*ff.*, 215*ff.*, 221*ff.*, 324, 349, *254-256*

Huguenots, 315 (*2*)

"Hypothesis," 71*ff.*, 79

359

A CATALOG OF SELECTED
DOVER BOOKS
IN ALL FIELDS OF INTEREST

A CATALOG OF SELECTED DOVER
BOOKS IN ALL FIELDS OF INTEREST

DRAWINGS OF REMBRANDT, edited by Seymour Slive. Updated Lippmann, Hofstede de Groot edition, with definitive scholarly apparatus. All portraits, biblical sketches, landscapes, nudes. Oriental figures, classical studies, together with selection of work by followers. 550 illustrations. Total of 630pp. 9⅛ × 12¼.
21485-0, 21486-9 Pa., Two-vol. set $29.90

GHOST AND HORROR STORIES OF AMBROSE BIERCE, Ambrose Bierce. 24 tales vividly imagined, strangely prophetic, and decades ahead of their time in technical skill: "The Damned Thing," "An Inhabitant of Carcosa," "The Eyes of the Panther," "Moxon's Master," and 20 more. 199pp. 5⅜ × 8½. 20767-6 Pa. $4.95

ETHICAL WRITINGS OF MAIMONIDES, Maimonides. Most significant ethical works of great medieval sage, newly translated for utmost precision, readability. Laws Concerning Character Traits, Eight Chapters, more. 192pp. 5⅜ × 8½.
24522-5 Pa. $4.50

THE EXPLORATION OF THE COLORADO RIVER AND ITS CANYONS, J. W. Powell. Full text of Powell's 1,000-mile expedition down the fabled Colorado in 1869. Superb account of terrain, geology, vegetation, Indians, famine, mutiny, treacherous rapids, mighty canyons, during exploration of last unknown part of continental U.S. 400pp. 5⅜ × 8½. 20094-9 Pa. $7.95

HISTORY OF PHILOSOPHY, Julián Marías. Clearest one-volume history on the market. Every major philosopher and dozens of others, to Existentialism and later. 505pp. 5⅜ × 8½. 21739-6 Pa. $9.95

ALL ABOUT LIGHTNING, Martin A. Uman. Highly readable non-technical survey of nature and causes of lightning, thunderstorms, ball lightning, St. Elmo's Fire, much more. Illustrated. 192pp. 5⅜ × 8½. 25237-X Pa. $5.95

SAILING ALONE AROUND THE WORLD, Captain Joshua Slocum. First man to sail around the world, alone, in small boat. One of great feats of seamanship told in delightful manner. 67 illustrations. 294pp. 5⅜ × 8½. 20326-3 Pa. $4.95

LETTERS AND NOTES ON THE MANNERS, CUSTOMS AND CONDITIONS OF THE NORTH AMERICAN INDIANS, George Catlin. Classic account of life among Plains Indians: ceremonies, hunt, warfare, etc. 312 plates. 572pp. of text. 6⅛ × 9¼. 22118-0, 22119-9, Pa. Two-vol. set $17.90

ALASKA: The Harriman Expedition, 1899, John Burroughs, John Muir, et al. Informative, engrossing accounts of two-month, 9,000-mile expedition. Native peoples, wildlife, forests, geography, salmon industry, glaciers, more. Profusely illustrated. 240 black-and-white line drawings. 124 black-and-white photographs. 3 maps. Index. 576pp. 5⅜ × 8½. 25109-8 Pa. $11.95

THE BOOK OF BEASTS: Being a Translation from a Latin Bestiary of the Twelfth Century, T. H. White. Wonderful catalog real and fanciful beasts: manticore, griffin, phoenix, amphivius, jaculus, many more. White's witty erudite commentary on scientific, historical aspects. Fascinating glimpse of medieval mind. Illustrated. 296pp. 5⅜ × 8¼. (Available in U.S. only) 24609-4 Pa. $6.95

FRANK LLOYD WRIGHT: ARCHITECTURE AND NATURE With 160 Illustrations, Donald Hoffmann. Profusely illustrated study of influence of nature—especially prairie—on Wright's designs for Fallingwater, Robie House, Guggenheim Museum, other masterpieces. 96pp. 9¼ × 10¾. 25098-9 Pa. $8.95

FRANK LLOYD WRIGHT'S FALLINGWATER, Donald Hoffmann. Wright's famous waterfall house: planning and construction of organic idea. History of site, owners, Wright's personal involvement. Photographs of various stages of building. Preface by Edgar Kaufmann, Jr. 100 illustrations. 112pp. 9¼ × 10. 23671-4 Pa. $8.95

YEARS WITH FRANK LLOYD WRIGHT: Apprentice to Genius, Edgar Tafel. Insightful memoir by a former apprentice presents a revealing portrait of Wright the man, the inspired teacher, the greatest American architect. 372 black-and-white illustrations. Preface. Index. vi + 228pp. 8¼ × 11. 24801-1 Pa. $10.95

THE STORY OF KING ARTHUR AND HIS KNIGHTS, Howard Pyle. Enchanting version of King Arthur fable has delighted generations with imaginative narratives of exciting adventures and unforgettable illustrations by the author. 41 illustrations. xviii + 313pp. 6⅛ × 9¼. 21445-1 Pa. $6.95

THE GODS OF THE EGYPTIANS, E. A. Wallis Budge. Thorough coverage of numerous gods of ancient Egypt by foremost Egyptologist. Information on evolution of cults, rites and gods; the cult of Osiris; the Book of the Dead and its rites; the sacred animals and birds; Heaven and Hell; and more. 956pp. 6⅛ × 9¼. 22055-9, 22056-7 Pa., Two-vol. set $21.90

A THEOLOGICO-POLITICAL TREATISE, Benedict Spinoza. Also contains unfinished *Political Treatise*. Great classic on religious liberty, theory of government on common consent. R. Elwes translation. Total of 421pp. 5⅜ × 8½. 20249-6 Pa. $7.95

INCIDENTS OF TRAVEL IN CENTRAL AMERICA, CHIAPAS, AND YUCATAN, John L. Stephens. Almost single-handed discovery of Maya culture; exploration of ruined cities, monuments, temples; customs of Indians. 115 drawings. 892pp. 5⅜ × 8½. 22404-X, 22405-8 Pa., Two-vol. set $15.90

LOS CAPRICHOS, Francisco Goya. 80 plates of wild, grotesque monsters and caricatures. Prado manuscript included. 183pp. 6⅞ × 9⅜. 22384-1 Pa. $5.95

AUTOBIOGRAPHY: The Story of My Experiments with Truth, Mohandas K. Gandhi. Not hagiography, but Gandhi in his own words. Boyhood, legal studies, purification, the growth of the Satyagraha (nonviolent protest) movement. Critical, inspiring work of the man who freed India. 480pp. 5⅜ × 8½. (Available in U.S. only) 24593-4 Pa. $6.95

HOW TO WRITE, Gertrude Stein. Gertrude Stein claimed anyone could understand her unconventional writing—here are clues to help. Fascinating improvisations, language experiments, explanations illuminate Stein's craft and the art of writing. Total of 414pp. 4⅝ × 6⅜. 23144-5 Pa. $6.95

ADVENTURES AT SEA IN THE GREAT AGE OF SAIL: Five Firsthand Narratives, edited by Elliot Snow. Rare true accounts of exploration, whaling, shipwreck, fierce natives, trade, shipboard life, more. 33 illustrations. Introduction. 353pp. 5⅜ × 8½. 25177-2 Pa. $8.95

THE HERBAL OR GENERAL HISTORY OF PLANTS, John Gerard. Classic descriptions of about 2,850 plants—with over 2,700 illustrations—includes Latin and English names, physical descriptions, varieties, time and place of growth, more. 2,706 illustrations. xlv + 1,678pp. 8½ × 12¼. 23147-X Cloth. $75.00

DOROTHY AND THE WIZARD IN OZ, L. Frank Baum. Dorothy and the Wizard visit the center of the Earth, where people are vegetables, glass houses grow and Oz characters reappear. Classic sequel to Wizard of Oz. 256pp. 5⅜ × 8. 24714-7 Pa. $5.95

SONGS OF EXPERIENCE: Facsimile Reproduction with 26 Plates in Full Color, William Blake. This facsimile of Blake's original "Illuminated Book" reproduces 26 full-color plates from a rare 1826 edition. Includes "The Tyger," "London," "Holy Thursday," and other immortal poems. 26 color plates. Printed text of poems. 48pp. 5¼ × 7. 24636-1 Pa. $3.95

SONGS OF INNOCENCE, William Blake. The first and most popular of Blake's famous "Illuminated Books," in a facsimile edition reproducing all 31 brightly colored plates. Additional printed text of each poem. 64pp. 5¼ × 7. 22764-2 Pa. $3.95

PRECIOUS STONES, Max Bauer. Classic, thorough study of diamonds, rubies, emeralds, garnets, etc.: physical character, occurrence, properties, use, similar topics. 20 plates, 8 in color. 94 figures. 659pp. 6⅛ × 9¼. 21910-0, 21911-9 Pa., Two-vol. set $15.90

ENCYCLOPEDIA OF VICTORIAN NEEDLEWORK, S. F. A. Caulfeild and Blanche Saward. Full, precise descriptions of stitches, techniques for dozens of needlecrafts—most exhaustive reference of its kind. Over 800 figures. Total of 679pp. 8¼ × 11. Two volumes. Vol. 1 22800-2 Pa. $11.95 Vol. 2 22801-0 Pa. $11.95

THE MARVELOUS LAND OF OZ, L. Frank Baum. Second Oz book, the Scarecrow and Tin Woodman are back with hero named Tip, Oz magic. 136 illustrations. 287pp. 5⅜ × 8½. 20692-0 Pa. $5.95

WILD FOWL DECOYS, Joel Barber. Basic book on the subject, by foremost authority and collector. Reveals history of decoy making and rigging, place in American culture, different kinds of decoys, how to make them, and how to use them. 140 plates. 156pp. 7⅞ × 10¾. 20011-6 Pa. $8.95

HISTORY OF LACE, Mrs. Bury Palliser. Definitive, profusely illustrated chronicle of lace from earliest times to late 19th century. Laces of Italy, Greece, England, France, Belgium, etc. Landmark of needlework scholarship. 266 illustrations. 672pp. 6⅛ × 9¼. 24742-2 Pa. $14.95

SUNDIALS, Albert Waugh. Far and away the best, most thorough coverage of ideas, mathematics concerned, types, construction, adjusting anywhere. Over 100 illustrations. 230pp. 5⅜ × 8½. 22947-5 Pa. $5.95

PICTURE HISTORY OF THE NORMANDIE: With 190 Illustrations, Frank O. Braynard. Full story of legendary French ocean liner: Art Deco interiors, design innovations, furnishings, celebrities, maiden voyage, tragic fire, much more. Extensive text. 144pp. 8⅜ × 11¾. 25257-4 Pa. $10.95

THE FIRST AMERICAN COOKBOOK: A Facsimile of "American Cookery," 1796, Amelia Simmons. Facsimile of the first American-written cookbook published in the United States contains authentic recipes for colonial favorites—pumpkin pudding, winter squash pudding, spruce beer, Indian slapjacks, and more. Introductory Essay and Glossary of colonial cooking terms. 80pp. 5⅜ × 8½. 24710-4 Pa $3.50

101 PUZZLES IN THOUGHT AND LOGIC, C. R. Wylie, Jr. Solve murders and robberies, find out which fishermen are liars, how a blind man could possibly identify a color—purely by your own reasoning! 107pp. 5⅜ × 8½. 20367-0 Pa. $2.50

ANCIENT EGYPTIAN MYTHS AND LEGENDS, Lewis Spence. Examines animism, totemism, fetishism, creation myths, deities, alchemy, art and magic, other topics. Over 50 illustrations. 432pp. 5⅜ × 8½. 26525-0 Pa. $8.95

ANTHROPOLOGY AND MODERN LIFE, Franz Boas. Great anthropologist's classic treatise on race and culture. Introduction by Ruth Bunzel. Only inexpensive paperback edition. 255pp. 5⅜ × 8½. 25245-0 Pa. $6.95

THE TALE OF PETER RABBIT, Beatrix Potter. The inimitable Peter's terrifying adventure in Mr. McGregor's garden, with all 27 wonderful, full-color Potter illustrations. 55pp. 4¼ × 5½. (Available in U.S. only) 22827-4 Pa. $1.75

THREE PROPHETIC SCIENCE FICTION NOVELS, H. G. Wells. *When the Sleeper Wakes, A Story of the Days to Come* and *The Time Machine* (full version). 335pp. 5⅜ × 8½. (Available in U.S. only) 20605-X Pa. $6.95

APICIUS COOKERY AND DINING IN IMPERIAL ROME, edited and translated by Joseph Dommers Vehling. Oldest known cookbook in existence offers readers a clear picture of what foods Romans ate, how they prepared them, etc. 49 illustrations. 301pp. 6⅛ × 9¼. 23563-7 Pa. $7.95

SHAKESPEARE LEXICON AND QUOTATION DICTIONARY, Alexander Schmidt. Full definitions, locations, shades of meaning of every word in plays and poems. More than 50,000 exact quotations. 1,485pp. 6½ × 9¼. 22726-X, 22727-8 Pa., Two-vol. set $31.90

THE WORLD'S GREAT SPEECHES, edited by Lewis Copeland and Lawrence W. Lamm. Vast collection of 278 speeches from Greeks to 1970. Powerful and effective models; unique look at history. 842pp. 5⅜ × 8½. 20468-5 Pa. $12.95

A CONCISE HISTORY OF PHOTOGRAPHY: Third Revised Edition, Helmut Gernsheim. Best one-volume history—camera obscura, photochemistry, daguerreotypes, evolution of cameras, film, more. Also artistic aspects—landscape, portraits, fine art, etc. 281 black-and-white photographs. 26 in color. 176pp. 8⅜ × 11¼. 25128-4 Pa. $13.95

THE DORÉ BIBLE ILLUSTRATIONS, Gustave Doré. 241 detailed plates from the Bible: the Creation scenes, Adam and Eve, Flood, Babylon, battle sequences, life of Jesus, etc. Each plate is accompanied by the verses from the King James version of the Bible. 241pp. 9 × 12. 23004-X Pa. $9.95

WANDERINGS IN WEST AFRICA, Richard F. Burton. Great Victorian scholar/adventurer's invaluable descriptions of African tribal rituals, fetishism, culture, art, much more. Fascinating 19th-century account. 624pp. 5⅜ × 8½. 26890-X Pa. $12.95

FLATLAND, E. A. Abbott. Intriguing and enormously popular science-fiction classic explores the complexities of trying to survive as a two-dimensional being in a three-dimensional world. Amusingly illustrated by the author. 16 illustrations. 103pp. 5⅜ × 8½. 20001-9 Pa. $2.50

THE HISTORY OF THE LEWIS AND CLARK EXPEDITION, Meriwether Lewis and William Clark, edited by Elliott Coues. Classic edition of Lewis and Clark's day-by-day journals that later became the basis for U.S. claims to Oregon and the West. Accurate and invaluable geographical, botanical, biological, meteorological and anthropological material. Total of 1,508pp. 5⅜ × 8½.
21268-8, 21269-6, 21270-X Pa. Three-vol. set $26.85

LANGUAGE, TRUTH AND LOGIC, Alfred J. Ayer. Famous, clear introduction to Vienna, Cambridge schools of Logical Positivism. Role of philosophy, elimination of metaphysics, nature of analysis, etc. 160pp. 5⅜ × 8½. (Available in U.S. and Canada only) 20010-8 Pa. $3.95

MATHEMATICS FOR THE NONMATHEMATICIAN, Morris Kline. Detailed, college-level treatment of mathematics in cultural and historical context, with numerous exercises. For liberal arts students. Preface. Recommended Reading Lists. Tables. Index. Numerous black-and-white figures. xvi + 641pp. 5⅜ × 8½.
24823-2 Pa. $11.95

HANDBOOK OF PICTORIAL SYMBOLS, Rudolph Modley. 3,250 signs and symbols, many systems in full; official or heavy commercial use. Arranged by subject. Most in Pictorial Archive series. 143pp. 8⅜ × 11. 23357-X Pa. $6.95

INCIDENTS OF TRAVEL IN YUCATAN, John L. Stephens. Classic (1843) exploration of jungles of Yucatan, looking for evidences of Maya civilization. Travel adventures, Mexican and Indian culture, etc. Total of 669pp. 5⅜ × 8½.
20926-1, 20927-X Pa., Two-vol. set $11.90